BESTSELLING BOOK SERIES

Linear Algebra For Dummies®

Values of Selected Trig Functions

When performing transformations such as rotations, you need the numerical values of the functions of some of the more commonly used angles.

Values of Selected Trigonometric Functions

Angle in Degrees	*Sine (sin)*	*Cosine (cos)*	*Tangent (tan)*
0	0	1	0
30	½	$\frac{\sqrt{3}}{2}$	$\frac{\sqrt{3}}{3}$
45	$\frac{\sqrt{2}}{2}$	$\frac{\sqrt{2}}{2}$	1
60	$\frac{\sqrt{3}}{2}$	½	$\sqrt{3}$
90	1	0	Undefined
180	0	–1	0
270	–1	0	Undefined

Vector Space Requirements

A set of elements is termed a vector space when particular requirements are met. Let a set consist of vectors **u**, **v**, and **w**. Also let k and l be real numbers, and consider the defined operations of ⊕ and ⊗. The set is a vector space if, under the operation of ⊕, you have

- **Closure: u ⊕ v** is in the set.
- **Commutativity: u ⊕ v = v ⊕ u.**
- **Associativity: u ⊕ (v ⊕ w) = (u ⊕ v) ⊕ w.**
- **An identity element 0: u ⊕ 0 = 0 ⊕ u = u** for any element **u.**
- **An inverse element –u: u ⊕ –u = –u ⊕ u = 0**

And if under the operation of ⊗, you have:

- **Closure:** *k* ⊗ **u** is in the set.
- **Distribution over a vector sum:** *k* ⊗ (**u** ⊕ **v**) = *k* ⊗ **u** ⊕ *k* ⊗ ⊗ **v**.
- **Distribution over a scalar sum:** (*k* + *l*) ⊗ **u** = *k* ⊗ **u** ⊕ *l* ⊗ **u.**
- **Associativity of a scalar product:** *k* ⊗ (*l* ⊗ u) = (*kl*) ⊗ **u.**
- **Multiplication by the scalar identity:** 1 ⊗ **u** = **u.**

Algebraic Properties

Some properties of algebraic expressions are also found when working with matrices, determinants, and vector spaces:

- **Commutative property of addition and multiplication:** $a + b = b + a$ and $ab = ba$
- **Associative property of addition and multiplication:** $a + (b + c) = (a + b) + c$ and $a(bc) = (ab)c$
- **Distributive property of multiplication over addition:** $a(b + c) = ab + ac$ and $(b + c)a = ba + ca$
- **Identities in addition and multiplication:** $0 + a = a + 0 = a$ and $1 \times a = a \times 1 = a$
- **Inverses in addition and multiplication:** $a + (-a) = (-a) + a = 0$ and $a \times {}^1\!/_a = {}^1\!/_a \times a = 1$ (where $a \neq 0$)

For Dummies: Bestselling Book Series for Beginners

Linear Algebra For Dummies®

Cheat Sheet

Calculator Commands

The following are all general instructions to apply to most available graphing calculators. You find more detailed instructions in Chapter 18.

To solve systems of equations by graphing:

1. **Write each equation in $y = mx + b$ form.**
2. **Insert equations in the *y*-menu.**
3. **Graph the lines.**
4. **Use the Intersection tool to get the answer.**

To add or subtract matrices:

1. **Insert the elements into the matrices A and B.**
2. **With a new screen, press [A] + [B] or [A] – [B], and press Enter.**

To multiply by a scalar:

1. **Insert the elements into the matrix A.**
2. **With a new screen, press the scalar and multiply: *k* * [A], and press Enter.**

To multiply two matrices together:

1. **Insert the elements into the matrices A and B.**
2. **With a new screen, press [A] * [B], and press Enter.**

To switch rows:

1. **Insert the elements into a matrix.**
2. **Use *row swap: rowSwap* ([matrix name], first row, second row), and press Enter.**

To add two rows together:

1. **Insert the elements into a matrix.**
2. **Use *row addition: row* +([matrix name], row to be added to target row, target row), and press Enter.**

To add the multiple of one row to another:

1. **Insert the elements into a matrix.**
2. **Use row *sum-of-multiple:* * *row* + (multiplier, [matrix name], row being multiplied, target row having multiple added to it), and press Enter.**

To multiply a row by a scalar:

1. **Insert the elements into a matrix.**
2. **Use *row multiple:* * *row* (multiplier, [matrix name], row), and press Enter.**

To create an echelon form:

1. **Insert the elements into a matrix.**
2. **Use *row-echelon form: ref* ([matrix name]) or the reduced row-echelon form: rref ([matrix name]), and press Enter.**

To raise a matrix to a power:

1. **Insert the elements into a matrix.**
2. **Use the caret operation with power, *p:* [matrix name] ^ *p*, and press Enter.**

To find inverses:

1. **Insert the elements into a matrix.**
2. **Use the reciprocal operation, x^{-1}: [matrix name]$^{-1}$, and press Enter.**

To solve systems of linear equations:

1. **Write each equation with the variables in the same order and the constant on the other side of the equation sign.**
2. **Create a matrix A, whose elements are the coefficients of the variables.**
3. **Create a matrix B, whose elements are the constants.**
4. **Press, A^{-1} * B, and press Enter.**

   The resulting vector has the values of the variables, in order.

Linear Algebra FOR DUMMIES®

by Mary Jane Sterling

Wiley Publishing, Inc.

Linear Algebra For Dummies®
Published by
Wiley Publishing, Inc.
111 River St.
Hoboken, NJ 07030-5774
www.wiley.com

For general information on our other products and services, please contact our Customer Care Department within the U.S. at 877-762-2974, outside the U.S. at 317-572-3993, or fax 317-572-4002.

For technical support, please visit www.wiley.com/techsupport.

Wiley also publishes its books in a variety of electronic formats. Some content that appears in print may not be available in electronic books.

Library of Congress Control Number: 2009927342

ISBN: 978-0-470-43090-3

Manufactured in the United States of America

10 9 8 7 6 5 4 3 2 1

About the Author

Mary Jane Sterling is the author of five other *For Dummies* titles (all published by Wiley): *Algebra For Dummies, Algebra II For Dummies, Trigonometry For Dummies, Math Word Problems For Dummies,* and *Business Math For Dummies.*

Mary Jane continues doing what she loves best: teaching mathematics. As much fun as the *For Dummies* books are to write, it's the interaction with students and colleagues that keeps her going. Well, there's also her husband, Ted; her children; Kiwanis; Heart of Illinois Aktion Club; fishing; and reading. She likes to keep busy!

Dedication

I dedicate this book to friends and colleagues, past and present, at Bradley University. Without their friendship, counsel, and support over these past 30 years, my teaching experience wouldn't have been quite so special and my writing opportunities wouldn't have been quite the same. It's been an interesting journey, and I thank all who have made it so.

Author's Acknowledgments

A big thank-you to Elizabeth Kuball, who has again agreed to see me through all the many victories and near-victories, trials and errors, misses and bull's-eyes — all involved in creating this book. Elizabeth does it all — project and copy editing. Her keen eye and consistent commentary are so much appreciated.

Also, a big thank-you to my technical editor, John Haverhals. I was especially pleased that he would agree to being sure that I got it right.

And, of course, a grateful thank-you to my acquisitions editor, Lindsay Lefevere, who found yet another interesting project for me.

Publisher's Acknowledgments

We're proud of this book; please send us your comments through our Dummies online registration form located at `http://dummies.custhelp.com`. For other comments, please contact our Customer Care Department within the U.S. at 877-762-2974, outside the U.S. at 317-572-3993, or fax 317-572-4002.

Some of the people who helped bring this book to market include the following:

Acquisitions, Editorial, and Media Development

Project Editor: Elizabeth Kuball

Acquisitions Editor: Lindsay Lefevere

Copy Editor: Elizabeth Kuball

Assistant Editor: Erin Calligan Mooney

Editorial Program Coordinator: Joe Niesen

Technical Editor: John S. Haverhals

Senior Editorial Manager: Jennifer Ehrlich

Editorial Supervisor and Reprint Editor: Carmen Krikorian

Editorial Assistants: Jennette ElNaggar, David Lutton

Cover Photos: PhotoAlto Agency

Cartoons: Rich Tennant (`www.the5thwave.com`)

Composition Services

Project Coordinator: Patrick Redmond

Layout and Graphics: Carl Byers, Reuben W. Davis, Mark Pinto

Proofreaders: Laura L. Bowman, Dwight Ramsey

Indexer: Christine Karpeles

Publishing and Editorial for Consumer Dummies

Diane Graves Steele, Vice President and Publisher, Consumer Dummies

Kristin Ferguson-Wagstaffe, Product Development Director, Consumer Dummies

Ensley Eikenburg, Associate Publisher, Travel

Kelly Regan, Editorial Director, Travel

Publishing for Technology Dummies

Andy Cummings, Vice President and Publisher, Dummies Technology/General User

Composition Services

Debbie Stailey, Director of Composition Services

Contents at a Glance

Table of Contents

Part III: Evaluating Determinants 173

Introduction

Linear algebra is usually the fledgling mathematician's first introduction to the *real* world of mathematics. "What?" you say. You're wondering what in tarnation you've been doing up to this point if it wasn't real mathematics. After all, you started with counting real numbers as a toddler and have worked your way through some really good stuff — probably even some calculus.

I'm not trying to diminish your accomplishments up to this point, but you've now ventured into that world of mathematics that sheds a new light on mathematical structure. All the tried-and-true rules and principles of arithmetic and algebra and trigonometry and geometry still apply, but linear algebra looks at those rules, dissects them, and helps you see them in depth.

You'll find, in linear algebra, that you can define your own set or grouping of objects — decide who gets to play the game by a particular, select criteria — and then determine who gets to stay in the group based on your standards. The operations involved in linear algebra are rather precise and somewhat limited. You don't have all the usual operations (such as addition, subtraction, multiplication, and division) to perform on the objects in your set, but that doesn't really impact the possibilities. You'll find new ways of looking at operations and use them in your investigations of linear algebra and the journeys into the different facets of the subject.

Linear algebra includes systems of equations, linear transformations, vectors and matrices, and determinants. You've probably seen most of these structures in different settings, but linear algebra ties them all together in such special ways.

About This Book

Linear algebra includes several different topics that can be investigated without really needing to spend time on the others. You really don't have to read this book from front to back (or even back to front!). You may be really, really interested in determinants and get a kick out of going through the chapters discussing them first. If you need a little help as you're reading the explanation on determinants, then I do refer you to the other places in the book where you find the information you may need. In fact, throughout this book, I send you scurrying to find more information on topics in other places. The layout of the book is logical and follows a plan, but my plan doesn't have to be your plan. Set your own route.

Conventions Used in This Book

You'll find the material in this book to be a helpful reference in your study of linear algebra. As I go through explanations, I use *italics* to introduce new terms. I define the words right then and there, but, if that isn't enough, you can refer to the glossary for more on that word and words close to it in meaning. Also, you'll find **boldfaced** text as I introduce a listing of characteristics or steps needed to perform a function.

What You're Not to Read

You don't have to read every word of this book to get the information you need. If you're in a hurry or you just want to get in and out, here are some pieces you can safely skip:

- **Sidebars:** Text in gray boxes are called sidebars. These contain interesting information, but they're not essential to understanding the topic at hand.
- **Text marked with the Technical Stuff icon:** For more on this icon, see "Icons Used in This Book," later in this Introduction.
- **The copyright page:** Unless you're the kind of person who reads the ingredients of every food you put in your mouth, you probably won't miss skipping this!

Foolish Assumptions

As I planned and wrote this book, I had to make a few assumptions about you and your familiarity with mathematics. I assume that you have a working knowledge of algebra and you've at least been exposed to geometry and trigonometry. No, you don't have to do any geometric proofs or measure any angles, but algebraic operations and grouping symbols are used in linear algebra, and I refer to geometric transformations such as rotations and reflections when working with the matrices. I do explain what's going on, but it helps if you have that background.

How This Book Is Organized

This book is divided into several different *parts*, and each part contains several chapters. Each chapter is also subdivided into sections, each with a unifying topic. It's all very organized and logical, so you should be able to go from section to section, chapter to chapter, and part to part with a firm sense of what you'll find when you get there.

The subject of linear algebra involves equations, matrices, and vectors, but you can't really separate them too much. Even though a particular section focuses on one or the other of the concepts, you find the other topics working their way in and getting included in the discussion.

Part I: Lining Up the Basics of Linear Algebra

In this part, you find several different approaches to organizing numbers and equations. The chapters on vectors and matrices show you rows and columns of numbers, all neatly arranged in an orderly fashion. You perform operations on the arranged numbers, sometimes with rather surprising results. The matrix structure allows for the many computations in linear algebra to be done more efficiently. Another basic topic is systems of equations. You find out how they're classified, and you see how to solve the equations algebraically or with matrices.

Part II: Relating Vectors and Linear Transformations

Part II is where you begin to see another dimension in the world of mathematics. You take nice, reasonable vectors and matrices and link them together with linear combinations. And, as if that weren't enough, you look at solutions of the vector equations and test for homogeneous systems. Don't get intimidated by all these big, impressive words and phrases I'm tossing around. I'm just giving you a hint as to what more you can do — some really interesting stuff, in fact.

Part III: Evaluating Determinants

A determinant is a function. You apply this function to a square matrix, and out pops the answer: a single number. The chapters in this part cover how to perform the determinant function on different sizes of matrices, how to change the matrices for more convenient computations, and what some of the applications of determinants are.

Part IV: Involving Vector Spaces

The chapters in this part get into the nitty-gritty details of vector spaces and their subspaces. You see how linear independence fits in with vector spaces. And, to top it all off, I tell you about eigenvalues and eigenvectors and how they interact with specific matrices.

Part V: The Part of Tens

The last three chapters are lists of ten items — with a few intriguing details for each item in the list. First, I list for you some of the many applications of matrices — some things that matrices are actually used for in the real world. The second chapter in this part deals with using your graphing calculator to work with matrices. Finally, I show you ten of the more commonly used Greek letters and what they stand for in mathematics and other sciences.

Icons Used in This Book

You undoubtedly see lots of interesting icons on the start-up screen of your computer. The icons are really helpful for quick entries and manipulations when performing the different tasks you need to do. What is very helpful with these icons is that they usually include some symbol that suggests what the particular program does. The same goes for the icons used in this book.

This icon alerts you to important information or rules needed to solve a problem or continue on with the explanation of the topic. The icon serves as a place marker so you can refer back to the item as you're reading through the material that follows. The information following the Remember icon is pretty much necessary for the mathematics involved in that section of the book.

The material following this icon is wonderful mathematics; it's closely related to the topic at hand, but it's not absolutely necessary for your understanding of the material. You can take it or leave it — whichever you prefer.

When you see this icon, you'll find something helpful or timesaving. It won't be earthshaking, but it'll keep you grounded.

The picture on this icon says it all. You should really pay attention when you see the Warning icon. I use it to alert you to a particularly serious pitfall or misconception. I don't use it too much, so you won't think I'm crying wolf when you do see it in a section.

Where to Go from Here

You really can't pick a bad place to dive into this book. If you're more interested in first testing the waters, you can start with vectors and matrices in Chapters 2 and 3, and see how they interact with one another. Another nice place to make a splash is in Chapter 4, where you discover different approaches to solving systems of equations. Then, again, diving right into transformations gives you more of a feel for how the current moves through linear algebra. In Chapter 8, you find linear transformations, but other types of transformations also make their way into the chapters in Part II. You may prefer to start out being a bit grounded with mathematical computations, so you can look at Chapter 9 on permutations, or look in Chapters 10 and 11, which explain how the determinants are evaluated. But if you're really into the high-diving aspects of linear algebra, then you need to go right to vector spaces in Part IV and look into eigenvalues and eigenvectors in Chapter 16. No matter what, you can change your venue at any time. Start or finish by diving or wading — there's no right or wrong way to approach this swimmin' subject.

Part I
Lining Up the Basics of Linear Algebra

In this part . . .

Welcome to L.A.! No, you're not in the sunny, rockin' state of California (okay, you *may* be), but regardless of where you live, you've entered the rollin' arena of linear algebra. Instead of the lights of Hollywood, I bring you the *de*lights of systems of equations. Instead of being mired in the La Brea Tar Pits, you get to *ad*mire matrices and vectors. Put on your shades, you're in for quite an adventure.

Chapter 1

Putting a Name to Linear Algebra

In This Chapter

- Aligning the algebra part of linear algebra with systems of equations
- Making waves with matrices and determinants
- Vindicating yourself with vectors
- Keeping an eye on eigenvalues and eigenvectors

The words *linear* and *algebra* don't always appear together. The word *linear* is an adjective used in many settings: *linear equations, linear regression, linear programming, linear technology,* and so on. The word *algebra,* of course, is familiar to all high school and most junior high students. When used together, the two words describe an area of mathematics in which some traditional algebraic symbols, operations, and manipulations are combined with vectors and matrices to create systems or structures that are used to branch out into further mathematical study or to use in practical applications in various fields of science and business.

The main elements of linear algebra are systems of linear equations, vectors and matrices, linear transformations, determinants, and vector spaces. Each of these topics takes on a life of its own, branching into its own special emphases and coming back full circle. And each of the main topics or areas is entwined with the others; it's a bit of a symbiotic relationship — the best of all worlds.

You can find the systems of linear equations in Chapter 4, vectors in Chapter 2, and matrices in Chapter 3. Of course, that's just the starting point for these topics. The uses and applications of these topics continue throughout the book. In Chapter 8, you get the big picture as far as linear transformations; determinants begin in Chapter 10, and vector spaces are launched in Chapter 13.

Solving Systems of Equations in Every Which Way but Loose

A *system of equations* is a grouping or listing of mathematical statements that are tied together for some reason. Equations may associate with one another because the equations all describe the relationships between two or more variables or unknowns. When studying systems of equations (see Chapter 4), you try to determine if the different equations or statements have any common solutions — sets of replacement values for the variables that make all the equations have the value of truth at the same time.

For example, the system of equations shown here consists of three different equations that are all true (the one side is equal to the other side) when $x = 1$ and $y = 2$.

$$\begin{cases} 2x + y = 4 \\ y^2 - x = 3 \\ \sqrt{x+3} = y \end{cases}$$

The only problem with the set of equations I've just shown you, as far as *linear* algebra is concerned, is that the second and third equations in the system are *not* linear.

A *linear equation* has the form $a_1x_1 + a_2x_2 + a_3x_3 + \ldots + a_nx_n = k$, where a_i is a real number, x_i is a variable, and k is some real constant.

Note that, in a linear equation, each of the variables has an exponent of exactly 1. Yes, I know that you don't see any exponents on the *x*s, but that's standard procedure — the 1s are assumed. In the system of equations I show you earlier, I used x and y for the variables instead of the subscripted *x*s. It's easier to write (or type) x, y, z, and so on when working with smaller systems than to use the subscripts on a single letter.

I next show you a system of *linear* equations. I'll use x, y, z, and w for the variables instead of x_1, x_2, x_3, and x_4.

$$\begin{cases} 2x + y - 3z + 4w = 11 \\ x + 3y - 2z + w = 5 \\ 3x + y + 2w = 13 \\ 4x + y + 5z - w = 17 \end{cases}$$

The system of four linear equations with four variables or unknowns *does* have a single solution. Each equation is true when $x = 1$, $y = 2$, $z = 3$, and $w = 4$. Now a caution: Not every system of linear equations has a solution. Some systems of equations have no solutions, and others have many or infinitely many solutions. What you find in Chapter 4 is how to determine which situation you have: none, one, or many solutions.

Systems of linear equations are used to describe the relationship between various entities. For example, you might own a candy store and want to create different selections or packages of candy. You want to set up a 1-pound box, a 2-pound box, a 3-pound box, and a diet-spoiler 4-pound box. Next I'm going to describe the contents of the different boxes. After reading through all the descriptions, you're going to have a greater appreciation for how nice and neat the corresponding equations are.

The four types of pieces of candy you're going to use are a nougat, a cream, a nut swirl, and a caramel. The 1-pounder is to contain three nougats, one cream, one nut swirl, and two caramels; the 2-pounder has three nougats, two creams, three nut swirls, and four caramels; the 3-pounder has four nougats, two creams, eight nut swirls, and four caramels; and the 4-pounder contains six nougats, five creams, eight nut swirls, and six caramels. What does each of these candies weigh?

Letting the weight of nougats be represented by x_1, the weight of creams be represented by x_2, the weight of nut swirls be represented by x_3, and the weight of caramels be represented by x_4, you have a system of equations looking like this:

$$\begin{cases} 3x_1 + 1x_2 + 1x_3 + 2x_4 = 16 \\ 3x_1 + 2x_2 + 3x_3 + 4x_4 = 32 \\ 4x_1 + 2x_2 + 8x_3 + 4x_4 = 48 \\ 6x_1 + 5x_2 + 8x_3 + 6x_4 = 64 \end{cases}$$

The pounds are turned to ounces in each case, and the solution of the system of linear equations is that $x_1 = 1$ ounce, $x_2 = 2$ ounces, $x_3 = 3$ ounces, and $x_4 = 4$ ounces. Yes, this is a very simplistic representation of a candy business, but it serves to show you how systems of linear equations are set up and how they work to solve complex problems. You solve such a system using algebraic methods or matrices. Refer to Chapter 4 if you want more information on how to deal with such a situation.

Systems of equations don't always have solutions. In fact, a single equation, all by itself, can have an infinite number of solutions. Consider the equation $2x + 3y = 8$. Using ordered pairs, (x,y), to represent the numbers you want, some of the solutions of the system are (1,2), (4,0), (-8,8), and (10,-4). But none of the solutions of the equation $2x + 3y = 8$ is also a solution of the equation $4x + 6y = 10$. You can try to find some matches, but there just aren't any. Some solutions of $4x + 6y = 10$ are (1,1), (4,−1), and (10,−5). Each equation has an infinite number of solutions, but no pairs of solutions match. So the system has no solution.

$$\begin{cases} 2x + 3y = 8 \\ 4x + 6y = 10 \end{cases}$$

Knowing that you don't have a solution is a very important bit of information, too.

Matchmaking by Arranging Data in Matrices

A matrix is a rectangular arrangement of numbers. Yes, all you see is a bunch of numbers — lined up row after row and column after column. Matrices are tools that eliminate all the fluff (such as those pesky variables) and set all the pertinent information in an organized logical order. (Matrices are introduced in Chapter 3, but you use them to solve systems of equations in Chapter 4.) When matrices are used for solving systems of equations, you find the coefficients of the variables included in a matrix and the variables left out. So how do you know what is what? You get organized, that's how.

Here's a system of four linear equations:

$$\begin{cases} x_1 - 2x_2 + 3x_3 - 6x_4 = 8 \\ x_1 \qquad - 10x_3 + x_4 = 11 \\ 4x_1 + x_2 \qquad - 4x_4 = 8 \\ 3x_1 + 5x_2 - 7x_3 \qquad = 6 \end{cases}$$

When working with this system of equations, you may use one matrix to represent all the coefficients of the variables.

$$A = \begin{bmatrix} 1 & -2 & 3 & -6 \\ 1 & 0 & -10 & 1 \\ 4 & 1 & 0 & -4 \\ 3 & 5 & -7 & 0 \end{bmatrix}$$

Notice that I placed a 0 where there was a missing term in an equation. If you're going to write down the coefficients only, you have to keep the terms in order according to the variable that they multiply and use *markers* or placeholders for missing terms. The coefficient matrix is so much easier to look at than the equation. But you have to follow the rules of order. And I named the matrix — nothing glamorous like *Angelina*, but something simple, like A.

When using coefficient matrices, you usually have them accompanied by two vectors. (A *vector* is just a one-dimensional matrix; it has one column and many rows or one row and many columns. See Chapters 2 and 3 for more on vectors.)

The vectors that correspond to this same system of equations are the vector of variables and the vector of constants. I name the vectors X and C.

$$X = \begin{bmatrix} x_1 \\ x_2 \\ x_3 \\ x_4 \end{bmatrix} \quad C = \begin{bmatrix} 8 \\ 11 \\ 8 \\ 6 \end{bmatrix}$$

Once in matrix and vector form, you can perform operations on the matrices and vectors individually or perform operations involving one operating on the other. All that good stuff is found beginning in Chapter 2.

Let me show you, though, a more practical application of matrices and why putting the numbers (coefficients) into a matrix is so handy. Consider an insurance agency that keeps track of the number of policies sold by the different agents each month. In my example, I'll keep the number of agents and policies small, and let you imagine how massive the matrices become with a large number of agents and different variations on policies.

At Pay-Off Insurance Agency, the agents are Amanda, Betty, Clark, and Dennis. In January, Amanda sold 15 auto insurance policies, 10 dwelling/home insurance policies, 5 whole-life insurance policies, 9 tenant insurance policies, and 1 health insurance policy. Betty sold . . . okay, this is already getting drawn out. I'm putting all the policies that the agents sold in January into a matrix.

$$\begin{array}{r} \\ \text{Amanda} \\ \text{Betty} \\ \text{Clark} \\ \text{Dennis} \end{array} \begin{array}{c} \begin{array}{ccccc} A & D & W & T & H \end{array} \\ \begin{bmatrix} 15 & 10 & 5 & 9 & 1 \\ 10 & 9 & 4 & 9 & 2 \\ 20 & 0 & 0 & 23 & 1 \\ 15 & 6 & 10 & 6 & 5 \end{bmatrix} \end{array}$$

If you were to put the number of policies from January, February, March, and so on in matrices, it's a simple task to perform *matrix addition* and get totals for the year. Also, the commissions to agents can be computed by performing *matrix multiplication*. For example, if the commissions on these policies are flat rates — say $110, $200, $600, $60, and $100, respectively, then you create a vector of the payouts and multiply.

$$\begin{array}{r} \\ \text{Amanda} \\ \text{Betty} \\ \text{Clark} \\ \text{Dennis} \end{array} \begin{array}{c} \begin{array}{ccccc} \text{A} & \text{D} & \text{W} & \text{T} & \text{H} \end{array} \\ \begin{bmatrix} 15 & 10 & 5 & 9 & 1 \\ 10 & 9 & 4 & 9 & 2 \\ 20 & 0 & 0 & 23 & 1 \\ 15 & 6 & 10 & 6 & 5 \end{bmatrix} \end{array} * \begin{bmatrix} 110 \\ 200 \\ 600 \\ 60 \\ 100 \end{bmatrix} = \begin{bmatrix} 7{,}290 \\ 6{,}040 \\ 3{,}680 \\ 9{,}710 \end{bmatrix} \begin{array}{l} \text{Amanda} \\ \text{Betty} \\ \text{Clark} \\ \text{Dennis} \end{array}$$

This matrix addition and matrix multiplication business is found in Chapter 3. Other processes for the insurance company that could be performed using matrices are figuring the percent increases or decreases of sales (of the whole company or individual salespersons) by performing operations on summary vectors, determining commissions by multiplying totals by their respective rates, setting percent increase goals, and so on. The possibilities are limited only by your lack of imagination, determination, or need.

Valuating Vector Spaces

In Part IV of this book, you find all sorts of good information and interesting mathematics all homing in on the topic of vector spaces. In other chapters, I describe and work with vectors. Sorry, but there's not really any separate chapter on *spaces* or space — I leave that to the astronomers. But the words *vector space* are really just a mathematical expression used to define a particular group of elements that exist under a particular set of conditions. (You can find information on the properties of vector spaces in Chapter 13.)

Think of a vector space in terms of a game of billiards. You have all the elements (the billiards balls) that are confined to the top of the table (well, they stay there if hit properly). Even when the billiard balls interact (bounce off one another), they stay somewhere on the tabletop. So the billiard balls are the elements of the vector space and the table top is that vector space. You have operations that cause actions on the table — hitting a ball with a cue stick or a ball being hit by another ball. And you have rules that govern how all the actions can occur. The actions keep the billiard balls on the table (in the vector space). Of course, a billiards game isn't nearly as exciting as a vector space, but I wanted to relate some real-life action to the confinement of elements and rules.

A vector space is linear algebra's version of a type of classification plan or design. Other areas in mathematics have similar entities (classifications and designs). The common theme of such designs is that they contain a set or grouping of objects that all have something in common. Certain properties are attached to the plan — properties that apply to all the members of the grouping. If all the members must abide by the rules, then you can make judgments or conclusions based on just a few of the members rather than having to investigate every single member (if that's even possible).

Vector spaces contain vectors, which really take on many different forms. The easiest form to show you is an actual vector, but the vectors may actually be matrices or polynomials. As long as these different forms follow the rules, then you have a vector space. (In Chapter 14, you see the rules when investigating the subspaces of vector spaces.)

The rules regulating a vector space are highly dependent on the operations that belong to that vector space. You find some new twists to some familiar operation notation. Instead of a simple plus sign, +, you find $\oplus$. And the multiplication symbol, ×, is replaced with $\otimes$. The new, revised symbols are used to alert you to the fact that you're not in Kansas anymore. With vector spaces, the operation of addition may be defined in a completely different way. For example, you may define the vector addition of two elements, x and y, to be $x \oplus y = 2x + y$. Does that rule work in a vector space? That's what you need to determine when studying vector spaces.

Determining Values with Determinants

A *determinant* is tied to a matrix, as you see in Chapter 10. You can think of a determinant as being an operation that's performed on a matrix. The determinant incorporates all the elements of a matrix into its grand plan. You have a few qualifications to meet, though, before performing the operation *determinant*.

Square matrices are the only candidates for having a determinant. Let me show you just a few examples of matrices and their determinants. The matrix A has a determinant |A| — which is also denoted det (A) — and so do matrices B and C.

$$A = \begin{bmatrix} 1 & 2 & 5 \\ -3 & 8 & 0 \\ 2 & -1 & 1 \end{bmatrix}, \det(A) = |A| = -51$$

$$B = \begin{bmatrix} -2 & 3 \\ 5 & -8 \end{bmatrix}, \det(B) = |B| = 1$$

$$C = \begin{bmatrix} 4 \end{bmatrix}, \det(C) = |C| = 4$$

The matrices A, B, and C go from a 3×3 matrix to a 2×2 matrix to a 1×1 matrix. The determinants of the respective matrices go from complicated to simple to compute. I give you all the gory details on computing determinants in Chapter 10, so I won't go into any of the computations here, but I do want to introduce you to the fact that these square matrices are connected, by a particular function, to single numbers.

All square matrices have determinants, but some of these determinants don't amount to much (the determinant equals 0). Having a determinant of 0 isn't a big problem to the matrix, but the value 0 causes problems with some of the applications of matrices and determinants. A common property that all these 0-determinant matrices have is that they don't have a multiplicative inverse.

For example, the matrix D, that I show you here, has a determinant of 0 and, consequently, no inverse.

$$D = \begin{bmatrix} 1 & 2 & 5 \\ -3 & 8 & -4 \\ -2 & 10 & 1 \end{bmatrix}, \det(D) = |D| = 0$$

Matrix D looks perfectly respectable on the surface, but, lurking beneath the surface, you have what could be a big problem when using the matrix to solve problems. You need to be aware of the consequences of the determinant being 0 and make arrangements or adjustments that allow you to proceed with the solution.

For example, determinants are used in Chapter 12 with Cramer's rule (for solving systems of equations). The values of the variables are ratios of different determinants computed from the coefficients in the equations. If the determinant in the denominator of the ratio is zero, then you're out of luck, and you need to pursue the solution using an alternate method.

Zeroing In on Eigenvalues and Eigenvectors

In Chapter 16, you see how eigenvalues and eigenvectors correspond to one another in terms of a particular matrix. Each eigenvalue has its related eigenvector. So what are these eigen-things?

First, the German word *eigen* means *own*. The word *own* is somewhat descriptive of what's going on with eigenvalues and eigenvectors. An eigenvalue is a number, called a *scalar* in this linear algebra setting. And an eigenvector is an $n \times 1$ vector. An eigenvalue and eigenvector are related to a particular $n \times n$ matrix.

For example, let me reach into the air and pluck out the number 13. Next, I take that number 13 and multiply it times a 2×1 vector. You'll see in Chapter 2 that multiplying a vector by a scalar just means to multiply each element in the vector by that number. For now, just trust me on this.

$$13 * \begin{bmatrix} 3 \\ 4 \end{bmatrix} = \begin{bmatrix} 39 \\ 52 \end{bmatrix}$$

That didn't seem too exciting, so let me up the ante and see if this next step does more for you. Again, though, you'll have to take my word for the multiplication step. I'm now going to multiply the same vector that just got multiplied by 13 by a matrix.

$$\begin{bmatrix} 1 & 9 \\ 12 & 4 \end{bmatrix} * \begin{bmatrix} 3 \\ 4 \end{bmatrix} = \begin{bmatrix} 39 \\ 52 \end{bmatrix}$$

The resulting vector is the same whether I multiply the vector by 13 or by the matrix. (You can find the hocus-pocus needed to do the multiplication in Chapter 3.) I just want to make a point here: Sometimes you can find a single number that will do the same job as a complete matrix. You can't just pluck the numbers out of the air the way I did. (I actually peeked.) Every matrix has its *own* set of *eigenvalues* (the numbers) and *eigenvectors* (that get multiplied by the eigenvalues). In Chapter 16, you see the full treatment — all the steps and procedures needed to discover these elusive entities.

Chapter 2

The Value of Involving Vectors

In This Chapter

- Relating two-dimensional vectors to all $n \times 1$ vectors
- Illustrating vector properties with rays drawn on axes
- Demonstrating operations performed on vectors
- Making magnitude meaningful
- Creating and measuring angles

The word *vector* has a very specific meaning in the world of mathematics and linear algebra. A vector is a special type of *matrix* (rectangular array of numbers). The vectors in this chapter are columns of numbers with brackets surrounding them. Two-space and three-space vectors are drawn on two axes and three axes to illustrate many of the properties, measurements, and operations involving vectors.

You may find the discussion of vectors to be both limiting and expanding — at the same time. Vectors seem limiting, because of the restrictive structure. But they're also expanding because of how the properties delineated for the simple vector are then carried through to larger groupings and more general types of number arrays.

As with any mathematical presentation, you find very specific meanings for otherwise everyday words (and some not so everyday). Keep track of the words and their meanings, and the whole picture will make sense. Lose track of a word, and you can fall back to the glossary or italicized definition.

Describing Vectors in the Plane

A *vector* is an ordered collection of numbers. Vectors containing two or three numbers are often represented by *rays* (a line segment with an arrow on one end and a point on the other end). Representing vectors as rays works with two or three numbers, but the ray loses its meaning when you deal with larger vectors and numbers. Larger vectors exist, but I can't draw you a nice picture of them. The properties that apply to smaller vectors also apply to larger vectors, so I introduce you to the vectors that have pictures to help make sense of the entire set.

When you create a vector, you write the numbers in a column surrounded by brackets. Vectors have names (no, not John Henry or William Jacob). The names of vectors are usually written as single, boldfaced, lowercase letters. You often see just one letter used for several vectors when the vectors are related to one another, and subscripts attached to distinguish one vector from another: $\mathbf{u}_1$, $\mathbf{u}_2$, $\mathbf{u}_3$, and so on.

Here, I show you four of my favorite vectors, named **u, v, w,** and **x:**

$$\mathbf{u} = \begin{bmatrix} 3 \\ -1 \end{bmatrix}, \mathbf{v} = \begin{bmatrix} 8 \\ 4 \\ 0 \end{bmatrix}, \mathbf{w} = \begin{bmatrix} -1 \\ 0 \\ 0 \\ 3 \end{bmatrix}, \mathbf{x} = \begin{bmatrix} 4 \\ 0 \\ -3 \\ 2 \\ 1 \end{bmatrix}$$

The size of a vector is determined by its rows or how many numbers it has. Technically, a vector is a *column matrix* (matrices are covered in great detail in Chapter 3), meaning that you have just one column and a certain number of rows. In this example, vector **u** is 2×1, **v** is 3×1, **w** is 4×1, and **x** is 5×1, meaning that **u** has two rows and one column, **v** has three rows and one column, and so on.

Homing in on vectors in the coordinate plane

Vectors that are 2×1 are said to belong to R^2, meaning that they belong to the set of *real* numbers that come paired two at a time. The capital R is used to emphasize that you're looking at *real* numbers. You also say that 2×1 vectors, or vectors in R^2, are a part of *two-space.*

Vectors in two-space are represented on the coordinate (x,y) plane by rays. In *standard position,* the ray representing a vector has its endpoint at the origin and its terminal point (or arrow) at the (x,y) coordinates designated by the column vector. The x coordinate is in the first row of the vector, and the y coordinate is in the second row. The following vectors are shown with their respective terminal points written as ordered pairs, (x,y):

$$\begin{bmatrix} 1 \\ 4 \end{bmatrix} \to (1,4), \begin{bmatrix} 2 \\ -3 \end{bmatrix} \to (2,-3), \begin{bmatrix} -5 \\ 3 \end{bmatrix} \to (-5,3),$$
$$\begin{bmatrix} -1 \\ -5 \end{bmatrix} \to (-1,-5), \begin{bmatrix} 0 \\ 4 \end{bmatrix} \to (0,4), \begin{bmatrix} -3 \\ 0 \end{bmatrix} \to (-3,0)$$

Displaying 2×1 vectors in standard position

In *standard position,* 2×1 vectors have their endpoints at the origin, at the point (0,0). The coordinate axes are used, with the horizontal x-axis and vertical y-axis. Figure 2-1 shows the six vectors listed in the preceding section, drawn in their standard positions. The coordinates of the terminal points are indicated on the graph.

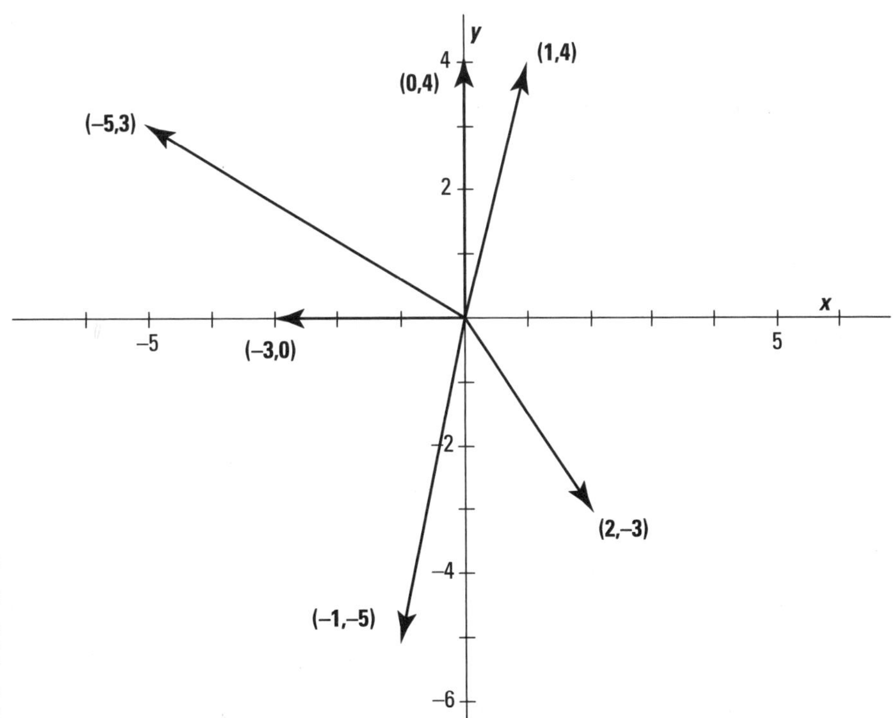

Figure 2-1: Vectors represented by rays.

Broadening your scope to vectors leaving their origin

You aren't limited to always drawing 2×1 vectors radiating from the origin. The following vector is just as correctly drawn by starting with the point (–1,4) as an endpoint, and then drawing the vector by moving two units to the right and three units down, ending up with the terminal point at (1,1).

$$\begin{bmatrix} 2 \\ -3 \end{bmatrix}$$

Figure 2-2 shows you the vector in this alternate position. (See "Adding and subtracting vectors," later in this chapter, for more on what's actually happening and how to compute the new endpoint.)

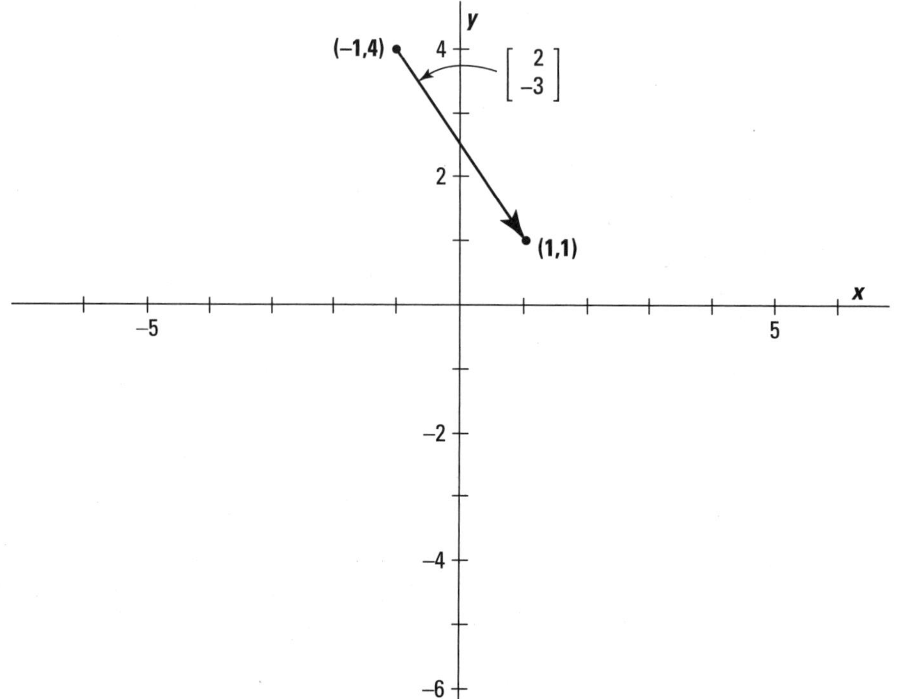

Figure 2-2: Drawing a vector in nonstandard position.

Notice, in Figures 2-1 and 2-2, that the following vector has the same length and points in the same direction with the same slant:

$$\begin{bmatrix} 2 \\ -3 \end{bmatrix}$$

The length of a vector is also called its *magnitude.* Both the length and the direction uniquely determine a vector and allow you to tell if one vector is *equal* to another vector.

Vectors can actually have any number of rows. You just can't draw pictures to illustrate the vectors that have more than three entries. Also, the applications for vectors involving hundreds of entries are rather limited and difficult to work with, except on computers. All the properties that apply to two-space and three-space vectors also apply to larger vectors; so you'll find that most of my examples use vectors of a more manageable size.

Adding a dimension with vectors out in space

Vectors in R^3 are said to be in *three-space.* The vectors representing three-space are column matrices with three entries or numbers in them. The R part of R^3 indicates that the vector involves real numbers.

Three-space vectors are represented by three-dimensional figures and arrows pointing to positions in space. Vectors in three-space are represented by column vectors with dimension 3×1. When they're drawn on the standard x, y, and z axes, the y-axis moves left and right, the x-axis seems to come off the page, and the z-axis moves up and down. Picture a vector drawn in three-space as being a diagonal drawn from one corner of a box to the opposite corner. A ray representing the following vector is shown in Figure 2-3 with the endpoint at the origin and the terminal point at (2,3,4).

$$\begin{bmatrix} 2 \\ 3 \\ 4 \end{bmatrix}$$

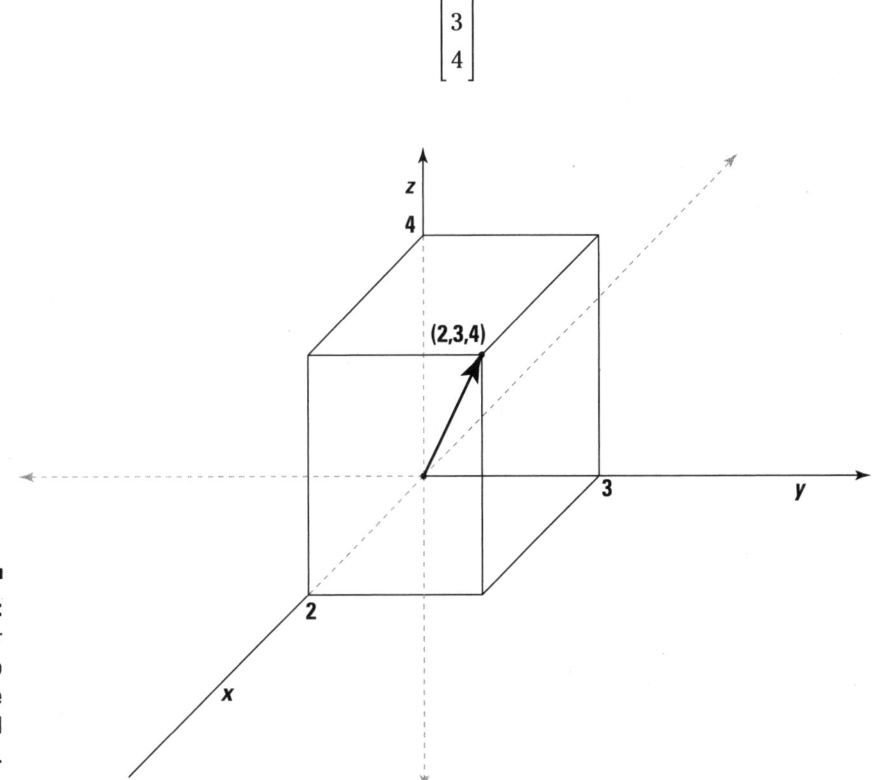

Figure 2-3: The vector seems to come off the page toward you.

Defining the Algebraic and Geometric Properties of Vectors

The two-space and three-space vectors are nice to refer to when defining the properties involving vectors. Vectors are groupings of numbers just waiting to have operations performed on them — ending up with predictable results. The different geometric transformations performed on vectors include rotations, reflections, expansions, and contractions. You find the rotations and reflections in Chapter 8, where larger matrices are also found. As far as operations on vectors, you add vectors together, subtract them, find their opposite, or multiply by a *scalar* (constant number). You can also find an *inner product* — multiplying each of the respective elements together.

Swooping in on scalar multiplication

Scalar multiplication is one of the two basic operations performed on vectors that preserves the original format. When you multiply a 2 × 1 vector by a scalar, your result is another 2 × 1 vector. You may not be all that startled by this revelation, but you really should appreciate the fact that the scalar maintains its original dimension. Such preservation isn't necessarily the rule in mathematics, as you'll see when doing inner products, later in this chapter.

Reading the recipe for multiplying by a scalar

A scalar is a real number — a constant value. Multiplying a vector by a scalar means that you multiply each element in the vector by the same constant value that appears just outside of and in front of the vector.

Multiplying vector **v** by scalar k, you multiply each element in the vector, v_i, by the scalar k:

$$k\begin{bmatrix} v_1 \\ v_2 \\ v_3 \\ \vdots \\ v_n \end{bmatrix} = \begin{bmatrix} kv_1 \\ kv_2 \\ kv_3 \\ \vdots \\ kv_n \end{bmatrix}$$

Here's an example:

$$5\begin{bmatrix} 2 \\ -3 \\ 0 \end{bmatrix} = \begin{bmatrix} 5\cdot 2 \\ 5(-3) \\ 5\cdot 0 \end{bmatrix} = \begin{bmatrix} 10 \\ -15 \\ 0 \end{bmatrix}$$

You started with a 3 × 1 vector and ended up with a 3 × 1 vector.

Opening your eyes to dilation and contraction of vectors

Vectors have operations that cause *dilations* (expansions) and *contractions* (shrinkages) of the original vector. Both operations of dilation and contraction are accomplished by multiplying the elements in the vector by a scalar.

If the scalar, k, that is multiplying a vector is greater than 1, then the result is a dilation of the original vector. If the scalar, k, is a number between 0 and 1, then the result is a contraction of the original vector. (Multiplying by exactly 1 is multiplying by the *identity* — it doesn't change the vector at all.)

For example, consider the following vector:

$$\begin{bmatrix} 2 \\ 4 \end{bmatrix}$$

If you multiply the vector by 3, you have a dilation; if you multiply it by $^1/_2$, you have a contraction.

$$3\begin{bmatrix} 2 \\ 4 \end{bmatrix} = \begin{bmatrix} 6 \\ 12 \end{bmatrix}, \quad \frac{1}{2}\begin{bmatrix} 2 \\ 4 \end{bmatrix} = \begin{bmatrix} 1 \\ 2 \end{bmatrix}$$

In Figure 2-4, you see the results of the dilation and contraction on the original vector.

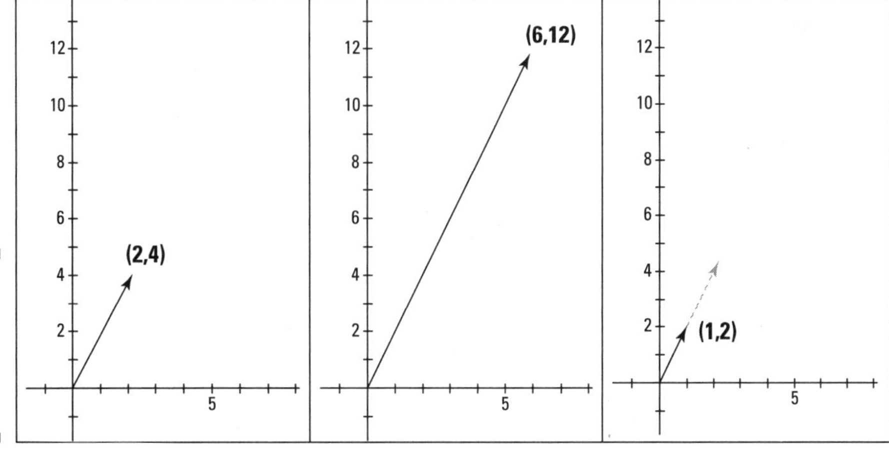

Figure 2-4: Dilating and contracting a vector in two-space.

As you can see, the length of the vector is affected by the scalar multiplication, but the direction or angle with the *axis* (slant) is not changed. You also may have wondered why I only multiplied by numbers greater than 0. The rule for contractions of vectors involves numbers between 0 and 1, nothing smaller. In the next section, I pursue the negative numbers and 0.

Multiplying by zero and getting something for your efforts

It probably comes as no surprise to you that multiplying a vector by the number 0 results in a vector that has all zeros for its entries. The illustration for multiplying by 0 in two-space is a single point or dot. Not unexpected.

But don't dismiss this all-zero vector. The vector created by having all zeros for entries is very important; it's called the *zero vector* (another big surprise). The zero vector is the identity for vector addition, just as the number 0 is the identity for the addition of real numbers. When you add the zero vector to another vector, the second vector keeps its identity — it doesn't change. (For more on adding vectors, see the "Adding and subtracting vectors" section, later in this chapter.)

$$0\begin{bmatrix}8\\-3\end{bmatrix}=\begin{bmatrix}0\\0\end{bmatrix},\ 0\begin{bmatrix}5\\2\\0\\-6\end{bmatrix}=\begin{bmatrix}0\\0\\0\\0\end{bmatrix}$$

Having a negative attitude about scalars

When multiplying a 2×1 or 3×1 vector by a negative scalar, you see two things happen:

- The size of the vector changes (except with –1).
- The direction of the vector reverses.

These properties hold for larger vectors, but I can't show you pictures of them. You'll see how multiplying by a negative scalar affects magnitude later in this chapter.

When you multiply a vector by –2, as shown with the following vector, each element in the vector changes and has a greater absolute value:

$$-2\begin{bmatrix}-2\\3\\5\end{bmatrix}=\begin{bmatrix}4\\-6\\-10\end{bmatrix}$$

In Figure 2-5, you see the original vector as a diagonal in a box moving upward and away from the page and the resulting vector in a larger box moving downward and toward you.

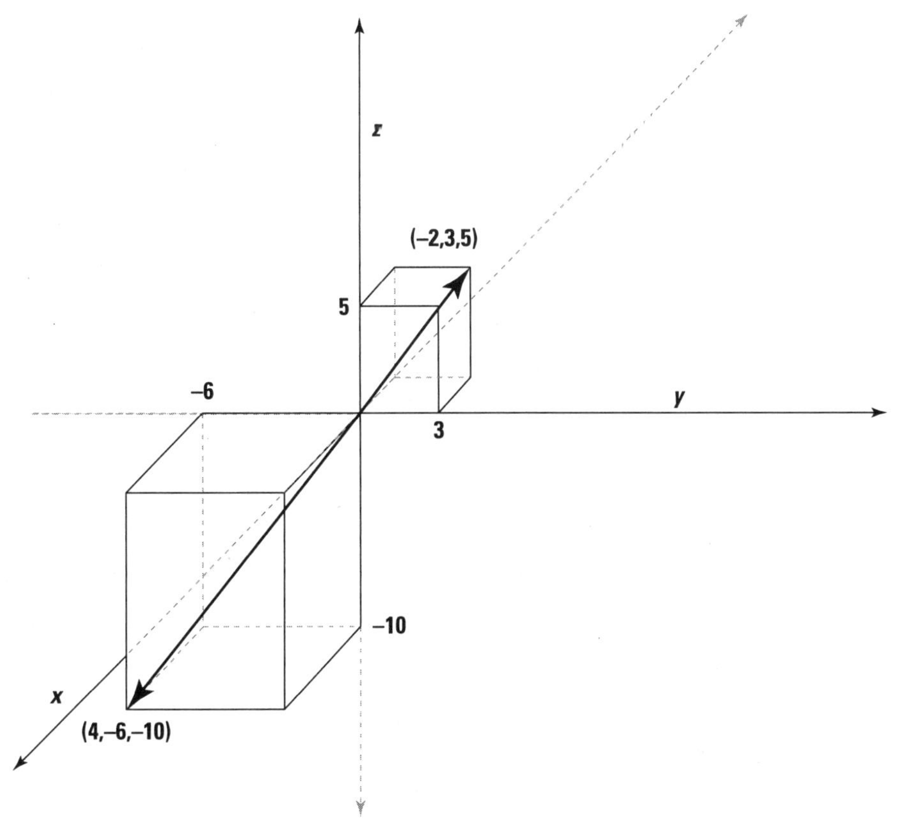

Figure 2-5: Multiplying a vector by a negative scalar.

Adding and subtracting vectors

Vectors are added to one another and subtracted from one another with just one stipulation: The vectors have to be the same size. The process of adding or subtracting vectors involves adding or subtracting the corresponding elements in the vectors, so you need to have a one-to-one match-up for the operations.

To add or subtract two $n \times 1$ vectors, you add or subtract the corresponding elements of the vectors:

$$\mathbf{u}+\mathbf{v}=\begin{bmatrix}u_1\\u_2\\u_3\\\vdots\\u_n\end{bmatrix}+\begin{bmatrix}v_1\\v_2\\v_3\\\vdots\\v_n\end{bmatrix}=\begin{bmatrix}u_1+v_1\\u_2+v_2\\u_3+v_3\\\vdots\\u_n+v_n\end{bmatrix} \text{ and } \mathbf{u}-\mathbf{v}=\begin{bmatrix}u_1\\u_2\\u_3\\\vdots\\u_n\end{bmatrix}-\begin{bmatrix}v_1\\v_2\\v_3\\\vdots\\v_n\end{bmatrix}=\begin{bmatrix}u_1-v_1\\u_2-v_2\\u_3-v_3\\\vdots\\u_n-v_n\end{bmatrix}$$

Illustrating vector addition

Look at what happens when you do the following vector addition:

$$\begin{bmatrix} 2 \\ 4 \end{bmatrix} + \begin{bmatrix} 3 \\ 1 \end{bmatrix} = \begin{bmatrix} 2+3 \\ 4+1 \end{bmatrix} = \begin{bmatrix} 5 \\ 5 \end{bmatrix}$$

Graphing the first vector,

$$\begin{bmatrix} 2 \\ 4 \end{bmatrix}$$

on the coordinate axes, and then adding the vector

$$\begin{bmatrix} 3 \\ 1 \end{bmatrix}$$

to the terminal point of the first vector, your end result is the vector obtained from vector addition. Figure 2-6 shows all three vectors.

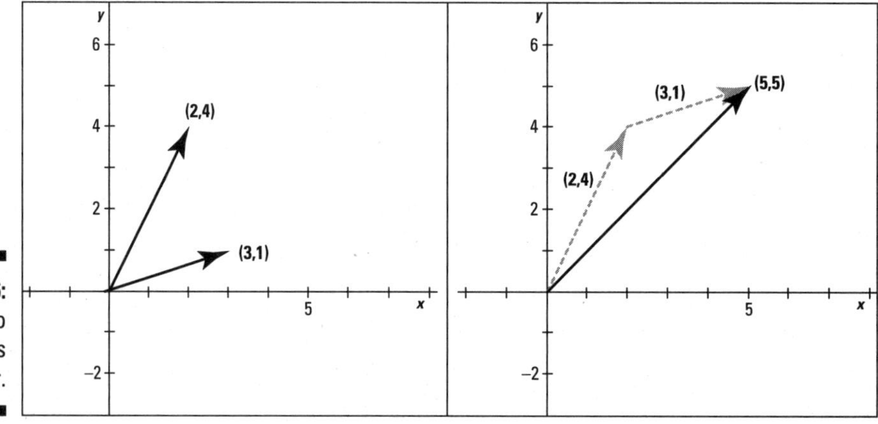

Figure 2-6: Adding two vectors together.

Adding nothing changes nothing

When you add the zero (identity) vector to another vector, you don't change anything:

$$\begin{bmatrix} 4 \\ 5 \\ 6 \end{bmatrix} + \begin{bmatrix} 0 \\ 0 \\ 0 \end{bmatrix} = \begin{bmatrix} 0 \\ 0 \\ 0 \end{bmatrix} + \begin{bmatrix} 4 \\ 5 \\ 6 \end{bmatrix} = \begin{bmatrix} 4 \\ 5 \\ 6 \end{bmatrix}$$

Whether the zero vector comes first or last in the addition problem, you maintain the original vector.

Revisiting with vector subtraction

Vector subtraction is just another way of saying that you're adding one vector to a second vector that's been multiplied by the scalar –1. So, if you want to change a subtraction problem to an addition problem (perhaps to change the order of the vectors in the operation), you rewrite the second vector in the problem in terms of its opposite.

For example, changing the following subtraction problem to an addition problem, and rewriting the order, you have:

$$\begin{bmatrix} 5 \\ -2 \\ 3 \end{bmatrix} - \begin{bmatrix} 3 \\ 1 \\ -5 \end{bmatrix} = \begin{bmatrix} 5 \\ -2 \\ 3 \end{bmatrix} + (-1)\begin{bmatrix} 3 \\ 1 \\ -5 \end{bmatrix} = \begin{bmatrix} 5 \\ -2 \\ 3 \end{bmatrix} + \begin{bmatrix} -3 \\ -1 \\ 5 \end{bmatrix} = \begin{bmatrix} -3 \\ -1 \\ 5 \end{bmatrix} + \begin{bmatrix} 5 \\ -2 \\ 3 \end{bmatrix}$$

Yes, of course the answers come out the same whether you subtract or change the second vector to its opposite. The maneuvers shown here are for the structure or order of the problem and are used in various applications of vectors.

Managing a Vector's Magnitude

The magnitude of a vector is also referred to as its *length* or *norm.* In two-space or three-space, you see a ray that has a particular length and can visualize where the magnitude's numerical value comes from. Vectors with more than three rows also have magnitude, and the computation is the same no matter what the size of the vector. I just can't show you a picture.

The *magnitude* of vector **v** is designated with two sets of vertical lines, $\|\mathbf{v}\|$, and the formula for computing the magnitude is

$$\|\mathbf{v}\| = \sqrt{v_1^2 + v_2^2 + \cdots + v_n^2}$$

where $v_1, v_2, \ldots, v_n$ are the elements of vector **v.**

So, for example, for the vector

$$\mathbf{v} = \begin{bmatrix} 3 \\ 2 \\ 4 \end{bmatrix}$$

the magnitude is found with:

$$\|\mathbf{v}\| = \sqrt{3^2 + 2^2 + 4^2} = \sqrt{9 + 4 + 16} = \sqrt{29} \approx 5.4$$

Think of this example in terms of a right-rectangular prism (a cardboard box). The box measures 3 x 2 x 4 feet. How long a rod can you fit in the box, diagonally? According to the formula for the magnitude of the vector whose numbers are the dimensions of the box, you can place a rod measuring about 5.4 feet into that box, from corner to corner.

Adjusting magnitude for scalar multiplication

The magnitude of a vector is determined by squaring each element in the vector, finding the sum of the squares, and then computing the square root of that sum. What happens to the magnitude of a vector, though, if you multiply it by a scalar? Can you predict the magnitude of the new vector without going through all the computation if you have the magnitude of the original vector?

Consider the following vector:

$$\mathbf{v} = \begin{bmatrix} 1 \\ -4 \\ -2 \\ 2 \end{bmatrix}$$

When you compute the magnitude, you get

$$\|\mathbf{v}\| = \sqrt{1^2 + (-4)^2 + (-2)^2 + 2^2} = \sqrt{1 + 16 + 4 + 4} = \sqrt{25} = 5$$

Now multiply the vector **v** by $k = 3$, $3\mathbf{v}$, and compute the magnitude of the resulting vector.

$$3\mathbf{v} = 3\begin{bmatrix} 1 \\ -4 \\ -2 \\ 2 \end{bmatrix} = \begin{bmatrix} 3 \\ -12 \\ -6 \\ 6 \end{bmatrix} \text{ and } \|3\mathbf{v}\| = \sqrt{3^2 + (-12)^2 + (-6)^2 + 6^2}$$

$$= \sqrt{9 + 144 + 36 + 36} = \sqrt{225} = 15$$

The magnitude of the new vector is three times that of the original. So it looks like all you have to do is multiply the original magnitude by the scalar to get the new magnitude.

Careful there! In mathematics, you need to be suspicious of results where someone gives you a bunch of numbers and declares that, because one example works, they all do. But in this case, it's true that, with just a bit of an adjustment to the rule, the magnitude is just a multiple of the original.

The magnitude of the product of a scalar, k, and a vector, $\mathbf{v}$, is equal to the absolute value of the scalar and the magnitude of the original vector, $|k|\cdot\|\mathbf{v}\|$.

And now, to show you that I'm not fooling, here's how the math works:

$$k\begin{bmatrix} v_1 \\ v_2 \\ v_3 \\ \vdots \\ v_n \end{bmatrix} = \begin{bmatrix} kv_1 \\ kv_2 \\ kv_3 \\ \vdots \\ kv_n \end{bmatrix} \text{ so } \|k\mathbf{v}\| = \sqrt{(kv_1)^2 + (kv_2)^2 + \cdots + (kv_n)^2}$$

$$\begin{aligned} &= \sqrt{k^2v_1^2 + k^2v_2^2 + \cdots + k^2v_n^2} \\ &= \sqrt{k^2\left(v_1^2 + v_2^2 + \cdots + v_n^2\right)} \\ &= \sqrt{k^2}\sqrt{\left(v_1^2 + v_2^2 + \cdots + v_n^2\right)} \\ &= |k|\sqrt{v_1^2 + v_2^2 + \cdots + v_n^2} \\ &= |k|\cdot\|\mathbf{v}\| \end{aligned}$$

The magnitude of $k\mathbf{v}$ is equal to the absolute value of k times the magnitude of the original vector, $\mathbf{v}$.

The square root of a square is equal to the absolute value of the number that's being squared:

$$\sqrt{k^2} = |k|$$

Using the absolute value takes care of the instances when k is a negative number.

So, if you multiply a vector by a negative number, the value of the magnitude of the resulting vector is still going to be a positive number.

For example, if

$$\mathbf{v} = \begin{bmatrix} -9 \\ 6 \\ -2 \end{bmatrix}$$

has the following magnitude

$$\|\mathbf{v}\| = \sqrt{(-9)^2 + 6^2 + (-2)^2} = \sqrt{81 + 36 + 4} = \sqrt{121} = 11$$

then multiplying the vector, **v**, by –8, the magnitude of the vector –8**v** is $|-8|(11) = 8(11) = 88$.

Making it all right with the triangle inequality

When dealing with the addition of vectors, a property arises involving the sum of the vectors. The theorem involving vectors, their magnitudes, and the sum of their magnitudes is called the *triangle inequality* or the *Cauchy-Schwarz inequality* (named for the mathematicians responsible).

For any vectors **u** and **v**, the following, which says that the magnitude of the sum of vectors is always less than or equal to the sum of the magnitudes of the vectors, holds:

$$\|\mathbf{u} + \mathbf{v}\| \leq \|\mathbf{u}\| + \|\mathbf{v}\|$$

Showing the inequality for what it is

In Figure 2-7, you see two vectors, **u** and **v**, with terminal points (x_1,y_1) and (x_2,y_2), respectively. The triangle inequality theorem says that the magnitude of the vector resulting from adding two vectors together is either smaller or sometimes the same as the sum of the magnitudes of the two vectors being added together.

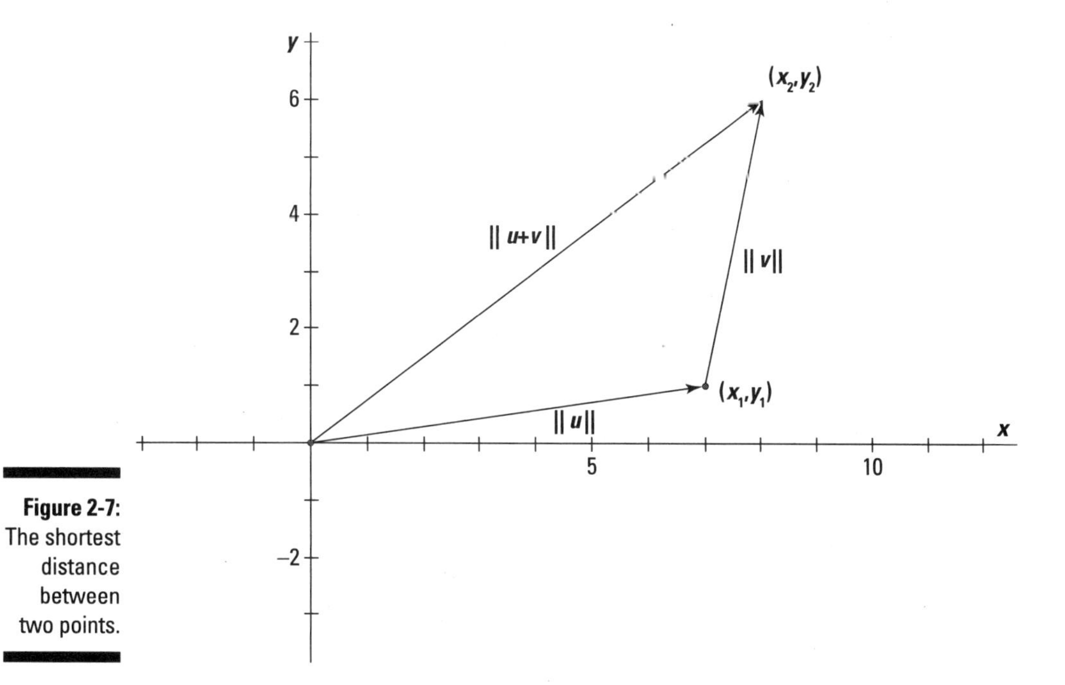

Figure 2-7: The shortest distance between two points.

For example, I illustrate the triangle inequality with the addition of the two vectors (and, just to make them *random,* I use the birthdates of my first two children):

$$\mathbf{u}+\mathbf{v}=\begin{bmatrix}1\\10\\79\end{bmatrix}+\begin{bmatrix}9\\19\\81\end{bmatrix}=\begin{bmatrix}10\\29\\160\end{bmatrix}$$

$$\left\|\mathbf{u}+\mathbf{v}\right\|=\left\|\begin{bmatrix}10\\29\\160\end{bmatrix}\right\|=\sqrt{10^2+29^2+160^2}=\sqrt{26{,}541}\approx 162.9$$

$$\left\|\mathbf{u}\right\|+\left\|\mathbf{v}\right\|=\left\|\begin{bmatrix}1\\10\\79\end{bmatrix}\right\|+\left\|\begin{bmatrix}9\\19\\81\end{bmatrix}\right\|=\sqrt{1^2+10^2+79^2}+\sqrt{9^2+19^2+81^2}$$
$$=\sqrt{6{,}342}+\sqrt{7{,}003}\approx 79.6+83.7=163.3$$

You see the magnitude of the sum first, $\|\mathbf{u}+\mathbf{v}\|$. Then I compare the magnitude to the sum of the two separate magnitudes. The sums are mighty close, but the magnitude of the sum is smaller, as expected.

Using the Cauchy-Schwarz inequality in averages

The *arithmetic mean* of two numbers is the value that most people think of when they hear *average.* You find the average of two numbers by adding them together and dividing by two. The *geometric mean* of two numbers is another number between the two such that there's a common ratio between (1) the first number and the mean and (2) the mean and the second number.

I'll explain the concept of geometric mean with an example. The geometric mean of 4 and 9 is 6, because the ratio $4/6$ is the same as the ratio $6/9$. Both ratios are equal to $2/3$. To find a geometric mean of two numbers, you just determine the square root of the product of the numbers.

The geometric mean of a and b is

$$\sqrt{ab}$$

while the arithmetic mean is

$$\frac{a+b}{2}$$

For an example of how the arithmetic and geometric means of two numbers compare, consider the two numbers 16 and 25. The arithmetic mean is the sum of the two numbers divided by two: $(16 + 25) \div 2 = 20.5$. The geometric mean is the square root of the product of the numbers.

$$\sqrt{16 \cdot 25} = \sqrt{400} = 20$$

In this example, the geometric mean is slightly smaller than the arithmetic mean.

In fact, the geometric mean is *never* larger than the arithmetic mean — the geometric mean is always smaller than, or the same as, the arithmetic mean. I show you why this is so by using two very carefully selected vectors, **u** and **v**, which have elements that illustrate my statement.

First, let

$$\mathbf{u} = \begin{bmatrix} \sqrt{a} \\ \sqrt{b} \end{bmatrix} \text{ and } \mathbf{v} = \begin{bmatrix} \sqrt{b} \\ \sqrt{a} \end{bmatrix}$$

Assume, also, that both a and b are positive numbers. You're probably wondering why I'm choosing such *awkward* entries for the vectors, but you'll soon see the reason for the choices.

Apply the triangle inequality to the two vectors:

$$\|\mathbf{u} + \mathbf{v}\| \le \|\mathbf{u}\| + \|\mathbf{v}\|$$

$$\left\| \begin{bmatrix} \sqrt{a} + \sqrt{b} \\ \sqrt{b} + \sqrt{a} \end{bmatrix} \right\| \le \left\| \begin{bmatrix} \sqrt{a} \\ \sqrt{b} \end{bmatrix} \right\| + \left\| \begin{bmatrix} \sqrt{b} \\ \sqrt{a} \end{bmatrix} \right\|$$

$$\sqrt{\left(\sqrt{a} + \sqrt{b}\right)^2 + \left(\sqrt{b} + \sqrt{a}\right)^2} \le \sqrt{\left(\sqrt{a}\right)^2 + \left(\sqrt{b}\right)^2} + \sqrt{\left(\sqrt{b}\right)^2 + \left(\sqrt{a}\right)^2}$$

$$\sqrt{2\left(\sqrt{a} + \sqrt{b}\right)^2} \le 2\sqrt{a + b}$$

You see that the beginning and ending inequalities are equivalent.

To get to the last step, I used the commutative property of addition on the left (changing the order) and found that I had two of the same term. On the right, I took the square roots of the squares (and, since both *a* and *b* are positive, I don't need absolute value) and found that the two terms are also alike.

Now I square both sides of the inequality, divide each side by 2, square the binomial, distribute the 2, and simplify by subtracting *a* and *b* from each side:

$$\left(\sqrt{2\left(\sqrt{a} + \sqrt{b}\right)^2}\right)^2 \le \left(2\sqrt{a + b}\right)^2$$

$$2\left(\sqrt{a} + \sqrt{b}\right)^2 \le 4\left(a + b\right)$$

$$\left(\sqrt{a} + \sqrt{b}\right)^2 \le 2\left(a + b\right)$$

$$a + 2\sqrt{ab} + b \le 2a + 2b$$

$$2\sqrt{ab} \le a + b$$

$$\sqrt{ab} \le \frac{a + b}{2}$$

See! The geometric mean of the two numbers, *a* and *b,* is less than or equal to the arithmetic mean of the same two numbers.

Getting an inside scoop with the inner product

The *inner product* of two vectors is also called its *dot product.* When **u** and **v** are both $n \times 1$ vectors, then the notation $\mathbf{u}^T\mathbf{v}$ indicates the *inner product* of **u** and **v.**

Augustin Louis Cauchy: A name found on many theorems

Augustin Louis Cauchy was a French mathematician, born in Paris in 1789, a few weeks after the fall of the Bastille. His birth, during a time of political upheaval, seemed to set the tone for the rest of his life.

Cauchy was home-schooled by his father, before going on to a university-level education. His family was often visited by mathematicians of the day — notably Joseph Louis Lagrange and Pierre-Simon Laplace — who encouraged the young prodigy to be exposed to languages, first, and then mathematics.

Cauchy's first profession was as a military engineer, but he "saw the light" and later abandoned engineering for mathematics. At a time when most jobs or positions for mathematicians were as professors at universities, Cauchy found it difficult to find such a position because of his outspoken religious and political stands. He even relinquished a position at the École Polytechnique because he wouldn't take the necessary oaths imposed by the French government.

At one point, Cauchy responded to a request from the then-deposed king, Charles X, to tutor his grandson. The experience wasn't particularly pleasant or successful — perhaps due to Cauchy's yelling and screaming at the prince. Cauchy was raised in a political environment and was rather political and opinionated. He was, more often than not, rather difficult in his dealings with other mathematicians. No one could take away his wonderful reputation, though, for his mathematical work and rigorous proofs, producing almost 800 mathematical papers. Cauchy was quick to publish his findings (unlike some mathematicians, who tended to sit on their discoveries), perhaps because of an advantage that he had as far as getting his work in print. He was married to Aloise de Bure, the close relative of a publisher.

You find the *inner product* of two $n \times 1$ vectors by multiplying their corresponding entries together and adding up all the products:

$$\mathbf{u} = \begin{bmatrix} u_1 \\ u_2 \\ \vdots \\ u_n \end{bmatrix}, \mathbf{v} = \begin{bmatrix} v_1 \\ v_2 \\ \vdots \\ v_n \end{bmatrix}$$

$$\mathbf{u}^T\mathbf{v} = \begin{bmatrix} u_1 & u_2 & \cdots & u_n \end{bmatrix} \begin{bmatrix} v_1 \\ v_2 \\ \vdots \\ v_n \end{bmatrix} = u_1v_1 + u_2v_2 + \cdots + u_nv_n$$

You probably noticed that the vector **u** was written horizontally, instead of vertically. The superscript T in the notation $\mathbf{u}^T\mathbf{v}$ means to *transpose* the vector **u,** to change its orientation. This is merely a notation technicality and tries to make it clear what's happening with the operation. You'll see more on *transposing* matrices in Chapter 3.

So, for example, if

$$\mathbf{u} = \begin{bmatrix} 4 \\ -2 \\ 0 \\ 1 \end{bmatrix} \text{ and } \mathbf{v} = \begin{bmatrix} -1 \\ -3 \\ 1 \\ 5 \end{bmatrix}$$

then the dot product is

$$\mathbf{u}^T\mathbf{v} = \begin{bmatrix} 4 & -2 & 0 & 1 \end{bmatrix} \begin{bmatrix} -1 \\ -3 \\ 1 \\ 5 \end{bmatrix} = 4(-1)+(-2)(-3)+0(1)+1(5) = -4+6+0+5 = 7$$

Also, you're probably wondering why in the world you'd ever want to compute an inner product of two vectors. The reason should become crystal clear in the following section.

Making it right with angles

Two lines, segments, or planes are said to be *orthogonal* if they're perpendicular to one another. Consider the two vectors shown in Figure 2-8, which are drawn perpendicular to one another and form a 90-degree angle (or right angle) where their endpoints meet. You can confirm that the rays forming the vectors are perpendicular to one another, using some basic algebra, because their slopes are negative reciprocals.

The slope of a line is determined by finding the difference between the y-coordinates of two points on the line and dividing that difference by the difference between the corresponding x-coordinates of the points on that line.

$$m = \frac{y_2 - y_1}{x_2 - x_1}$$

And, further, two lines are perpendicular (form a right angle) if the product of their slopes is –1. (Then they're negative reciprocals of one another.)

$$m_1 \cdot m_2 = -1 \text{ or } m_1 = -\frac{1}{m_2}$$

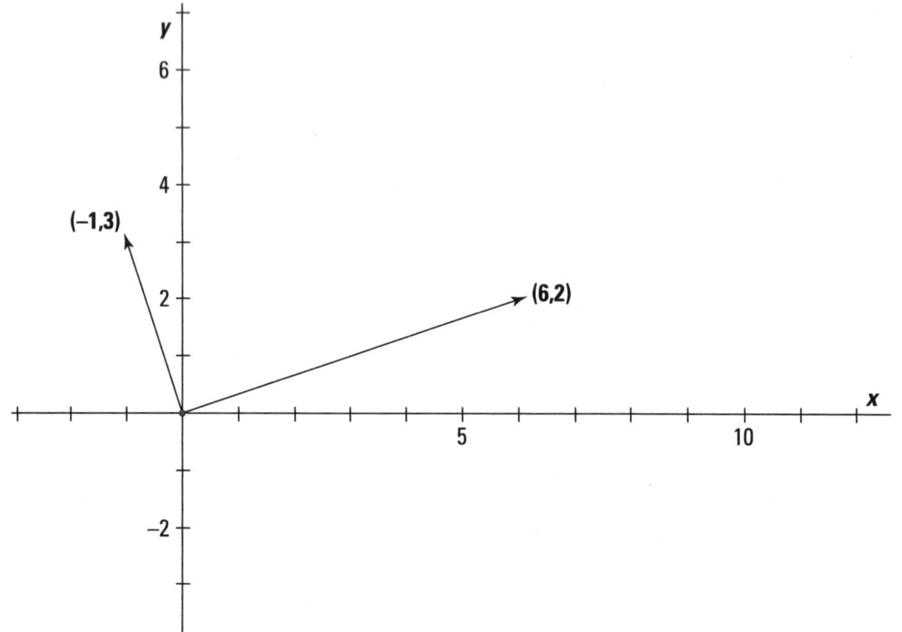

Figure 2-8: The two rays form a right angle.

So what does this have to do with vectors and their orthogonality? (Sure, that's your question!) Read on.

Determining orthogonality (right angles)

When dealing with vectors, you determine if they're perpendicular to one another by finding their inner product. If the inner product of vectors **u** and **v** is equal to 0, then the vectors are perpendicular.

Referring to the two vectors in Figure 2-8, you have

$$\mathbf{u} = \begin{bmatrix} 6 \\ 2 \end{bmatrix}, \mathbf{v} = \begin{bmatrix} -1 \\ 3 \end{bmatrix}$$

Now, finding their inner product,

$$\mathbf{u}^T\mathbf{v} = \begin{bmatrix} 6 & 2 \end{bmatrix}\begin{bmatrix} -1 \\ 3 \end{bmatrix} = 6(-1) + 2(3) = -6 + 6 = 0$$

Since the inner product is equal to 0, the rays must be perpendicular to one another and form a right angle.

Determining whether two vectors are perpendicular is just fine and dandy, but you may be even more interested in what the measure of the angle between two rays is when they're *not* perpendicular (see the following section).

Finding the angle between two vectors in two-space

A very nice relationship between the inner product of two vectors, the magnitude of the vectors, and the angle between the vectors is that:

$$\mathbf{u} \cdot \mathbf{v} = \|\mathbf{u}\| \cdot \|\mathbf{v}\| \cos\theta$$

where **u** and **v** are two vectors in two-space, $\mathbf{u} \cdot \mathbf{v}$ is the inner product of the two vectors, $\|\mathbf{u}\|$ and $\|\mathbf{v}\|$ are the respective magnitudes of the vectors, and $\cos\theta$ is the measure (counterclockwise) of the angle formed between the two vectors.

So you can solve for the measure of angle θ by first finding the cosine:

$$\frac{\mathbf{u} \cdot \mathbf{v}}{\|\mathbf{u}\| \cdot \|\mathbf{v}\|} = \cos\theta$$

and then determining the angle θ.

In Figure 2-9, you see the two vectors whose terminal points are (2,6) and (–1,5). The angle between the two vectors is the angle θ.

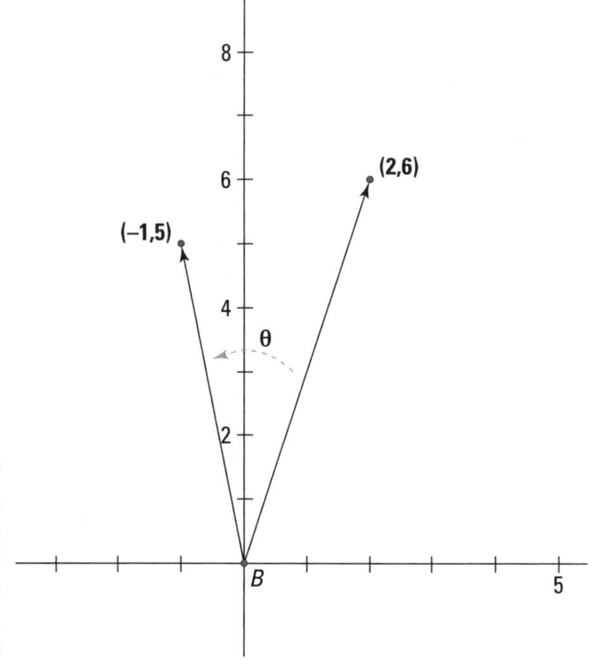

Figure 2-9: The angle is measured counter-clockwise.

Find the inner product of the two vectors and the magnitudes of each. Then put the numbers in their respective places in the formula:

$$\mathbf{u} = \begin{bmatrix} 2 \\ 6 \end{bmatrix}, \mathbf{v} = \begin{bmatrix} -1 \\ 5 \end{bmatrix}$$

$$\mathbf{u} \cdot \mathbf{v} = \begin{bmatrix} 2 & 6 \end{bmatrix} \begin{bmatrix} -1 \\ 5 \end{bmatrix} = 2(-1) + 6(5) = 28$$

$$\|\mathbf{u}\| = \sqrt{2^2 + 6^2} = \sqrt{40}, \quad \|\mathbf{v}\| = \sqrt{(-1)^2 + 5^2} = \sqrt{26}$$

$$\frac{\mathbf{u} \cdot \mathbf{v}}{\|\mathbf{u}\| \cdot \|\mathbf{v}\|} = \frac{28}{\sqrt{40} \cdot \sqrt{26}} = \frac{28}{\sqrt{1{,}040}} \approx 0.8682$$

Using either a calculator or table of trigonometric functions, you find that the angle whose cosine is closest to 0.8682 is an angle of about 29.75 degrees. The angle formed by the two vectors is close to a 30-degree angle.

Chapter 3

Mastering Matrices and Matrix Algebra

In This Chapter

- Making note of matrix vocabulary and notation
- Operating on matrices every which way
- Categorizing matrices by size and format
- Recognizing properties of matrices and matrix operations
- Finding inverses of invertible matrices

Matrices are essentially rectangular arrays of numbers. The reason that you'd ever even consider putting numbers in rectangular arrays is to give some order or arrangement to them so that they can be studied or used in an application. Matrices have their own arithmetic. What you think of when you hear *multiplication* has just a slight resemblance to matrix multiplication. When working with matrices, you'll find just enough familiarity to let you feel as though you haven't left the world of mathematics as you know it — and you'll find just enough differences to let you see the possibilities beyond. Matrix algebra has identities, inverses, and operations. You'll find all these good things and more in this chapter.

Getting Down and Dirty with Matrix Basics

A matrix is made up of some rows and columns of numbers — a rectangular array of numbers. You have the same number of numbers in each row and the same number of numbers in each column. The number of rows and columns in a matrix does not have to be the same. A vector is a matrix with just one column and one or more rows; a vector is also called a *column vector.* (Turn to Chapter 2 for all you'd ever want to know about vectors.)

Becoming familiar with matrix notation

I show you here four matrices, A, B, C, and D. Matrices are generally named so you can distinguish one matrix from another in a discussion or text. Nice, simple, capital letters are usually the names of choice for matrices:

$$A = \begin{bmatrix} 2 & 3 \\ 6 & 0 \end{bmatrix} \quad B = \begin{bmatrix} 1 & -1 & 2 & 0 & 0 & 3 \\ -4 & 8 & 3 & 2 & 0 & 5 \\ 2 & 0 & 1 & 0 & 20 & 4 \\ -3 & 6 & 8 & 5 & 9 & 0 \end{bmatrix}$$

$$C = \begin{bmatrix} 1 \\ 2 \\ 3 \\ 4 \end{bmatrix} \quad D = \begin{bmatrix} 3 & 5 & 6 \end{bmatrix}$$

Matrix A has two rows and two columns, and Matrix B has four rows and six columns. The rectangular arrays of numbers are surrounded by a bracket to indicate that this is a mathematical structure called a matrix. The different positions or values in a matrix are called *elements*. The elements themselves are named with lowercase letters with subscripts. The subscripts are the *index* of the element. The element a_{12} is in matrix A and is the number in the first row and second column. Looking at matrix A, you see that $a_{12} = 3$. A general notation for the elements in a matrix A is a_{ij} where i represents the row and j represents the column. In matrix B, you refer to the elements with b_{ij}. In matrix B, the element $b_{35} = 20$.

Sometimes a rule or pattern is used to construct a particular matrix. The rule may be as simple as "Every element is four times the first subscript plus three," or "Every element is the sum of the subscripts in its index." To avoid confusion or misrepresentation, you use symbols to describe a rule for constructing the elements of a matrix.

For example, if you want to construct a matrix with three rows and four columns in which each element is the sum of the digits in its index, you write:

A is a 3×4 matrix where $a_{ij} = i + j$.

I explain the $\times$ sign in the next section. And here's matrix A:

$$A = \begin{bmatrix} 1+1 & 1+2 & 1+3 & 1+4 \\ 2+1 & 2+2 & 2+3 & 2+4 \\ 3+1 & 3+2 & 3+3 & 3+4 \end{bmatrix} = \begin{bmatrix} 2 & 3 & 4 & 5 \\ 3 & 4 & 5 & 6 \\ 4 & 5 & 6 & 7 \end{bmatrix}$$

And now I show you matrix B which is a 3×3 matrix where $b_{ij} = i^2 - 2j$.

$$B = \begin{bmatrix} 1^2 - 2(1) & 1^2 - 2(2) & 1^2 - 2(3) \\ 2^2 - 2(1) & 2^2 - 2(2) & 2^2 - 2(3) \\ 3^2 - 2(1) & 3^2 - 2(2) & 3^2 - 2(3) \end{bmatrix} = \begin{bmatrix} -1 & -3 & -5 \\ 2 & 0 & -2 \\ 7 & 5 & 3 \end{bmatrix}$$

Defining dimension

Matrices come in all sizes or *dimensions.* The dimension gives the number of rows, followed by a multiplication sign, followed by the number of columns. Matrix A is a 2×2 matrix, because it has 2 rows and 2 columns. Matrix B has 4 rows and 6 columns, so it's dimension is 4×6. Matrix C is a *column* matrix, because it has just one column; its dimension is 4×1. And D is a *row* matrix with dimension 1×3.

Determining the dimension of a matrix is important when performing operations involving more than one matrix. When adding or subtracting matrices, the two matrices need to have the same dimension. When multiplying matrices, the number of columns in the first matrix has to match the number of columns in the second matrix. You find more on adding, subtracting, multiplying, dividing, and finding inverses of matrices later in this chapter. And each operation requires paying attention to dimension.

Putting Matrix Operations on the Schedule

Matrix operations are special operations defined specifically for matrices. When you do *matrix addition,* you use the traditional process of addition of numbers, but the operation has special requirements and specific rules. *Matrix multiplication* is actually a combination of multiplication and addition. The operations and rules aren't difficult; you just have to follow the rules for the particular processes carefully.

Adding and subtracting matrices

Adding and subtracting matrices requires that the two matrices involved have the same dimension. The process of adding or subtracting two matrices involves performing the respective operation on each pair of corresponding elements.

To add or subtract two $m \times n$ matrices, you add (or subtract) each corresponding element in the matrices.

$$\mathrm{A} + \mathrm{B} = \begin{bmatrix} a_{11}+b_{11} & a_{12}+b_{12} & \cdots & a_{1n}+b_{1n} \\ a_{21}+b_{21} & a_{22}+b_{22} & \cdots & a_{2n}+b_{2n} \\ \vdots & \vdots & \ddots & \vdots \\ a_{m1}+b_{m1} & a_{m2}+b_{m2} & \cdots & a_{mn}+b_{mn} \end{bmatrix}$$

$$\mathrm{A} - \mathrm{B} = \begin{bmatrix} a_{11}-b_{11} & a_{12}-b_{12} & \cdots & a_{1n}-b_{1n} \\ a_{21}-b_{21} & a_{22}-b_{22} & \cdots & a_{2n}-b_{2n} \\ \vdots & \vdots & \ddots & \vdots \\ a_{m1}-b_{m1} & a_{m2}-b_{m2} & \cdots & a_{mn}-b_{mn} \end{bmatrix}$$

For example, if matrix J represents the number of life insurance policies, car insurance policies, and homeowner's insurance policies sold by eight different agents in January, and matrix F represents the number of polices sold by those same agents in February, then J + F represents the total number of polices for each agent. The *matrices* (rectangular arrangements) always have the same type of policy in each column and the same agents in each row.

$$\begin{matrix} & \begin{matrix} \text{LI} & \text{CI} & \text{HI} \end{matrix} & & \begin{matrix} \text{LI} & \text{CI} & \text{HI} \end{matrix} & & \begin{matrix} \text{LI} & \text{CI} & \text{HI} \end{matrix} \\ \mathrm{J} + \mathrm{F} = & \begin{bmatrix} 12 & 23 & 14 \\ 20 & 29 & 38 \\ 3 & 6 & 10 \\ 15 & 12 & 2 \\ 90 & 5 & 16 \\ 40 & 40 & 40 \\ 0 & 0 & 83 \\ 16 & 26 & 39 \end{bmatrix} & + & \begin{bmatrix} 13 & 33 & 22 \\ 10 & 19 & 8 \\ 30 & 0 & 20 \\ 0 & 0 & 0 \\ 60 & 15 & 26 \\ 30 & 40 & 50 \\ 0 & 0 & 69 \\ 12 & 48 & 11 \end{bmatrix} & = & \begin{bmatrix} 25 & 56 & 36 \\ 30 & 48 & 46 \\ 33 & 6 & 30 \\ 15 & 12 & 2 \\ 150 & 20 & 42 \\ 70 & 80 & 90 \\ 0 & 0 & 152 \\ 28 & 74 & 50 \end{bmatrix} \end{matrix}$$

The rectangular array allows the sales manager to quickly observe any trends or patterns or problems with the production of the salespersons.

Matrix addition is commutative. *Commutativity* means that you can reverse the order, add F + J, and get the same results. Matrix subtraction, however, is *not* commutative. The same way that 4 – 1 doesn't equal 1 – 4, the subtraction J – F isn't the same as F – J.

Scaling the heights with scalar multiplication

Multiplying two matrices together takes some doing — perhaps like climbing the Matterhorn. But *scalar multiplication* is a piece of cake — more like riding the tram to the top of yonder hill. I don't mean to imply that scalar multiplication isn't just as important. You can't have matrix arithmetic without scalar multiplication. I just wanted to set you straight before proceeding.

Multiplying a matrix A by a *scalar* (constant number), *k*, means to multiply every element in matrix A by the number *k* out in front of the matrix.

$$kA = k\begin{bmatrix} a_{11} & a_{12} & \cdots & a_{1n} \\ a_{21} & a_{22} & \cdots & a_{2n} \\ \vdots & \vdots & \ddots & \vdots \\ a_{m1} & a_{m2} & \cdots & a_{mn} \end{bmatrix} = \begin{bmatrix} ka_{11} & ka_{12} & \cdots & ka_{1n} \\ ka_{21} & ka_{22} & \cdots & ka_{2n} \\ \vdots & \vdots & \ddots & \vdots \\ ka_{m1} & ka_{m2} & \cdots & ka_{mn} \end{bmatrix}$$

So, multiplying some matrix A by –4,

$$A = \begin{bmatrix} 2 & -3 & -4 & 0 \\ -3 & 1 & -1 & 5 \\ 4 & 0 & -6 & -7 \end{bmatrix}$$

$$-4A = -4\begin{bmatrix} 2 & -3 & -4 & 0 \\ -3 & 1 & -1 & 5 \\ 4 & 0 & -6 & -7 \end{bmatrix} = \begin{bmatrix} -8 & 12 & 16 & 0 \\ 12 & -4 & 4 & -20 \\ -16 & 0 & 24 & 28 \end{bmatrix}$$

Making matrix multiplication work

Matrix multiplication actually involves two different operations: multiplication and addition. Elements in the respective matrices are aligned carefully, multiplied, added, and then the grand sum is placed carefully into the resulting matrix. Matrix multiplication is only performed when the two matrices involved meet very specific standards. You can't just multiply any two matrices together. Quite often, you can't even multiply a matrix times itself.

To distinguish *matrix multiplication* from the multiplication of real numbers, you'll see the asterisk, * , whenever two matrices are multiplied together. So multiplying matrix A by matrix B is designated A * B.

Determining which matrices and what order

To perform matrix multiplication on two matrices, the number of columns in the first matrix must be the same as the number of rows in the second matrix.

To perform matrix multiplication on $m \times n$ matrix A and $p \times q$ matrix B, it is necessary that $n = p$. Furthermore, the resulting matrix A * B has dimension $m \times q$.

For example, if matrix A has dimension 3×4 and matrix B has dimension 4×7, then the product A * B has dimension 3×7. But you can't multiply the matrices in the reverse order. The product B * A cannot be performed, because matrix B has seven columns and matrix A has three rows.

Multiplying two matrices

The process used when multiplying two matrices together is to add up a bunch of products. Each element in the new matrix created by matrix multiplication is the sum of all the products of the elements in a row of the first matrix times a column in the second matrix. For example, if you're finding the element a_{23} in the product matrix A, you get that element by multiplying all the elements in the second row of the first matrix times the elements in the third column of the second matrix and then adding up the products.

Let me show you an example before giving the rule symbolically. The example involves multiplying the 2×3 matrix K times the 3×4 matrix L.

$$K = \begin{bmatrix} 2 & -3 & 0 \\ 1 & 4 & -4 \end{bmatrix}, \; L = \begin{bmatrix} 1 & 2 & 3 & 4 \\ 5 & 6 & 7 & 6 \\ 5 & 4 & 3 & 2 \end{bmatrix}$$

The number of columns in matrix K is 3, as is the number of rows in matrix L. The resulting matrix, K * L, has dimension 2×4.

$$K * L = \begin{bmatrix} a_{11} & a_{12} & a_{13} & a_{14} \\ a_{21} & a_{22} & a_{23} & a_{24} \end{bmatrix} = A$$

Now, to find each element in the product matrix, K * L, you multiply the elements in the rows of K by the elements in the columns of L and add up the products:

- The element a_{11} is found by multiplying the elements in the first row of K by the elements in the first column of L: $2(1) + (-3)(5) + 0(5) = 2 - 15 + 0 = -13$

- The element a_{12} is found by multiplying the elements in the first row of K by the elements in the second column of L: 2(2) + (–3)(6) + 0(4) = 4 – 18 + 0 = –14
- The element a_{13} is found by multiplying the elements in the first row of K by the elements in the third column of L: 2(3) + (–3)(7) + 0(3) = 6 – 21 + 0 = –15
- The element a_{14} is found by multiplying the elements in the first row of K by the elements in the fourth column of L: 2(4) + (–3)(6) + 0(2) = 8 – 18 + 0 = –10
- The element a_{21} is found by multiplying the elements in the second row of K by the elements in the first column of L: 1(1) + 4(5) + (–4)(5) = 1 + 20 – 20 = 1
- The element a_{22} is found by multiplying the elements in the second row of K by the elements in the second column of L: 1(2) + 4(6) + (–4)(4) = 2 + 24 – 16 = 10
- The element a_{23} is found by multiplying the elements in the second row of K by the elements in the third column of L: 1(3) + 4(7) + (–4)(3) = 3 + 28 – 12 = 19
- The element a_{24} is found by multiplying the elements in the second row of K by the elements in the fourth column of L: 1(4) + 4(6) + (–4)(2) = 4 + 24 – 8 = 20

Placing the elements in their correct positions,

$$K * L = \begin{bmatrix} -13 & -14 & -15 & -10 \\ 1 & 10 & 19 & 20 \end{bmatrix}$$

To perform the matrix multiplication, A * B, where matrix A has dimension $m \times n$ and matrix B has dimension $n \times p$, the elements in the resulting $m \times p$ matrix A * B are:

$$A * B = \begin{bmatrix} a_{11}b_{11} + a_{12}b_{21} + \cdots + a_{1m}b_{m1} & a_{11}b_{12} + a_{12}b_{22} + \cdots a_{1m}b_{m2} & \cdots \\ a_{21}b_{11} + a_{22}b_{21} + \cdots + a_{2m}b_{m1} & a_{21}b_{12} + a_{22}b_{22} + \cdots a_{2m}b_{m2} & \cdots \\ \vdots & \vdots & \ddots \end{bmatrix}$$

In general, matrix multiplication is not commutative. Even when you have two square matrices (the number of rows and number of columns are the same), their product usually is not the same when the matrices are reversed. For example, matrices C and D are both 3 × 3 matrices. I compute C * D and then D * C.

$$C = \begin{bmatrix} 4 & -3 & 2 \\ 0 & 1 & -1 \\ 5 & 4 & 0 \end{bmatrix} \text{ and } D = \begin{bmatrix} 2 & 2 & -5 \\ 3 & 1 & 0 \\ -1 & 1 & 4 \end{bmatrix}$$

$$C^*D = \begin{bmatrix} 4\cdot2+(-3)3+2(-1) & 4\cdot2+(-3)1+2(1) & 4(-5)+(-3)0+2(4) \\ 0\cdot2+1\cdot3+(-1)(-1) & 0\cdot2+1\cdot1+(-1)1 & 0(-5)+1\cdot0+(-1)4 \\ 5\cdot2+4\cdot3+0(-1) & 5\cdot2+4\cdot1+0\cdot1 & 5(-5)+4\cdot0+0\cdot4 \end{bmatrix}$$

$$= \begin{bmatrix} -3 & 7 & -12 \\ 4 & 0 & -4 \\ 22 & 14 & -25 \end{bmatrix}$$

$$D^*C = \begin{bmatrix} 2\cdot4+2\cdot0+(-5)5 & 2(-3)+2\cdot1+(-5)4 & 2\cdot2+2(-1)+(-5)0 \\ 3\cdot4+1\cdot0+0\cdot5 & 3(-3)+1\cdot1+0\cdot4 & 3\cdot2+1(-1)+0\cdot0 \\ -1\cdot4+1\cdot0+4\cdot5 & -1(-3)+1\cdot1+4\cdot4 & -1\cdot2+1(-1)+4\cdot0 \end{bmatrix}$$

$$= \begin{bmatrix} -17 & -24 & 2 \\ 12 & -8 & 5 \\ 16 & 20 & -3 \end{bmatrix}$$

So, even though, in arithmetic, multiplication is commutative, this is not the case with matrix multiplication. Having said that, I have to tell you that there are cases where matrix multiplication *is* commutative. Here are the special cases in which matrix multiplication is commutative:

- When multiplying by the multiplicative identity on a square matrix
- When multiplying by the inverse of the matrix — if the matrix does, indeed, have an inverse

I refer to commutativity again, later, in the "Identifying with identity matrices," "Tackling the properties under multiplication," and "Establishing the properties of an invertible matrix" sections.

Putting Labels to the Types of Matrices

Matrices are traditionally named using capital letters. So you have matrices A, B, C, and so on. Matrices are also identified by their structure or elements; you identity matrices by their characteristics just as you identify people by their height or age or country of origin. Matrices can be square, identity, triangular, singular — or not.

Identifying with identity matrices

The two different types of identity matrices are somewhat related to the two identity numbers in arithmetic. The additive identity in arithmetic is 0. If you add 0 to a number, you don't change the number — the number keeps its identity. The same idea works for the multiplicative identity: The multiplicative identity in arithmetic is 1. You multiply any number by 1, and the number keeps its original identity.

Zeroing in on the additive identity

The additive identity for matrices is the *zero matrix.* A zero matrix has elements that are all zero. How convenient! But the zero matrix takes on many shapes and sizes. The additive identity for a 3×4 matrix is a 3×4 zero matrix. Matrices are added together only when they have the same dimension. When adding numbers in arithmetic, you have just one 0. But in matrix addition, you have more than one 0 — in fact, you have an infinite number of them (technically). Sorta neat.

In addition to having many additive identities — one for each size matrix — you also have commutativity of addition when using the zero matrix. Addition is commutative, anyway, so extending commutativity to the zero matrix should come as no surprise.

Matrices A, B, and C are added to their zero matrices:

$$A = \begin{bmatrix} 4 & 5 \\ 0 & -3 \end{bmatrix} \quad B = \begin{bmatrix} 1 \\ -2 \\ 3 \\ 7 \end{bmatrix} \quad C = \begin{bmatrix} 1 & 0 & 3 & 0 \\ 0 & 2 & 0 & 4 \\ 5 & 0 & 6 & 7 \end{bmatrix}$$

$$A + 0 = \begin{bmatrix} 4 & 5 \\ 0 & -3 \end{bmatrix} + \begin{bmatrix} 0 & 0 \\ 0 & 0 \end{bmatrix} = \begin{bmatrix} 4 & 5 \\ 0 & -3 \end{bmatrix} = A$$

$$0 + B = \begin{bmatrix} 0 \\ 0 \\ 0 \\ 0 \end{bmatrix} + \begin{bmatrix} 1 \\ -2 \\ 3 \\ 7 \end{bmatrix} = \begin{bmatrix} 1 \\ -2 \\ 3 \\ 7 \end{bmatrix} = B$$

$$C + 0 = \begin{bmatrix} 1 & 0 & 3 & 0 \\ 0 & 2 & 0 & 4 \\ 5 & 0 & 6 & 7 \end{bmatrix} + \begin{bmatrix} 0 & 0 & 0 & 0 \\ 0 & 0 & 0 & 0 \\ 0 & 0 & 0 & 0 \end{bmatrix} = \begin{bmatrix} 1 & 0 & 3 & 0 \\ 0 & 2 & 0 & 4 \\ 5 & 0 & 6 & 7 \end{bmatrix} = C$$

Marching in step with the multiplicative identity

The multiplicative identity for matrices has one thing in common with the additive identity and, then, one big difference. The common trait of the multiplicative identity is that the multiplicative identity also comes in many sizes; the difference is that the multiplicative identity comes in only one *shape:* a square. A square matrix is $n \times n$; the number of rows and number of columns is the same.

The multiplicative identity is a square matrix, and the elements on the main diagonal running from the top left to the bottom right are 1s. All the rest of the elements in the matrix are 0s. Here, I show you three identity matrices with dimensions 2×2, 3×3, and 4×4, respectively.

$$I_2 = \begin{bmatrix} 1 & 0 \\ 0 & 1 \end{bmatrix} \quad I_3 = \begin{bmatrix} 1 & 0 & 0 \\ 0 & 1 & 0 \\ 0 & 0 & 1 \end{bmatrix} \quad I_4 = \begin{bmatrix} 1 & 0 & 0 & 0 \\ 0 & 1 & 0 & 0 \\ 0 & 0 & 1 & 0 \\ 0 & 0 & 0 & 1 \end{bmatrix}$$

When you multiply a matrix times an identity matrix, the original matrix stays the same — it keeps its identity. Of course, you have to have the correct match-up of columns and rows.

For example, let me show you matrix D being multiplied by identity matrices. Matrix D is a 3×2 matrix. When multiplying D * I, the identity matrix, I, has to be a 2×2 matrix. When multiplying I * D, the identity matrix has to be a 3×3 matrix.

$$D = \begin{bmatrix} 3 & 4 \\ 2 & -1 \\ -5 & 6 \end{bmatrix}$$

$$D * I = \begin{bmatrix} 3 & 4 \\ 2 & -1 \\ -5 & 6 \end{bmatrix} * \begin{bmatrix} 1 & 0 \\ 0 & 1 \end{bmatrix} = \begin{bmatrix} 3 & 4 \\ 2 & -1 \\ -5 & 6 \end{bmatrix}$$

$$I * D = \begin{bmatrix} 1 & 0 & 0 \\ 0 & 1 & 0 \\ 0 & 0 & 1 \end{bmatrix} * \begin{bmatrix} 3 & 4 \\ 2 & -1 \\ -5 & 6 \end{bmatrix} = \begin{bmatrix} 3 & 4 \\ 2 & -1 \\ -5 & 6 \end{bmatrix}$$

The size of the identity matrix is pretty much dictated by the dimension of the matrix being multiplied and the order of the multiplication.

In the "Multiplying two matrices" section, earlier in this chapter, I mention that multiplication, in general, is *not* commutative. The exception to that rule is when a *square* matrix is multiplied by its identity matrix.

For example, consider a 2×2 matrix being multiplied by a 2×2 identity matrix:

$$\begin{bmatrix} 1 & 0 \\ 0 & 1 \end{bmatrix} * \begin{bmatrix} 4 & -1 \\ 2 & 3 \end{bmatrix} = \begin{bmatrix} 4 & -1 \\ 2 & 3 \end{bmatrix} = \begin{bmatrix} 4 & 1 \\ 2 & 3 \end{bmatrix} * \begin{bmatrix} 1 & 0 \\ 0 & 1 \end{bmatrix}$$

Whether the identity matrix comes first or second in the multiplication, the original matrix keeps its identity. You have commutativity of multiplication in this special case.

Triangulating with triangular and diagonal matrices

Matrices labeled as being *triangular* or *diagonal* have the common characteristic that they're *square* matrices. A triangular matrix is either *upper triangular* or *lower triangular.* The best way to define or describe these matrices is to show you what they look like, first.

$$A = \begin{bmatrix} 4 & 3 & 5 \\ 0 & 1 & -4 \\ 0 & 0 & 2 \end{bmatrix} \quad B = \begin{bmatrix} 7 & 0 & 0 & 0 \\ 7 & 8 & 0 & 0 \\ 5 & 2 & 3 & 0 \\ 5 & -4 & -9 & 1 \end{bmatrix} \quad C = \begin{bmatrix} 4 & 0 & 0 & 0 \\ 0 & 3 & 0 & 0 \\ 0 & 0 & 1 & 0 \\ 0 & 0 & 0 & 2 \end{bmatrix}$$

Matrix A is an *upper triangular* matrix; all the elements *below* the main diagonal (the diagonal running from upper left to lower right) are 0s. Matrix B is a *lower triangular* matrix; all the elements *above* the main diagonal are 0s. And matrix C is a *diagonal* matrix, because all the entries above and below the main diagonal are 0s.

Triangular and diagonal matrices are desirable and sought-after in matrix applications. You'll find the triangular and diagonal matrices used when solving systems of equations and computing determinants (see Chapter 11).

Doubling it up with singular and non-singular matrices

The classification as singular or non-singular matrices applies to just square matrices. Square matrices get quite a workout in linear algebra, and this is just another example.

A square matrix is *singular* if it has a multiplicative inverse; a matrix is *non-singular* if it does *not* have a multiplicative inverse.

Two numbers are multiplicative inverses if their product is 1 (the multiplicative identity). For example, 4 and $^1/_4$ are multiplicative inverses. The numbers $^3/_5$ and $^5/_3$ are multiplicative inverses. The number 0 has no multiplicative inverse, because there's no number you can multiply 0 by to get a result of 1.

When a matrix has a multiplicative inverse, the product of the matrix and its inverse is equal to an identity matrix (multiplicative identity). And, furthermore, you can multiply the two matrices involved in either order (commutativity) and still get the identity. Later in this chapter, in "Investigating the Inverse of a Matrix," you see what a matrix's inverse looks like and how to find it.

Connecting It All with Matrix Algebra

Arithmetic and matrix algebra have many similarities and many differences. To begin with, the components in arithmetic and matrix algebra are completely different. In arithmetic, you have numbers like 4, 7, and 0. In matrix algebra, you have rectangular arrays of numbers surrounded by a bracket.

In "Making matrix multiplication work" and "Identifying with identity matrices," earlier in this chapter, I point out when the commutative property applies to matrices. In this section, I get down-and-dirty and discuss all the properties you find when working with matrix algebra. You need to know if a particular operation or property applies so that you can take advantage of the property when doing computations.

Delineating the properties under addition

Matrix addition requires that the matrices involved have the same dimension. Once the dimension question is settled, then you have two very nice properties applied to addition: commutativity and associativity.

Commutativity

Matrix addition is *commutative;* if you add matrix A to matrix B, you get the same result as adding matrix B to matrix A:

$$A + B = B + A$$

In the "Adding and subtracting matrices" section, earlier in this chapter, you see how the corresponding elements are added together to accomplish matrix addition. Because adding real numbers is commutative, the sums occurring in matrix addition are also the same when added in the reverse order.

$$\begin{bmatrix} 5 & 3 \\ 0 & 1 \end{bmatrix} + \begin{bmatrix} 4 & -1 \\ 2 & 3 \end{bmatrix} = \begin{bmatrix} 9 & 2 \\ 2 & 4 \end{bmatrix} = \begin{bmatrix} 4 & -1 \\ 2 & 3 \end{bmatrix} + \begin{bmatrix} 5 & 3 \\ 0 & 1 \end{bmatrix}$$

Associativity

Matrix addition is *associative.* When adding three matrices together, you get the same result if you add the sum of the first two to the third as you do when you add the first to the sum of the second two:

$$(A + B) + C = A + (B + C)$$

Note that the order doesn't change, just the groupings. The following equations are equivalent:

$$\left(\begin{bmatrix} 5 & 3 \\ 0 & 1 \end{bmatrix} + \begin{bmatrix} 4 & -1 \\ 2 & 3 \end{bmatrix}\right) + \begin{bmatrix} 9 & 8 \\ -6 & 2 \end{bmatrix} = \begin{bmatrix} 4 & -1 \\ 2 & 3 \end{bmatrix} + \left(\begin{bmatrix} 5 & 3 \\ 0 & 1 \end{bmatrix} + \begin{bmatrix} 9 & 8 \\ -6 & 2 \end{bmatrix}\right)$$

$$\begin{bmatrix} 9 & 2 \\ 2 & 4 \end{bmatrix} + \begin{bmatrix} 9 & 8 \\ -6 & 2 \end{bmatrix} = \begin{bmatrix} 4 & -1 \\ 2 & 3 \end{bmatrix} + \begin{bmatrix} 14 & 11 \\ -6 & 3 \end{bmatrix}$$

$$\begin{bmatrix} 18 & 10 \\ -4 & 6 \end{bmatrix} = \begin{bmatrix} 18 & 10 \\ -4 & 6 \end{bmatrix}$$

Tackling the properties under multiplication

When multiplying numbers together, you have just one process or operation to consider. When you multiply 6×4, you know from your times tables that the answer is 24. You don't have any other type of multiplication that gives you a different result. Multiplication of matrices goes in one of two different directions. Matrix multiplication is described in detail in "Making matrix multiplication work," earlier in this chapter. And another type of multiplication, *scalar multiplication,* is covered in "Scaling the heights with scalar multiplication," earlier in this chapter.

The multiplication processes or operations involving matrices have very specific properties associated with them. You find some properties closely related to real number multiplication and other properties that are very special to just matrices:

- **A * B ≠ B * A:** Matrix multiplication, as a rule, is *not* commutative. The exceptions involve multiplying a matrix by its inverse or by the multiplicative identity. Or, if you want to consider a rather unproductive exercise, you could multiply a square matrix by a square zero matrix of the same size and consider the process to be commutative. (You do get the same answer.)

- **kA = Ak:** Scalar multiplication is commutative — you can put the scalar in front of the matrix or behind it and get the same answer:

$$5\begin{bmatrix}4 & -1\\2 & 3\end{bmatrix} = \begin{bmatrix}4 & -1\\2 & 3\end{bmatrix}5$$
$$\begin{bmatrix}5\cdot 4 & 5(-1)\\5\cdot 2 & 5\cdot 3\end{bmatrix} = \begin{bmatrix}4\cdot 5 & -1\cdot 5\\2\cdot 5 & 3\cdot 5\end{bmatrix}$$

- **(A * B) * C = A * (B * C):** Matrix multiplication is associative when you deal with three matrices. Multiplying the product of two matrices times the third gives you the same result as multiplying the first matrix times the product of the last two. The following equations are equivalent:

$$\left(\begin{bmatrix}5 & 3\\0 & 1\end{bmatrix} * \begin{bmatrix}4 & -1\\2 & 3\end{bmatrix}\right) * \begin{bmatrix}9 & 8\\-6 & 2\end{bmatrix} = \begin{bmatrix}5 & 3\\0 & 1\end{bmatrix} * \left(\begin{bmatrix}4 & -1\\2 & 3\end{bmatrix} * \begin{bmatrix}9 & 8\\-6 & 2\end{bmatrix}\right)$$
$$\begin{bmatrix}26 & 4\\2 & 3\end{bmatrix} * \begin{bmatrix}9 & 8\\-6 & 2\end{bmatrix} = \begin{bmatrix}5 & 3\\0 & 1\end{bmatrix} * \begin{bmatrix}42 & 30\\0 & 22\end{bmatrix}$$
$$\begin{bmatrix}210 & 216\\0 & 22\end{bmatrix} = \begin{bmatrix}210 & 216\\0 & 22\end{bmatrix}$$

- **(kA) * B = k (A * B):** Multiplying a matrix by a scalar and then that result times a second matrix gives the same result as you get if you first multiply the two matrices together and then multiply by the scalar. This could be called a *mixed associative* rule, mixing the scalar and the matrices. The following equations are equivalent:

$$\left(3\begin{bmatrix}4 & -1\\2 & 3\end{bmatrix}\right) * \begin{bmatrix}9 & 8\\-6 & 2\end{bmatrix} = 3\left(\begin{bmatrix}4 & -1\\2 & 3\end{bmatrix} * \begin{bmatrix}9 & 8\\-6 & 2\end{bmatrix}\right)$$
$$\begin{bmatrix}12 & -3\\6 & 9\end{bmatrix} * \begin{bmatrix}9 & 8\\-6 & 2\end{bmatrix} = 3\begin{bmatrix}42 & 30\\0 & 22\end{bmatrix}$$
$$\begin{bmatrix}126 & 90\\0 & 66\end{bmatrix} = \begin{bmatrix}126 & 90\\0 & 66\end{bmatrix}$$

- **(kl)A = k(lA):** Another *mixed associative* rule allows you to associate the two scalars and multiply the product times a matrix, or, if you prefer, you can multiply one of the scalars times the matrix and then that product times the other scalar.

Distributing the wealth using matrix multiplication and addition

The distributive property in arithmetic involves multiplying the sum of two numbers by another number. The distributive rule is written: $a(b + c) = ab + ac$. The distributive rule states that you get the same result if you multiply a number times the sum of two other numbers as you do if you first multiply each of the numbers involved in the sum by the third number before adding the results together. The distributive rule works in matrix algebra, too:

$$A * (B + C) = A * B + A * C$$

Adding two matrices together and then multiplying the sum by a third matrix gives you the same result as multiplying each of the matrices by the third matrix before finding the sum. The following equations are equivalent:

$$\begin{bmatrix} 4 & -1 \\ 2 & 3 \end{bmatrix} * \left(\begin{bmatrix} 5 & 3 \\ 0 & 1 \end{bmatrix} + \begin{bmatrix} 9 & 8 \\ -6 & 2 \end{bmatrix} \right) = \left(\begin{bmatrix} 4 & -1 \\ 2 & 3 \end{bmatrix} * \begin{bmatrix} 5 & 3 \\ 0 & 1 \end{bmatrix} \right) + \left(\begin{bmatrix} 4 & -1 \\ 2 & 3 \end{bmatrix} * \begin{bmatrix} 9 & 8 \\ -6 & 2 \end{bmatrix} \right)$$

$$\begin{bmatrix} 4 & -1 \\ 2 & 3 \end{bmatrix} * \begin{bmatrix} 14 & 11 \\ -6 & 3 \end{bmatrix} = \begin{bmatrix} 20 & 11 \\ 10 & 9 \end{bmatrix} + \begin{bmatrix} 42 & 30 \\ 0 & 22 \end{bmatrix}$$

$$\begin{bmatrix} 62 & 41 \\ 10 & 31 \end{bmatrix} = \begin{bmatrix} 62 & 41 \\ 10 & 31 \end{bmatrix}$$

Changing the order in the distribution also creates a true statement:

$$(B + C) * A = B * A + C * A$$

Here are three other variations on the distributive rule (which incorporate the use of scalars):

$$k(A + B) = kA + kB$$

$$(k + l)A = kA + lA$$

$$(A + B)k = Ak + Bk$$

Transposing a matrix

Transposing a matrix is like telling all the elements in the matrix to switch places. When you perform a *matrix transpose,* the element in row 1, column 4 moves to row 4, column 1. The effect is that the rows become columns, and the columns become rows. A 3 × 5 matrix becomes a 5 × 3 matrix when you perform a matrix transpose.

Performing a matrix transpose on matrix A, the notation is A^T. When A is changed to A^T, each a_{ij} in A becomes a_{ij} in A^T:

$$A = \begin{bmatrix} 1 & 2 & 3 & 4 & 5 \\ 6 & 7 & 8 & 9 & 10 \\ 11 & 12 & 13 & 14 & 15 \end{bmatrix} \quad A^T = \begin{bmatrix} 1 & 6 & 11 \\ 2 & 7 & 12 \\ 3 & 8 & 13 \\ 4 & 9 & 14 \\ 5 & 10 & 15 \end{bmatrix}$$

The transpose of the sum of two matrices is equal to the sum of the two transposes:

$$(A + B)^T = A^T + B^T$$

The transpose of the product of two matrices is equal to the product of the two transposes in the *opposite order:*

$$(A * B)^T = B^T * A^T$$

Implied in this rule is that multiplication is possible (the dimensions fit). Here are two matrices, A and B, their product, the transpose of their product, and the product of their transposes (in reverse order).

And matrix multiplication is not generally commutative, so multiplying $A^T * B^T$ does not give you the same result as multiplying in the reverse order.

$$A = \begin{bmatrix} 1 & 2 \\ 3 & 4 \\ 5 & 6 \end{bmatrix}, B = \begin{bmatrix} 1 & 0 & 2 \\ 4 & 5 & 6 \end{bmatrix}, A{*}B = \begin{bmatrix} 9 & 10 & 14 \\ 19 & 20 & 30 \\ 29 & 30 & 46 \end{bmatrix}, (A{*}B)^T = \begin{bmatrix} 9 & 19 & 29 \\ 10 & 20 & 30 \\ 14 & 30 & 46 \end{bmatrix},$$

$$A^T = \begin{bmatrix} 1 & 3 & 5 \\ 2 & 4 & 6 \end{bmatrix}, B^T = \begin{bmatrix} 1 & 4 \\ 0 & 5 \\ 2 & 6 \end{bmatrix}, \text{and } B^T{*}A^T = \begin{bmatrix} 9 & 19 & 29 \\ 10 & 20 & 30 \\ 14 & 30 & 46 \end{bmatrix}$$

Zeroing in on zero matrices

The zero matrix has all zeros, and some interesting properties arise out of this situation.

When you add a matrix to the scalar multiple created by multiplying by –1, you get the zero matrix:

$$A + (-1)A = 0$$

AB = 0 does *not* require that A = 0 or B = 0. This rule is in direct contradiction to the zero property of arithmetic, which says that in order for the product of two numbers to be 0, at least one of them has to be a 0. In the world of matrices, the zero property says that you can get a 0 matrix without either matrix in the product being a zero matrix. For example, here are matrices A and B, neither of which is a zero matrix:

$$A = \begin{bmatrix} 1 & 0 \\ 1 & 0 \end{bmatrix},\ B = \begin{bmatrix} 0 & 0 \\ 1 & 1 \end{bmatrix},\ \text{and } A*B = \begin{bmatrix} 0 & 0 \\ 0 & 0 \end{bmatrix}$$

Establishing the properties of an invertible matrix

An *invertible matrix* is a matrix that has an inverse. "That's a big help!" you say. Okay, let me try again. An *invertible matrix* is a square matrix. If matrix A is invertible, then there's also a matrix A^{-1} where, when you multiply the two matrices, $A * A^{-1}$, you get an identity matrix of the same dimension as A and A^{-1}.

For example, matrix B is invertible, because $B * B^{-1} = I$.

$$B = \begin{bmatrix} 3 & 7 \\ 2 & 5 \end{bmatrix} \quad B^{-1} = \begin{bmatrix} 5 & -7 \\ -2 & 3 \end{bmatrix}$$

$$B*B^{-1} = \begin{bmatrix} 3 & 7 \\ 2 & 5 \end{bmatrix} * \begin{bmatrix} 5 & -7 \\ -2 & 3 \end{bmatrix} = \begin{bmatrix} 1 & 0 \\ 0 & 1 \end{bmatrix}$$

Also, as I mention in "Marching in step with the multiplicative identity," earlier in this chapter, when you multiply a matrix and its inverse together, the order doesn't matter. Either order of multiplication produces the identity matrix:

$$B = \begin{bmatrix} 3 & 7 \\ 2 & 5 \end{bmatrix} \quad B^{-1} = \begin{bmatrix} 5 & -7 \\ -2 & 3 \end{bmatrix}$$

$$B^{-1} * B = \begin{bmatrix} 5 & -7 \\ -2 & 3 \end{bmatrix} * \begin{bmatrix} 3 & 7 \\ 2 & 5 \end{bmatrix} = \begin{bmatrix} 1 & 0 \\ 0 & 1 \end{bmatrix}$$

Not all matrices are invertible. For example, a matrix with a row of 0s or a column of 0s isn't invertible — it has no inverse. To show you why a matrix with a row or column of 0s has no inverse, consider this multiplication problem involving two 3×3 matrices, the first with a row of 0s.

Wassily Wassilyovitch Leontief: The friend of industrial planners

Wassily Leontief is probably best known in the world of mathematics for his work on *input-output* models — how inputs from one industry affect production of outputs for another industry, and how this is all kept in balance. Input-output models are best investigated using matrices. Input-output models allow for approximate predictions for the change in demand for inputs resulting from a change in demand for the finished product.

Leontief was born in 1905 or 1906, either in Germany or Russia, depending on the source. He entered the University of Leningrad in 1921 and graduated in 1924, earning a Learned Economist degree (somewhat like today's master's degree); he earned a PhD in Economics in 1928. Leontief's career included employment at the University of Kiel, advisor to the Ministry of Railroads in China, working for the U.S. National Bureau of Economic Research as consultant at the Office of Strategic Services during World War II, and professor of economics at Harvard University. Leontief's Nobel Prize in Economics was awarded in 1973 for his work on input-output tables.

$$\begin{bmatrix} a_{11} & a_{12} & a_{13} \\ a_{21} & a_{22} & a_{23} \\ 0 & 0 & 0 \end{bmatrix} * \begin{bmatrix} b_{11} & b_{12} & b_{13} \\ b_{21} & b_{22} & b_{23} \\ b_{31} & b_{32} & b_{33} \end{bmatrix}$$

$$= \begin{bmatrix} a_{11}b_{11} + a_{12}b_{21} + a_{13}b_{31} & a_{11}b_{12} + a_{12}b_{22} + a_{13}b_{32} & a_{11}b_{13} + a_{12}b_{23} + a_{13}b_{33} \\ a_{21}b_{11} + a_{22}b_{21} + a_{23}b_{31} & a_{21}b_{12} + a_{22}b_{22} + a_{23}b_{32} & a_{21}b_{13} + a_{22}b_{23} + a_{23}b_{33} \\ 0 \bullet b_{11} + 0 \bullet b_{21} + 0 \bullet b_{31} & 0 \bullet b_{12} + 0 \bullet b_{22} + 0 \bullet b_{32} & 0 \bullet b_{13} + 0 \bullet b_{23} + 0 \bullet b_{33} \end{bmatrix}$$

The bottom row in the product of the two matrices is all 0s. The bottom row can never have a 1 in the last position, so such a matrix can never be the identity matrix. When you have a matrix with a row of 0s or a column of 0s, there's no matrix multiplier for it that can be the matrix's inverse; such matrices cannot have inverses.

Investigating the Inverse of a Matrix

Many square matrices have inverses. When a matrix and its inverse are multiplied together, in either order, the result is an identity matrix. Matrix inverses are used to solve problems involving systems of equations and to perform matrix *division.*

Quickly quelling the 2 × 2 inverse

In general, if a matrix has an inverse, that inverse can be found using a method called *row reduction*. (You can find a complete description of that process later in the "Finding Inverses using row reduction" section.) Happily, though, a much quicker and neater process is available for 2 × 2 matrices. To find the inverse (if it exists) of a 2 × 2 matrix, you switch some elements, negate some elements, and divide all the newly positioned elements by a number created from the original elements.

It's easier to just show you than it is to try to describe this in words. For starters, consider a general 2 × 2 matrix M with the elements *a, b, c,* and *d.*

$$\mathrm{M} = \begin{bmatrix} a & b \\ c & d \end{bmatrix}$$

To find the inverse of matrix M, you first compute the number: $ad - bc$. This number is the difference between the two cross-products of the elements in the matrix M. Next, you reverse the elements *a* and *d,* and then you negate (change to the opposite) the elements *b* and *c.* Now divide each adjusted element by the number computed from the cross-products:

$$\mathrm{M}^{-1} = \begin{bmatrix} \dfrac{d}{ad - bc} & \dfrac{-b}{ad - bc} \\ \dfrac{-c}{ad - bc} & \dfrac{a}{ad - bc} \end{bmatrix}$$

Because you're dividing by the number obtained from $ad - bc$, it's essential that the difference $ad - bc$ not be equal to 0.

For example, to find the inverse of the following matrix A,

$$\mathrm{A} = \begin{bmatrix} 4 & -7 \\ 2 & -3 \end{bmatrix}$$

you first compute the number that will be dividing each element: $(4)(-3) - (-7)(2) = -12 + 14 = 2$. The divisor is 2.

Now reverse the positions of the 4 and –3, and change the –7 to 7 and the 2 to –2. Divide each element by 2.

$$A^{-1} = \begin{bmatrix} \frac{-3}{2} & \frac{7}{2} \\ \frac{-2}{2} & \frac{4}{2} \end{bmatrix} = \begin{bmatrix} -\frac{3}{2} & \frac{7}{2} \\ -1 & 2 \end{bmatrix}$$

When you multiply the original matrix A times the inverse A^{-1}, the result is the 2×2 identity matrix.

$$\begin{aligned} A{*}A^{-1} &= \begin{bmatrix} 4 & -7 \\ 2 & -3 \end{bmatrix} * \begin{bmatrix} -\frac{3}{2} & \frac{7}{2} \\ -1 & 2 \end{bmatrix} \\ &= \begin{bmatrix} 4\left(-\frac{3}{2}\right)+7 & 4\left(\frac{7}{2}\right)-14 \\ 2\left(-\frac{3}{2}\right)+(-3)(-1) & 2\left(\frac{7}{2}\right)+(-3)(2) \end{bmatrix} \\ &= \begin{bmatrix} -6+7 & 14-14 \\ -3+3 & 7-6 \end{bmatrix} = \begin{bmatrix} 1 & 0 \\ 0 & 1 \end{bmatrix} \end{aligned}$$

The quick rule for finding the inverse of a 2×2 matrix gives a good opportunity to show you why some matrices do not have inverses. If the number $ad - bc$ (obtained by finding the difference of the cross-products) is equal to 0, then you'd have to divide by 0. That's an impossible situation — no numbers have 0 in their denominator. For example, the matrix B has no inverse. The cross-products are both 24, and their difference is 0.

$$B = \begin{bmatrix} 8 & 2 \\ 12 & 3 \end{bmatrix}$$

Some matrices, such as matrix B, have no inverse.

Finding inverses using row reduction

The quick rule for finding inverses of 2×2 matrices is the desirable method for matrices of that dimension. For larger square matrices, the inverses are found by performing additions and multiplications in the form of *row operations.* Before showing you how to find the inverse of a 3×3 matrix, let me first describe the different operations possible under row reduction.

Rendering row reduction rules

Row operations are performed with the goal of changing the format into something more convenient. The convenient part depends on the particular application, so *convenient* in one case may be different from *convenient* in another case. In any case, row reductions change the first matrix to another matrix that is *equivalent* to the first.

The row operations are

1. **Interchange any two rows.**
2. **Multiply all the elements in a row by a real number (not 0).**
3. **Add multiples of the elements of one row to the elements of another row.**

Consider row operations on matrix C:

$$C = \begin{bmatrix} 3 & 4 & -2 & 8 \\ 5 & -6 & 1 & -3 \\ 0 & 5 & 20 & 30 \\ 1 & 6 & 3 & 3 \end{bmatrix}$$

1. **Interchange rows 1 and 4.**

   The notation $R_1 \leftrightarrow R_4$ means to replace one row with the other.

$$R_1 \leftrightarrow R_4 \begin{bmatrix} 1 & 6 & 3 & 3 \\ 5 & -6 & 1 & -3 \\ 0 & 5 & 20 & 30 \\ 3 & 4 & -2 & 8 \end{bmatrix}$$

2. **Multiply all the elements in row 3 by $^1/_5$.**

   The notation $^1/_5\, R_3 \rightarrow R_3$ is read, "One-fifth of each element in row 3 becomes the new row 3."

$$\frac{1}{5}R_3 \rightarrow R_3 \begin{bmatrix} 1 & 6 & 3 & 3 \\ 5 & -6 & 1 & -3 \\ 0 & 1 & 4 & 6 \\ 3 & 4 & -2 & 8 \end{bmatrix}$$

3. **Multiply all the elements in row 1 by –5, and add them to row 2.**

   The notation $-5R_1 + R_2 \rightarrow R_2$ is read, "Negative 5 times each element in row 1 added to the elements in row 2 produces a new row 2." Note that the elements in row 1 do not change; you just use the multiples of the elements in the operation to create a new row 2.

$$-5R_1 + R_2 \rightarrow R_2 \begin{bmatrix} 1 & 6 & 3 & 3 \\ 0 & -36 & -14 & -18 \\ 0 & 1 & 4 & 6 \\ 3 & 4 & -2 & 8 \end{bmatrix}$$

You may not see any point to my row operations, but each one has a real application to working with matrices. The importance of the operations is evident when solving for an inverse matrix and, also, when working with systems of equations in Chapter 4.

Stepping through inverse matrix steps

When you have a square matrix whose dimensions are greater than 2×2, you solve for the inverse of that matrix (if it exists) by creating a double-width matrix with the original on the left and an identity matrix on the right. Then you go through row operations to transform the left-hand matrix into an identity matrix. When you're finished, the matrix on the right is the original matrix's inverse.

For example, finding the inverse for matrix D, you put a 3×3 identity matrix next to the elements in D.

$$D = \begin{bmatrix} 1 & 2 & 0 \\ -1 & 3 & 3 \\ 0 & 1 & 1 \end{bmatrix} \quad \left[\begin{array}{ccc|ccc} 1 & 2 & 0 & 1 & 0 & 0 \\ -1 & 3 & 3 & 0 & 1 & 0 \\ 0 & 1 & 1 & 0 & 0 & 1 \end{array}\right]$$

You want to change the 3×3 matrix on the left to the identity matrix (a matrix with a diagonal of 1s and the rest 0s). You must use proper row operations. Note that the element in row 1, column 1, is already a 1, so you concentrate on the elements below that 1. The element in row 3 is already a 0, so you just have to change the –1 in row 2 to a 0. The row operation you use is to add row 1 to row 2, creating a new row 2. You don't have to multiply row 1 by anything (technically, you're multiplying it by 1).

$$R_1 + R_2 \rightarrow R_2 \left[\begin{array}{ccc|ccc} 1 & 2 & 0 & 1 & 0 & 0 \\ 0 & 5 & 3 & 1 & 1 & 0 \\ 0 & 1 & 1 & 0 & 0 & 1 \end{array}\right]$$

Whatever you do to the elements in the left-hand matrix should also be done to the right-hand matrix. You're concentrating on changing the left-hand matrix to an identity matrix, and the elements on the other side just have to "come along for the ride." The result of the row operations is that you now have a 0 for element d_{21}. In fact, both the elements under the 1 are 0s.

Now move to the second column. You want a 1 in row 2, column 2, and 0s above and below that 1. First multiply row 2 by $^1/_5$. Every element, all the way across, gets multiplied by that fraction.

$$\frac{1}{5}R_2 \rightarrow R_2 \left[\begin{array}{ccc|ccc} 1 & 2 & 0 & 1 & 0 & 0 \\ 0 & 1 & \frac{3}{5} & \frac{1}{5} & \frac{1}{5} & 0 \\ 0 & 1 & 1 & 0 & 0 & 1 \end{array}\right]$$

To get the 0s above and below the 1 in column 2, you multiply row 2 by –2 and add it to row 1 to get a new row 1. And you multiply row 2 by –1 and add it to row 3 to get a new row 3. Note that, in both cases, row 2 doesn't change; you're just adding a multiple of the row to another row.

$$-2R_2 + R_1 \rightarrow R_1 \left[\begin{array}{ccc|ccc} 1 & 0 & -\frac{6}{5} & \frac{3}{5} & -\frac{2}{5} & 0 \\ 0 & 1 & \frac{3}{5} & \frac{1}{5} & \frac{1}{5} & 0 \\ 0 & 1 & 1 & 0 & 0 & 1 \end{array}\right]$$

$$-1R_2 + R_3 \rightarrow R_3 \left[\begin{array}{ccc|ccc} 1 & 0 & -\frac{6}{5} & \frac{3}{5} & -\frac{2}{5} & 0 \\ 0 & 1 & \frac{3}{5} & \frac{1}{5} & \frac{1}{5} & 0 \\ 0 & 0 & \frac{2}{5} & -\frac{1}{5} & -\frac{1}{5} & 1 \end{array}\right]$$

Your first two columns in the left-hand matrix look good. You have 1s on the main diagonal and 0s above and below the 1s. You need a 1 in row 3, column 3, so multiply row 3 by $^5/_2$.

$$\frac{5}{2}R_3 \rightarrow R_3 \left[\begin{array}{ccc|ccc} 1 & 0 & -\frac{6}{5} & \frac{3}{5} & -\frac{2}{5} & 0 \\ 0 & 1 & \frac{3}{5} & \frac{1}{5} & \frac{1}{5} & 0 \\ 0 & 0 & 1 & -\frac{1}{2} & -\frac{1}{2} & \frac{5}{2} \end{array}\right]$$

Now create 0s above the 1 in column 3 by multiplying row 3 by $^6/_5$ and adding it to row 1, and then multiplying row 3 by $-^3/_5$ and adding it to row 2.

$$\frac{6}{5}R_3 + R_1 \rightarrow R_1 \left[\begin{array}{ccc|ccc} 1 & 0 & 0 & 0 & -1 & 3 \\ 0 & 1 & \frac{3}{5} & \frac{1}{5} & \frac{1}{5} & 0 \\ 0 & 0 & 1 & -\frac{1}{2} & -\frac{1}{2} & \frac{5}{2} \end{array}\right]$$

$$-\frac{3}{5}R_3 + R_2 \rightarrow R_2 \left[\begin{array}{ccc|ccc} 1 & 0 & 0 & 0 & -1 & 3 \\ 0 & 1 & 0 & \frac{1}{2} & \frac{1}{2} & -\frac{3}{2} \\ 0 & 0 & 1 & -\frac{1}{2} & -\frac{1}{2} & \frac{5}{2} \end{array}\right]$$

You now have the identity matrix on the left and the inverse of matrix D on the right.

$$D = \begin{bmatrix} 1 & 2 & 0 \\ -1 & 3 & 3 \\ 0 & 1 & 1 \end{bmatrix} \quad D^{-1} = \begin{bmatrix} 0 & -1 & 3 \\ \frac{1}{2} & \frac{1}{2} & -\frac{3}{2} \\ -\frac{1}{2} & -\frac{1}{2} & \frac{5}{2} \end{bmatrix}$$

$$D{*}D^{-1} = \begin{bmatrix} 1 & 2 & 0 \\ -1 & 3 & 3 \\ 0 & 1 & 1 \end{bmatrix} * \begin{bmatrix} 0 & -1 & 3 \\ \frac{1}{2} & \frac{1}{2} & -\frac{3}{2} \\ -\frac{1}{2} & -\frac{1}{2} & \frac{5}{2} \end{bmatrix} = \begin{bmatrix} 1 & 0 & 0 \\ 0 & 1 & 0 \\ 0 & 0 & 1 \end{bmatrix}$$

And, of course, $D * D^{-1}$ also equals the identity matrix.

Chapter 4

Getting Systematic with Systems of Equations

In This Chapter

- Solving systems of equations algebraically
- Visualizing solutions with graphs
- Maximizing the capabilities of matrices to solve systems
- Reconciling the possibilities for solutions of a system of equations

Systems of equations show up in applications of just about every mathematical subject. You solve systems of equations in first-year algebra just to get the feel for how systems behave and how to best approach finding a solution.

Systems of equations are used to represent different countable entities and how the different entities relate to one another. The systems may consist of all linear equations, or there could even be nonlinear equations. (The nonlinear equations belong in another setting, though — not in this book.) In this chapter, you see how to quickly and efficiently solve a simple system algebraically. Then you find out how to make matrices work for you when the system of equations gets too large and cumbersome.

Single solutions of systems of equations are written by listing the numerical value of each variable or as an ordered pair, triple, quadruple, and so on. To write multiple solutions, I show you how to introduce a parameter and write a nice, tidy rule.

Investigating Solutions for Systems

A system of equations has a *solution* when you have at least one set of numbers that replaces the variables in the equations and makes each equation read as a true statement. A system of equations may have more than one solution. Systems with more than one solution generally have an infinite number of solutions, and those solutions are written as an expression or rule in terms of a parameter.

Recognizing the characteristics of having just one solution

A system of equations is *consistent* if it has at least one solution. A system with *exactly* one solution has two characteristics that distinguish it from other systems: The system consists of linear equations, and you have as many *independent* equations as there are variables. The two characteristics don't guarantee a solution, but they must be in place for you to have that single solution.

The first prerequisite for a single solution is that the system must contain linear equations. A linear equation is of the form: $y = a_1x_1 + a_2x_2 + a_3x_3 + \ldots + a_nx_n$ where the a multipliers are real numbers and the x factors are variables. A *linear* system consists of variables with exponents or powers of 1.

The following system of linear equations has a single solution, $x = 2$, $y = 3$, and $z = 1$:

$$\begin{cases} 2x + 3y + z = 14 \\ 4x - y + 5z = 10 \\ x + 2y - 3z = 5 \end{cases}$$

The solution of the system is also written as the ordered triple (2,3,1). When writing this system, I used the consecutive letters *x, y,* and *z,* just to avoid subscripts. When systems contain a large number of variables, it's more efficient to use subscripted variables such as x_0, x_1, x_2, and so on. None of the variables are raised to a higher, negative, or fractional power.

The second requirement for a single solution is that the system of equations have as many independent equations as there are variables in the system. If the system contains three variables, then it must contain at least three independent equations. So, what's with this *independent* business? When equations are *independent,* that just means that none of the equations is a multiple or combination of multiples of the other equations in the system.

For example, the following system has two equations and two unknowns or variables:

$$\begin{cases} 3x_1 - 2x_2 = 16 \\ x_1 - 5x_2 = 14 \end{cases}$$

Including the equation $-4x_1 - 6x_2 = -4$ in the system increases the number of equations in the system but introduces an equation that isn't independent. The new equation is obtained by multiplying the top equation by –2, multiplying the bottom equation by 2, and adding the two multiples together.

Writing expressions for infinite solutions

Some systems of equations have many solutions — an infinite number, in fact. When systems have multiple solutions, you'll find a pattern or rule that predicts what those solutions are. For example, the following system has solutions: (3,1,1), (8,5,2), (13,9,3), (–2,–3,0), and many, many more.

$$\begin{cases} -x + y + z = -1 \\ -x + 2y - 3z = -4 \\ 3x - 2y - 7z = 0 \end{cases}$$

The pattern in the solutions is that x is always two less than five times z, and y is always three less than four times z. An efficient way to write the solutions as a rule is to use an ordered triple and a *parameter* (a variable used as a base for the rule). Starting with the ordered triple (x, y, z), replace the z with the parameter k. That makes the first equation read $-x + y + k = -1$ and the second equation read $-x + 2y - 3k = -4$. Solving for y in this system of two equations gives you $y = 4k - 3$. Substituting this value for y in the third equation gives you $x = 5k - 2$. The ordered triple is now written $(5k - 2, 4k - 3, k)$. You find a new solution by choosing a value for k and solving for x and y using the rules. For example, if you let k be equal to 4, then $x = 5(4) - 2 = 18$ and $y = 4(4) - 3 = 13$; the solution in this case is (18, 13, 4).

Graphing systems of two or three equations

A picture is worth a thousand words. Sounds good, and you do see several pictures or graphs in this section, but I can't keep from adding some words — maybe not a thousand, but plenty.

Systems of two equations and two variables are represented in a graph by two lines. When the system has a single solution, you see two lines intersecting at a single point — the solution of the system. When two lines have no solution at all, the graph of the system has two parallel lines that never touch. And systems that have an infinite number of solutions have two lines that completely overlap one another — you actually just see one line.

Systems of three equations are a bit harder to draw or graph. An equation with three variables is represented by a plane — a flat surface. You draw these planes in three-space and have to imagine an axis coming off the page toward you. Three planes intersect in a single point, or they may intersect in a single line. But sometimes planes don't have any common plane or line at all.

Graphing two lines for two equations

Systems of equations with two variables have a solution, many solutions, or no solutions.

The graph of the system of equations $ax + by = c$ and $dx + ey = f$, where x and y are variables and a, b, c, d, e, and f are real numbers, has:

- Intersecting lines (exactly one solution) when the system is *consistent* and the equations are *independent*
- Parallel lines (no solution) when the system is *inconsistent*
- Coincident lines (infinite solutions) when the system is *consistent* and the equations are *dependent*

A consistent and independent system of two equations has a single point of intersection. The equations are independent because one equation isn't a multiple of the other. In Figure 4-1, you see the graphs of the two lines in the following system with the solution shown as the point of intersection:

$$\begin{cases} x + y = 5 \\ x - y = -3 \end{cases}$$

A consistent and dependent system of two equations has an infinite number of points of intersection. In other words, every point on one of the lines is a point on the other line. The graph of such a system is just a single line (or two lines on top of one another). An example of such a system is:

$$\begin{cases} 2x + 4y = 6 \\ x + 2y = 3 \end{cases}$$

You may have noticed that the first equation is twice the second equation. When one equation is a nonzero multiple of the other, their graphs are the same, and, of course, they have the same solutions.

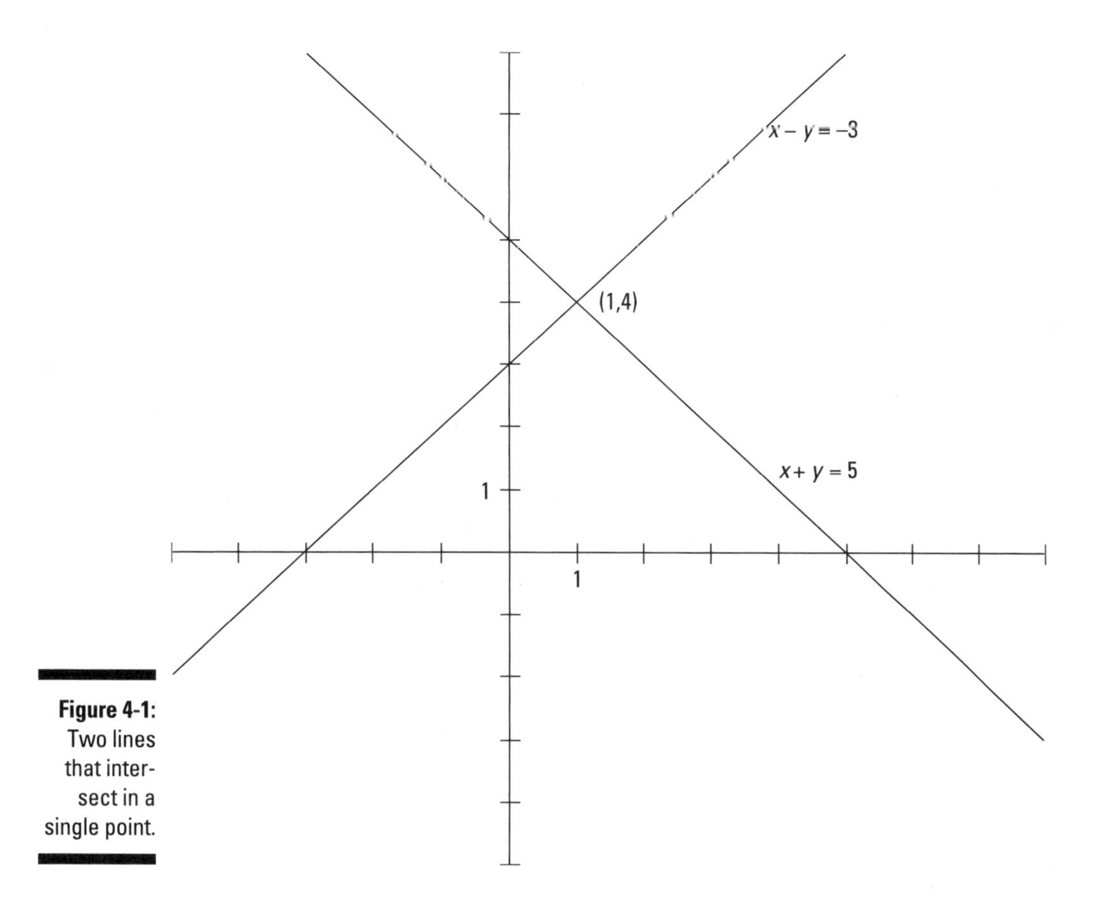

Figure 4-1: Two lines that intersect in a single point.

The last situation involving systems of equations represented by lines is that involving inconsistent systems. You find more on inconsistent systems in the "Dealing with Inconsistent Systems and No Solution" section, later in this chapter; there, I also show you the graph of parallel lines.

Taking a leap with three planes

A plane is a flat surface. In mathematics, a plane goes on forever and ever. It's a bit difficult to draw "forever," so drawing a close-up view shows the general slant and direction of the plane, and you get to project from there. The equation of a plane has the form $ax + by + cz = 0$, where x, y, and z are the variables (and the names of the three axes), and a, b, and c are real numbers. The more interesting situations involving planes are when they all meet in just one point — sort of like two walls and the floor meeting in a corner — or when they all meet along a straight line, like the extra-wide blades of a paddle wheel.

In Figure 4-2, you see an illustration of three planes meeting — the front, bottom, and left side of a box. The planes meet in a single point, the corner.

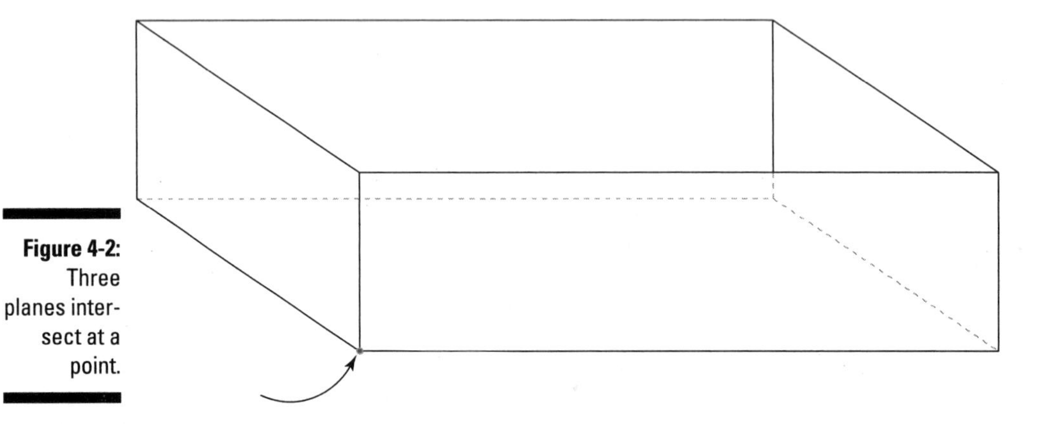

Figure 4-2: Three planes intersect at a point.

In Figure 4-3, you see three planes all sharing the same straight line — sort of like the axle of a paddle wheel.

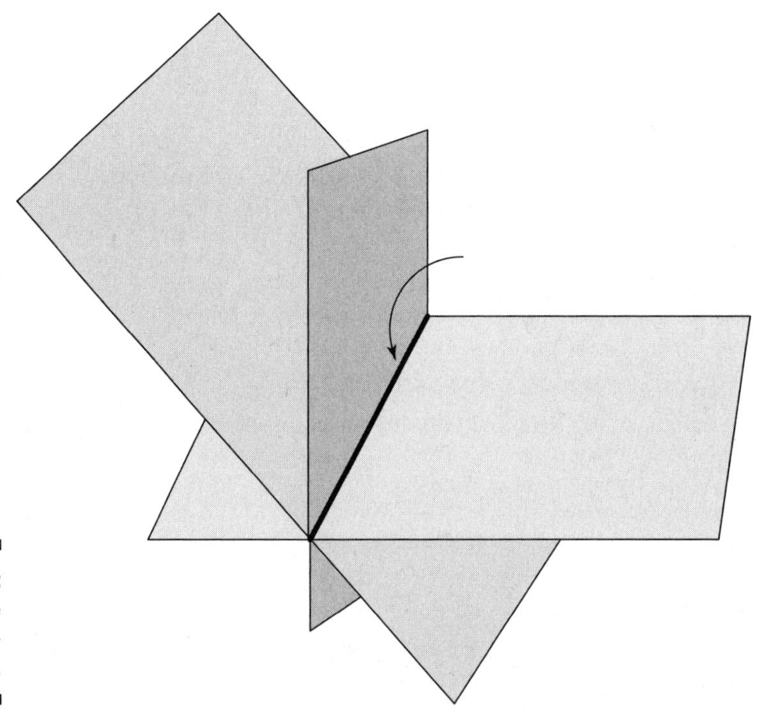

Figure 4-3: Three planes sharing a line.

Dealing with Inconsistent Systems and No Solution

An *inconsistent* system of two equations has two lines that never intersect. The lines are parallel, so the system has no solution. Figure 4-4 illustrates the following system:

$$\begin{cases} 2x + \ \ y = 5 \\ 6x + 3y = 1 \end{cases}$$

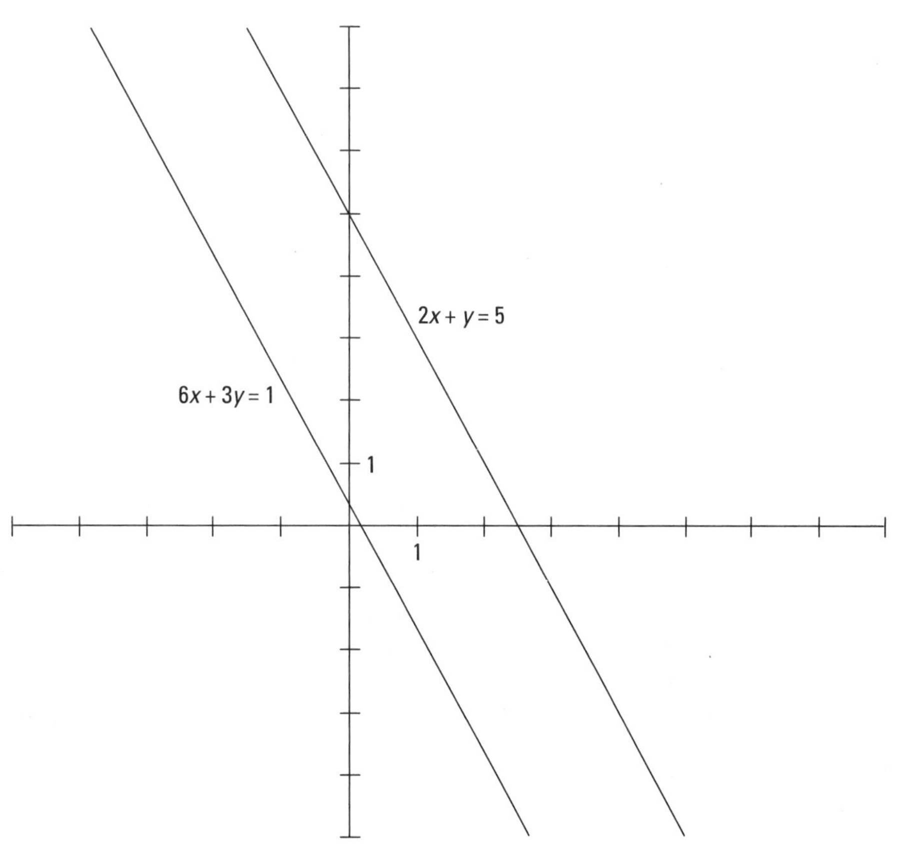

Figure 4-4: Distinct parallel lines never intersect.

The system is inconsistent because the lines have the same slope but different *y*-intercepts. The first equation, $2x + y = 5$, can be rewritten in the slope-intercept form as $y = -2x + 5$. With the equation written in the slope-intercept form, you see that the slope of the line is –2, and the line crosses the y-axis at (0,5).

The *slope-intercept form* of a linear equation in two variables is $y = mx + b$, where m is the slope of the line and b is the y-intercept.

The second equation, $6x + 3y = 1$, is rewritten as $3y = -6x + 1$ and then, dividing by 3, as $y = -2x + {}^1/_3$. This line also has a slope of –2, but its y-intercept is $(0,{}^1/_3)$. The two lines are parallel "and ne'er the twain shall meet" (whatever a *twain* is).

A logical leap from parallel lines is into the world of parallel planes. (Sounds like a science-fiction novel.) Planes that have the same slant but different intersections of the axes will never intersect. Figure 4-5 shows parallel planes.

Figure 4-5: These three planes are always the same distance apart.

Solving Systems Algebraically

Solving a system of equations means that you find the common solution or solutions that make each of the equations into a true statement at the same time. Finding the solutions for systems is sometimes a rather quick procedure, if the equations are few and the numbers in the equations cooperative. You have to assume the worst, though, and be prepared to use all the tools at your command to solve the systems — and be pleasantly surprised when they happen to work out without too much ado.

The tools or methods for solving systems of equations algebraically are

- **Elimination:** You eliminate one of the variables in the system to make it a simpler system, while preserving the original relationship between the variables.
- **Back substitution:** You use back substitution after you've found a value for one of the variables and then solve for the values of the other variables.

Starting with a system of two equations

A system of two equations and two unknowns is best represented by two lines. The point of intersection of the lines corresponds to the solution of the system. You see how this works in the earlier section "Graphing two lines for two equations."

To solve the following system of equations, I first use elimination to eliminate the *y*s and then back-substitute to find the value of *y* after finding *x*.

$$\begin{cases} 2x + 3y = 23 \\ 4x - 3y = 1 \end{cases}$$

Add the two equations together. Then divide each side of the equation by 6:

$$\begin{aligned} 2x + 3y &= 23 \\ \underline{4x - 3y} &= \underline{\ \ 1} \\ 6x \qquad &= 24 \\ x = \frac{24}{6} &= 4 \end{aligned}$$

You find that the value of $x = 4$. Now back-substitute into one of the equations to get the value of y. Using the first equation, and letting x be equal to 4, you get

$$\begin{aligned} 2(4) + 3y &= 23 \\ 8 + 3y &= 23 \\ 3y &= 15 \\ y &= 5 \end{aligned}$$

So the one solution for the system is $x = 4$ and $y = 5$, which is written (4,5) as an *ordered pair.*

The previous system of equations had a very convenient set of terms: the two *y*-terms that were opposites of one another. More likely than not, you won't have that very nice situation when solving the systems. The next system has a more common disjointedness between the numbers.

$$\begin{cases} 3x + 5y = 12 \\ 4x - 3y = -13 \end{cases}$$

Neither set of coefficients on the variables is the same, opposite of one another, or even the multiple of one another; you always hope for one of these convenient possibilities. My choice, now, would be to take advantage of the opposite signs on the *y*s, so I won't have to multiply by a negative number. To solve this system, multiply the terms in the first equation by 3 and the terms in the second equation by 5. These steps create 15*y* and –15*y* in the two equations. Adding the equations together, you eliminate the *y*s.

$$\begin{aligned} 9x + 15y &= \ \ 36 \\ 20x - 15y &= -65 \\ \hline 29x \qquad &= -29 \\ x = \frac{-29}{29} &= -1 \end{aligned}$$

Now, armed with the fact that $x = -1$, substitute back into the first equation to solve for *y*. You get $3(-1) + 5y = 12$, $5y = 15$, and $y = 3$. The solution is $(-1,3)$.

Extending the procedure to more than two equations

When solving systems of equations by elimination and back-substitution, you need to have a plan and be systematic when working through the steps. First, you identify a particular variable to be eliminated; then you perform multiplications and additions to eliminate that variable from all the equations. If more than one variable still appears in the equations after performing an elimination, you go through the process again.

The following operations may be performed (preserving the original relationship between the variables), giving you the solution to the original equation:

- Interchange two equations (change their order).
- Multiply or divide each term in an equation by a constant number ($\neq 0$).
- Add two equations together, combining like terms.
- Add the multiple of one equation to another equation.

The following system has four variables and four equations:

$$\begin{cases} 2x + 3y - 4z + w = 15 \\ x - 2y + 3z - 2w = -3 \\ 3x + 5y + z - w = 20 \\ 4x + y - z + w = 5 \end{cases}$$

My choice is to eliminate the *w*s first. This choice is somewhat arbitrary, but it's also considerate; I want to take advantage of the three coefficients that are 1 or –1, and the coefficient –2.

Add row 2 to two times row 1; add row 1 to row 3; and add row 4 to row 3. I'm using row reduction (row operation) notation to show what the steps are. For example, $2R_1 + R_2 \rightarrow 5x + 4y - 5z = 27$ means: "Adding two times row 1 plus row 2 gives you this new equation, $5x + 4y - 5z = 27$." The notation used here is similar to that used when writing row operation notation for matrices (see Chapter 3). The difference between the notation for matrices and the notation for equations is that, with matrices, I indicate a row number after the arrow. With equations, I show the resulting equation after the arrow.

$$\begin{cases} 2x + 3y - 4z + w = 15 \\ x - 2y + 3z - 2w = -3 \\ 3x + 5y + z - w = 20 \\ 4x + y - z + w = 5 \end{cases} \qquad \begin{aligned} 2R_1 + R_2 &\rightarrow 5x + 4y - 5z = 27 \\ R_1 + R_3 &\rightarrow 5x + 8y - 3z = 35 \\ R_4 + R_3 &\rightarrow 7x + 6y = 25 \end{aligned}$$

Now I have a system of three equations with three unknowns. Because the last equation doesn't have a *z* term, I choose to eliminate the *z* in the other two equations. Multiply the first equation by 3 and the second equation by –5. Then add those two equations together:

$$\begin{cases} 5x + 4y - 5z = 27 \\ 5x + 8y - 3z = 35 \\ 7x + 6y = 25 \end{cases} \qquad \begin{aligned} 3R_1 &\rightarrow -15x + 12y - 15z = 81 \\ -5R_2 &\rightarrow \underline{-25x - 40y + 15z = -175} \\ 3R_1 + (-5R_2) &\rightarrow -10x - 28y = -94 \end{aligned}$$

The new system of equations consists of the sum from the previous step and the equation that had no *z* in it. The new system has just *x*s and *y*s for variables. Multiply the first equation by 14 and the second equation by 3 to create coefficients for the *y* terms that are opposites. Then add the new equations together:

$$\begin{aligned} 7x + 6y &= 25 \\ -10x - 28y &= -94 \end{aligned} \qquad \begin{aligned} 14R_1 &\rightarrow 98x + 84y = 350 \\ 3R_2 &\rightarrow \underline{-30x - 84y = -282} \\ 14R_1 + 3R_2 &\rightarrow 68x = 68 \end{aligned}$$

Dividing each side of $68x = 68$ by 68, you get $x = 1$. Then take the fact that $x = 1$ and substitute into $7x + 6y = 25$ to get $7(1) + 6y = 25$, which tells you that $y = 3$. Now put the 1 and 3 into $5x + 4y - 5z = 27$ to get $5(1) + 4(3) - 5z = 27$, which becomes $17 - 5z = 27$, or $z = -2$. Similarly, with the values of *x, y,* and *z* in any of the original equations, you get $w = -4$. The solution, written as an *ordered quadruple,* is (1, 3, –2, –4).

Aren't you glad that there's another choice of procedure to use when solving these equations? Not that you don't just love these algebraic adventures. But I do have an alternative to offer for solving large systems of equations. You just need some information on matrices and how they work in order to use the method effectively. For the scoop on matrices, see Chapter 3; for the method of solving equations, see "Instituting inverses to solve systems" and "Introducing augmented matrices," later in this chapter.

Revisiting Systems of Equations Using Matrices

Solving systems of equations algebraically is the traditional method and what has been used by mathematicians since such concepts were developed or discovered. With the advent of computers and handheld calculators, matrix algebra and matrix applications are much preferred when dealing with large systems of equations. It's still easier to solve smaller systems by hand, but the larger systems almost beg to have matrices used for efficient and accurate solutions.

George Bernard Dantzig

If you've ever seen the movie *Good Will Hunting,* then you've already had a glimpse into the legend that is George Dantzig. The opening scene of *Good Will Hunting* is an altered version of what actually happened to Dantzig while he was a graduate student at the University of California, Berkeley.

As the story goes, Dantzig arrived late to class one day and saw that two problems were written on the board. He dutifully copied down the problems, assuming that they were a homework assignment. Dantzig thought that the problems were a little more difficult than those usually assigned, but he turned them in a few days later.

What Dantzig didn't know was that the professor had written down those two problems as examples of famous unsolved statistics problems. Several weeks later, Dantzig's professor excitedly told him that one of the solutions was being submitted for publication.

George Dantzig is regarded as one of the founders of linear programming, a method of optimization and strategy using matrices. The applications of linear programming were explored during World War II, when military supplies and personnel had to be moved efficiently. Linear programming involves systems of equations and inequalities and matrices. The solutions had to be found, in those early days, without the benefit of electronic computers. Dantzig devised the simplex method to help find solutions to the problems. Nowadays, you find the applications of linear programming in the scheduling of flight crews for airlines, shipping merchandise to retailers, oil refinery planning, and more.

Instituting inverses to solve systems

When a system of equations happens to meet a very strict set of criteria, you can use a square matrix and its inverse to solve the system for the single solution. Even though it may sound as if you have to jump through too many hoops to use this method, don't discount it out of hand. Many applications meet the requirements of having a single solution and also having as many equations as there are unknowns. Yes, those are the requirements: one solution and enough equations for variables.

To solve a system of n equations with n unknowns using an $n \times n$ matrix and its inverse, do the following:

1. **Write each of the equations in the system with the variables in the same order and the constant on the other side of the equal sign from the variables.**
2. **Construct a *coefficient matrix* (a square matrix whose elements are the coefficients of the variables).**
3. **Write the *constant matrix* (an $n \times 1$ column matrix) using the constants in the equations.**
4. **Find the inverse of the coefficient matrix.**
5. **Multiply the inverse of the coefficient matrix times the constant matrix.**

   The resulting matrix shows the values of the variables in the solution of the system.

Here's an example. (I show the steps using the same numbering as given in the instructions.)

Solve the system of equations:

$$\begin{cases} 0 = 3 + x + y + z \\ 4x + 5y + 6 = 0 \\ y = 3z + 5 \end{cases}$$

1. **Writing the system of equations with the variables in alphabetical order and the constants on the other side of the equal sign, I get**

$$\begin{cases} -x - y - z = 3 \\ 4x + 5y \quad = 3 \\ \quad y - 3z = 5 \end{cases}$$

2. **Constructing a coefficient matrix using 0s for the missing variables, I get**

$$A = \begin{bmatrix} -1 & -1 & -1 \\ 4 & 5 & 0 \\ 0 & 1 & -3 \end{bmatrix}$$

3. **Writing the constant matrix, I get**

$$B = \begin{bmatrix} 3 \\ -6 \\ 5 \end{bmatrix}$$

4. **Finding the inverse of the coefficient matrix, I get**

$$A^{-1} = \begin{bmatrix} 15 & 4 & -5 \\ -12 & -3 & 4 \\ -4 & -1 & 1 \end{bmatrix}$$

For instructions on how to find an inverse matrix, turn to Chapter 3.

If the coefficient matrix doesn't have an inverse, then the system does not have a single solution.

5. **Multiplying the inverse of the coefficient matrix times the constant matrix, I get**

$$A^{-1} * B = \begin{bmatrix} 15 & 4 & -5 \\ -12 & -3 & 4 \\ -4 & -1 & 1 \end{bmatrix} * \begin{bmatrix} 3 \\ -6 \\ 5 \end{bmatrix} = \begin{bmatrix} -4 \\ 2 \\ -1 \end{bmatrix}$$

The resulting matrix has the values of x, y, and z in order from top to bottom. So $x = -4$, $y = 2$, and $z = -1$.

Introducing augmented matrices

Not all matrices have inverses. Some square matrices are *non-singular.* But matrices and row operations are still the way to go when solving systems, because you deal only with the coefficients (numbers). You don't need to involve the variables until writing out the solution; the matrix's rectangular arrangement keeps the different components in order so that the variables are recognizable by their position. Using an augmented matrix and performing row operations, you reduce the original system of equations to a simpler form. Either you come out with a single solution, or you're able to write solutions in terms of a parameter — many solutions.

Setting up an augmented matrix

Writing a system of equations as an augmented matrix requires that all the variables are in the same order in each equation. You're going to be recording only the coefficients, so you have to be sure that each number in the matrix represents the correct variable.

Unlike the method of multiplying a coefficient matrix by its inverse, as shown in the "Instituting inverses to solve systems" section, earlier in this chapter, you incorporate the constants into a single, augmented matrix, drawing a dotted line down through the matrix to separate the coefficients from the constants. Here is a system of equations and its augmented matrix:

$$\begin{cases} 2x - y - 4z = 3 \\ 4x + 5y - z = -6 \\ 3y - 3z = 50 \end{cases}$$

$$\left[\begin{array}{ccc|c} 2 & -1 & -4 & 3 \\ 4 & 5 & -1 & -6 \\ 0 & 3 & -3 & 50 \end{array}\right]$$

Aiming for the echelon form

When a matrix is in the *reduced row echelon form,* it has the following characteristics:

- If the matrix has any rows whose elements are all 0s, those rows are at the bottom of the matrix.
- Reading from left to right, the first element of a row that is *not* a 0 is always a 1. This first 1 is called the *leading* 1 of its row.
- In any row, the leading 1 is always to the right of and below any leading 1 in a row above.
- If a column contains a leading 1, then all the entries above and below that leading 1 are 0s.

The rules for the reduced row echelon form are pretty strict and have to be followed exactly. A not-so-strictly defined form for a matrix is the *row echelon form* (as opposed to *reduced* row echelon form). The difference between the two forms is that the row echelon form doesn't require 0s above and below each leading 1. All the other rules apply, though.

Matrix R is in reduced row echelon form and matrix E is in row echelon form.

$$R = \begin{bmatrix} 1 & 0 & 0 & 3 & 4 \\ 0 & 1 & 0 & 0 & 2 \\ 0 & 0 & 1 & -1 & 1 \\ 0 & 0 & 0 & 0 & 0 \end{bmatrix} \quad E = \begin{bmatrix} 1 & 3 & 0 & 2 & -3 \\ 0 & 1 & 4 & 0 & 3 \\ 0 & 0 & 1 & 1 & 4 \\ 0 & 0 & 0 & 0 & 0 \end{bmatrix}$$

Each echelon form has its place in solving systems of equations and linear algebra manipulations. Sometimes you get to choose, and other times you're directed one way or the other.

Creating equivalent matrices

In Chapter 3, you find the three different row operations that can be performed on matrices resulting in an equivalent matrix. I use these rules, again, to change an augmented matrix to an echelon form. Briefly, those operations are

- Interchanging two rows
- Multiplying a row by a nonzero number
- Adding a multiple of one row to another row

Refer to Chapter 3 for more explanation of the row operations and the notation employed to indicate their use.

Solving a system of equations using an augmented matrix and working toward an echelon form

After writing a system of equations as an augmented matrix, you perform row operations until the matrix is in either row echelon form or reduced row echelon form. Taking the extra steps necessary to put the matrix in reduced row echelon form allows you to read the solution directly from the matrix without further back-solving or algebraic manipulations.

For example, consider the following system of equations and its corresponding augmented matrix:

$$\begin{cases} -x - y - z = 3 \\ 4x + 5y = -6 \\ y - 3z = 5 \end{cases}$$

$$\left[\begin{array}{ccc|c} -1 & -1 & -1 & 3 \\ 4 & 5 & 0 & -6 \\ 0 & 1 & -3 & 5 \end{array}\right]$$

To put the augmented matrix in reduced row echelon form, start by multiplying the first row by –1:

$$-1R_1 \rightarrow R_1 \left[\begin{array}{ccc|c} 1 & 1 & 1 & -3 \\ 4 & 5 & 0 & -6 \\ 0 & 1 & -3 & 5 \end{array}\right]$$

You need to have all 0s under the leading 1 in the first row. Row 3 already has a 0 in that first position, so you only have to deal with the 4 in row 2. Multiply row 1 by –4 and add it to row 2 so that the first element in row 2 becomes a 0.

$$-4R_1 + R_2 \rightarrow R_2 \left[\begin{array}{ccc|c} 1 & 1 & 1 & -3 \\ 0 & 1 & -4 & 6 \\ 0 & 1 & -3 & 5 \end{array}\right]$$

You already have a 1 in the second row, second column, so you don't have to make any adjustments to create a leading 1. But you do need to create 0s above and below that 1. So you multiply –1 by row 2 and add it to row 1; then do the same to row 3:

$$-1R_2 + R_1 \rightarrow R_1 \left[\begin{array}{ccc|c} 1 & 0 & 5 & -9 \\ 0 & 1 & -4 & 6 \\ 0 & 1 & -3 & 5 \end{array}\right]$$

$$-1R_2 + R_3 \rightarrow R_3 \left[\begin{array}{ccc|c} 1 & 0 & 5 & -9 \\ 0 & 1 & -4 & 6 \\ 0 & 0 & 1 & -1 \end{array}\right]$$

You could stop right now. The matrix is in row echelon form. Using this row echelon form and back-solving, you first write the last row as the equation $z = -1$. Then back-substitute the value for z into the row above, $y - 4z = 6$; you get $y + 4 = 6$, or $y = 2$. Now put the –1 for the z and the 2 for the y into the top equation (if there is a y), $x + 5z = -9$; you get $x - 5 = -9$, or $x = -4$.

If, however, you decide to finish what you started and create the reduced row echelon form, bring your attention to that last row. Again, a nice thing has happened: The element in row 3, column 3, is a 1. You create 0s above the 1 by multiplying –5 by row 3 and adding it to row 1, and then multiplying 4 times row 3 and adding it to row 2.

$$-5R_3 + R_1 \rightarrow R_1 \left[\begin{array}{ccc|c} 1 & 0 & 0 & -4 \\ 0 & 1 & -4 & 6 \\ 0 & 0 & 1 & -1 \end{array}\right]$$

$$4R_3 + R_2 \rightarrow R_2 \left[\begin{array}{ccc|c} 1 & 0 & 0 & -4 \\ 0 & 1 & 0 & 2 \\ 0 & 0 & 1 & -1 \end{array}\right]$$

You read directly from the last column that $x = -4$, $y = 2$, and $z = -1$.

Writing parametric solutions from augmented matrices

A system of equations does not always have a single solution. When the system contains an equation that is a linear combination of other equations, then the system is *dependent* and has more than one solution. You recognize that the system has more than one solution when the row echelon form has one or more rows of 0s.

For example, I found the following system, which has an infinite number of solutions. To show that the system has multiple solutions, I write the system as an augmented matrix to get started.

$$\begin{cases} x - y - z = 1 \\ 3x - 2y - 7z = 0 \\ 5x - 4y - 9z = 2 \end{cases} \left[\begin{array}{ccc|c} 1 & -1 & -1 & 1 \\ 3 & -2 & -7 & 0 \\ 5 & -4 & -9 & 2 \end{array}\right]$$

Now I perform row operations to change the augmented matrix to reduced row echelon form.

$$\begin{array}{l} -3R_1 + R_2 \rightarrow R_2 \\ -5R_1 + R_3 \rightarrow R_3 \end{array} \left[\begin{array}{ccc|c} 1 & -1 & -1 & 1 \\ 0 & 1 & -4 & -3 \\ 0 & 1 & -4 & -3 \end{array}\right]$$

$$-1R_2 + R_3 \rightarrow R_3 \left[\begin{array}{ccc|c} 1 & -1 & -1 & 1 \\ 0 & 1 & -4 & -3 \\ 0 & 0 & 0 & 0 \end{array}\right]$$

$$R_2 + R_1 \rightarrow R_1 \left[\begin{array}{ccc|c} 1 & 0 & -5 & -2 \\ 0 & 1 & -4 & -3 \\ 0 & 0 & 0 & 0 \end{array}\right]$$

I now have an augmented matrix with a row of 0s at the bottom, so I write the first two rows as their equivalent equations in x, y, and z. Then I solve for x and y in the two equations.

$$\begin{cases} x - 5z = -2 \\ y - 4z = -3 \end{cases} \text{ or } \begin{cases} x = 5z - 2 \\ y = 4z - 3 \end{cases}$$

Both x and y are dependent on what z is. So I choose the parameter k to represent some real number. I let $z = k$, which makes $x = 5k - 2$ and $y = 4k - 3$. The solution is also written $(5k - 2, 4k - 3, k)$ as an ordered triple. Once a real number is selected for k, then a solution is created from the ordered triple. For example, if $k = 1$, then a solution for the system is $(3,1,1)$.

Part II
Relating Vectors and Linear Transformations

In this part . . .

Tracking an LIV? Doesn't everyone have his own favorite Lunar Interplanetary Vehicle? If not, then you'll be happy to keep your feet on the ground with these chapters on linearly independent vectors — and other linear launchings.

Chapter 5

Lining Up Linear Combinations

In This Chapter

- Combining vectors with linear combinations
- Recognizing vectors in a span
- Determining sets of vectors to span R^2 and R^3

Vectors are $n \times 1$ matrices that can have various operations performed upon them. Two special types of vectors are those for which $n = 2$ and $n = 3$ — you can draw pictures that serve as graphical representations of the vectors to aid you in understanding vector properties. Using both scalar multiplication and matrix addition, a *linear combination* uses vectors in a set to create a new vector that's the same dimension as the vectors involved in the operations.

The questions arising from creating linear combinations on sets of vectors range from how extensive the resulting vectors are to whether a particular vector may be produced from a given set of vectors. In this chapter, I answer these questions and explore the methods used to answer them.

Defining Linear Combinations of Vectors

A linear *equation*, such as $4x + 3y + (-5z) + 6w = 7$, is made up of products and sums. The variables are multiplied by coefficients, and the products are then added together. A linear *combination* of vectors is also the result of products and sums. Multipliers are called *scalars* to signify that scalar multiplication is being performed, and the sums of the products result in new vectors that have the same dimension as those being multiplied.

Writing vectors as sums of other vectors

A linear combination of vectors is written $\mathbf{y} = c_1\mathbf{v}_1 + c_2\mathbf{v}_2 + c_3\mathbf{v}_3 + \ldots + c_k\mathbf{v}_k$ where $\mathbf{v}_1, \mathbf{v}_2, \mathbf{v}_3, \ldots, \mathbf{v}_k$ are vectors and c_i is a real coefficient called a *scalar*.

Given a set of vectors with the same dimensions, many different linear combinations may be formed. And, given a vector, you can determine if it was formed from a linear combination of a particular set of vectors.

Here I show you three vectors and a linear combination. Note that the column vectors all have the dimension 3×1. The operations involved in the linear combination are *scalar multiplication* and *vector addition*. (Refer to Chapter 2 if you want to know more about vector operations.)

Here are the three vectors:

$$\mathbf{v}_1 = \begin{bmatrix} 3 \\ -2 \\ 0 \end{bmatrix}, \mathbf{v}_2 = \begin{bmatrix} 4 \\ 0 \\ -4 \end{bmatrix}, \mathbf{v}_3 = \begin{bmatrix} 8 \\ -1 \\ -3 \end{bmatrix}$$

And here's the linear combination: $\mathbf{y} = 3\mathbf{v}_1 + 4\mathbf{v}_2 + 2\mathbf{v}_3$

Now, applying the rule defined by the linear combination,

$$\mathbf{y} = 3\begin{bmatrix} 3 \\ -2 \\ 0 \end{bmatrix} + 4\begin{bmatrix} 4 \\ 0 \\ -4 \end{bmatrix} + 2\begin{bmatrix} 8 \\ -1 \\ -3 \end{bmatrix} = \begin{bmatrix} 9 \\ -6 \\ 0 \end{bmatrix} + \begin{bmatrix} 16 \\ 0 \\ -16 \end{bmatrix} + \begin{bmatrix} 16 \\ -2 \\ -6 \end{bmatrix} = \begin{bmatrix} 41 \\ -8 \\ -22 \end{bmatrix}$$

The final result is another 3×1 vector.

Consider, next, the linear combination, $\mathbf{y} = -4\mathbf{v}_1 - 13\mathbf{v}_2 + 8\mathbf{v}_3$ and the vector resulting from applying the operations on the vectors $\mathbf{v}_1$, $\mathbf{v}_2$, and $\mathbf{v}_3$.

$$\mathbf{y} = -4\begin{bmatrix} 3 \\ -2 \\ 0 \end{bmatrix} - 13\begin{bmatrix} 4 \\ 0 \\ -4 \end{bmatrix} + 8\begin{bmatrix} 8 \\ -1 \\ -3 \end{bmatrix} = \begin{bmatrix} -12 \\ 8 \\ 0 \end{bmatrix} + \begin{bmatrix} -52 \\ 0 \\ 52 \end{bmatrix} + \begin{bmatrix} 64 \\ -8 \\ -24 \end{bmatrix} = \begin{bmatrix} 0 \\ 0 \\ 28 \end{bmatrix}$$

In this case, the resulting vector has 0s for the first two elements. I was trying to come up with a linear combination of the three vectors that makes all the elements 0s in the resulting vector, and I couldn't do it. In Chapters 6 and 7, you see how to determine if a result of all 0s is possible and just how grand it is to have all 0s for the result. For now, I'm just showing you the processes and procedures for working with the combinations.

Determining whether a vector belongs

When working with a set of vectors, the linear combinations of those vectors are numerous. If you have no restrictions on the values of the scalars, then you have an infinite number of possibilities for the resulting vectors. What you want to determine, though, is whether a particular vector is the result of some particular linear combination of a given set of vectors.

For example, if you have the set of vectors

$$\mathbf{v}_1 = \begin{bmatrix} 3 \\ -2 \\ 0 \end{bmatrix}, \mathbf{v}_2 = \begin{bmatrix} 4 \\ 0 \\ -4 \end{bmatrix}, \mathbf{v}_3 = \begin{bmatrix} 8 \\ -1 \\ -3 \end{bmatrix}$$

and you want to determine if there's some way you can produce the vector

$$\begin{bmatrix} 4 \\ 5 \\ -1 \end{bmatrix}$$

as a linear combination of $\mathbf{v}_1$, $\mathbf{v}_2$, and $\mathbf{v}_3$, you investigate the various linear combinations that may produce the elements in the vector.

KISS: Keeping it simple, silly

The nicest situation to have occur when looking for an appropriate linear combination is when all but one of the scalars is 0, which happens when the desired vector is just a multiple of one of the vectors in the set.

For example, what if the vector you want to create with a linear combination of a set of vectors is

$$\begin{bmatrix} -8 \\ 1 \\ 3 \end{bmatrix}$$

Using the previous set of vectors, $\mathbf{v}_1$, $\mathbf{v}_2$, and $\mathbf{v}_3$, you see that multiplying –1 times the vector $\mathbf{v}_3$ results in your desired vector, so you write the linear combination $0\mathbf{v}_1 + 0\mathbf{v}_2 - 1\mathbf{v}_3$ to create that vector.

Wising up for a solution

When a desired vector isn't a simple multiple of just one of the vectors in the set you're working from, then you resort to another method — solving for the individual scalars that produce the vector.

The scalars you seek are the multipliers of the vectors in the set under consideration. Again, using the vectors $\mathbf{v}_1$, $\mathbf{v}_2$, and $\mathbf{v}_3$, you write the equation $x_1\mathbf{v}_1 + x_2\mathbf{v}_2 + x_3\mathbf{v}_3 = \mathbf{b}$, where x_i is a scalar and $\mathbf{b}$ is the target vector.

$$x_1\begin{bmatrix}3\\-2\\0\end{bmatrix}+x_2\begin{bmatrix}4\\0\\-4\end{bmatrix}+x_3\begin{bmatrix}8\\-1\\-3\end{bmatrix}=\begin{bmatrix}4\\5\\-1\end{bmatrix}$$

Multiplying each vector by its respective scalar, you get

$$\begin{bmatrix}3x_1\\-2x_1\\0\end{bmatrix}+\begin{bmatrix}4x_2\\0\\-4x_2\end{bmatrix}+\begin{bmatrix}8x_3\\-x_3\\-3x_3\end{bmatrix}=\begin{bmatrix}4\\5\\-1\end{bmatrix}$$

Now, to solve for the values of the scalars that make the equation true, rewrite the vector equation as a system of linear equations.

$$\begin{cases}3x_1+4x_2+8x_3=4\\-2x_1\qquad\quad-x_3=5\\\qquad-4x_2-3x_3=-1\end{cases}$$

The system is solved by creating an augmented matrix where each column of the matrix corresponds to one of the vectors in the vector equation.

$$\left[\begin{array}{ccc|c}3&4&8&4\\-2&0&-1&5\\0&-4&-3&-1\end{array}\right]$$
$$\begin{array}{cccc}\uparrow&\uparrow&\uparrow&\uparrow\\\mathbf{v}_1&\mathbf{v}_2&\mathbf{v}_3&\mathbf{b}\end{array}$$

Now, perform the row operations needed to produce a *reduced row echelon form* (see Chapter 4 for details on these steps):

$$\left[\begin{array}{ccc|c} 3 & 4 & 8 & 4 \\ -2 & 0 & -1 & 5 \\ 0 & -4 & -3 & -1 \end{array}\right]$$

$$R_1 + R_2 \rightarrow R_1 \left[\begin{array}{ccc|c} 1 & 4 & 7 & 9 \\ -2 & 0 & 1 & 5 \\ 0 & -4 & -3 & -1 \end{array}\right]$$

$$2R_1 + R_2 \rightarrow R_2 \left[\begin{array}{ccc|c} 1 & 4 & 7 & 9 \\ 0 & 8 & 13 & 23 \\ 0 & -4 & -3 & -1 \end{array}\right]$$

$$R_2 \leftrightarrow R_3 \left[\begin{array}{ccc|c} 1 & 4 & 7 & 9 \\ 0 & -4 & -3 & -1 \\ 0 & 8 & 13 & 23 \end{array}\right]$$

$$\begin{array}{c} R_2 + R_1 \rightarrow R_1 \\ 2R_2 + R_3 \rightarrow R_3 \end{array} \left[\begin{array}{ccc|c} 1 & 0 & 4 & 8 \\ 0 & -4 & -3 & -1 \\ 0 & 0 & 7 & 21 \end{array}\right]$$

$$\begin{array}{c} -\frac{1}{4}R_2 \rightarrow R_2 \\ \frac{1}{7}R_3 \rightarrow R_3 \end{array} \left[\begin{array}{ccc|c} 1 & 0 & 4 & 8 \\ 0 & 1 & \frac{3}{4} & \frac{1}{4} \\ 0 & 0 & 1 & 3 \end{array}\right]$$

$$\begin{array}{c} -4R_3 + R_1 \rightarrow R_1 \\ -\frac{3}{4}R_3 + R_2 \rightarrow R_2 \end{array} \rightarrow \left[\begin{array}{ccc|c} 1 & 0 & 0 & -4 \\ 0 & 1 & 0 & -2 \\ 0 & 0 & 1 & 3 \end{array}\right]$$

The solution of the system of equations is $x_1 = -4$, $x_2 = -2$, $x_3 = 3$. The solutions correspond to the scalars needed for the vector equation.

$$x_1\mathbf{v}_1 + x_2\mathbf{v}_2 + x_3\mathbf{v}_3 = \mathbf{b}$$

$$-4\begin{bmatrix} 3 \\ -2 \\ 0 \end{bmatrix} - 2\begin{bmatrix} 4 \\ 0 \\ -4 \end{bmatrix} + 3\begin{bmatrix} 8 \\ -1 \\ -3 \end{bmatrix} = \begin{bmatrix} -12 \\ 8 \\ 0 \end{bmatrix} + \begin{bmatrix} -8 \\ 0 \\ 8 \end{bmatrix} + \begin{bmatrix} 24 \\ -3 \\ -9 \end{bmatrix} = \begin{bmatrix} 4 \\ 5 \\ -1 \end{bmatrix}$$

Recognizing when there's no combination possible

Not every vector you choose is going to turn out to be a linear combination of a particular set of vectors. But when you begin the process of trying to determine the necessary scalars, you don't know that a linear combination isn't possible. You'll find out that there's no solution after performing some row operations and noting a discrepancy or impossible situation.

For example, consider the following set of vectors and the target vector, **b:**

$$\mathbf{v}_1 = \begin{bmatrix} 3 \\ -2 \\ 1 \end{bmatrix}, \mathbf{v}_2 = \begin{bmatrix} 4 \\ 0 \\ 12 \end{bmatrix}, \mathbf{v}_3 = \begin{bmatrix} 8 \\ -1 \\ 20 \end{bmatrix}, \mathbf{b} = \begin{bmatrix} 8 \\ -7 \\ 4 \end{bmatrix}$$

You want to solve the vector equation formed by multiplying the vectors in the set by scalars and setting them equal to the target vector.

$$x_1 \begin{bmatrix} 3 \\ -2 \\ 1 \end{bmatrix} + x_2 \begin{bmatrix} 4 \\ 0 \\ 12 \end{bmatrix} + x_3 \begin{bmatrix} 8 \\ -1 \\ 20 \end{bmatrix} = \begin{bmatrix} 8 \\ -7 \\ 4 \end{bmatrix}$$

Creating the augmented matrix for the system of equations and solving for the scalars, you have:

$$\left[\begin{array}{ccc|c} 3 & 4 & 8 & 8 \\ -2 & 0 & -1 & -7 \\ 1 & 12 & 20 & 4 \end{array}\right]$$

$$R_1 + R_2 \to R_1 \left[\begin{array}{ccc|c} 1 & 4 & 7 & 1 \\ -2 & 0 & -1 & -7 \\ 1 & 12 & 20 & 4 \end{array}\right]$$

$$\begin{array}{r} 2R_1 + R_2 \to R_2 \\ -1R_1 + R_3 \to R_3 \end{array} \left[\begin{array}{ccc|c} 1 & 0 & 4 & 8 \\ 0 & 8 & 13 & -5 \\ 0 & 8 & 13 & 3 \end{array}\right]$$

$$-1R_2 + R_3 \to R_3 \to \left[\begin{array}{ccc|c} 1 & 0 & 4 & 8 \\ 0 & 8 & 13 & -5 \\ 0 & 0 & 0 & 8 \end{array}\right]$$

After performing some row operations, you find that the last row of the matrix has 0s and an 8. The corresponding equation is $0x_1 + 0x_2 + 0x_3 = 8$, or $0 + 0 + 0 = 8$. The equation makes no sense — it can't be true. So there's no solution to the system of equations and no set of scalars that provide the target vector. The vector **b** is not one of the linear combinations possible from the chosen set of vectors.

Searching for patterns in linear combinations

Many different vectors can be written as linear combinations of a given set of vectors. Conversely, you can find a set of vectors to use in writing a particular target vector.

Finding a vector set for a target vector

For example, if you want to create the vector

$$\begin{bmatrix} 4 \\ 1 \\ 3 \\ 2 \end{bmatrix}$$

you could use the set of vectors

$$\left\{ \begin{bmatrix} 1 \\ 0 \\ 1 \\ 0 \end{bmatrix}, \begin{bmatrix} 1 \\ 0 \\ 0 \\ 1 \end{bmatrix}, \begin{bmatrix} 0 \\ 1 \\ 1 \\ 0 \end{bmatrix} \right\}$$

and the linear combination

$$2\begin{bmatrix} 1 \\ 0 \\ 1 \\ 0 \end{bmatrix} + 2\begin{bmatrix} 1 \\ 0 \\ 0 \\ 1 \end{bmatrix} + 1\begin{bmatrix} 0 \\ 1 \\ 1 \\ 0 \end{bmatrix} = \begin{bmatrix} 4 \\ 1 \\ 3 \\ 2 \end{bmatrix}$$

The vector set and linear combination shown here are in no way unique; you can find many different combinations and many different vector sets to use in creating the particular vector. Note, though, that my set of vectors is somewhat special, because the elements are all either 0 or 1. (You see more vectors with those two elements later in this chapter and in Chapters 6 and 7.)

Generalizing a pattern and writing a vector set

Sets of vectors can be described by listing all the vectors in the set or by recognizing and writing a rule for a pattern. When a set of vectors is very large or even has an infinite number of members, a pattern and generalized rule is preferable to describe all those members, if this is possible.

Consider the following vectors:

$$\left\{ \begin{bmatrix} 1 \\ 4 \\ 5 \\ 6 \end{bmatrix}, \begin{bmatrix} 3 \\ 12 \\ 7 \\ 10 \end{bmatrix}, \begin{bmatrix} -2 \\ -8 \\ 3 \\ 1 \end{bmatrix}, \begin{bmatrix} 0 \\ 0 \\ -11 \\ -11 \end{bmatrix} \right\}$$

One possibility for describing the vectors in this set is with a rule in terms of two real numbers, a and b, as shown here:

$$\left\{ \begin{bmatrix} 1 \\ 4 \\ 5 \\ 6 \end{bmatrix}, \begin{bmatrix} 3 \\ 12 \\ 7 \\ 10 \end{bmatrix}, \begin{bmatrix} -2 \\ -8 \\ 3 \\ 1 \end{bmatrix}, \begin{bmatrix} 0 \\ 0 \\ -11 \\ -11 \end{bmatrix}, \ldots, \begin{bmatrix} a \\ 4a \\ b \\ a+b \end{bmatrix} \right\}$$

Two elements, the first and third, determine the values of the other two elements.

This rule is just one possibility for a pattern in the set of vectors. A list of four vectors isn't really all that large. When you have only a few vectors to work with, you have to proceed with caution before applying that pattern or rule to some specific application.

The rule shows how to construct the vectors using the one vector and its elements. An alternative to using the one vector is to use two vectors and a linear combination:

$$a \begin{bmatrix} 1 \\ 4 \\ 0 \\ 1 \end{bmatrix} + b \begin{bmatrix} 0 \\ 0 \\ 1 \\ 1 \end{bmatrix}$$

So many choices!

Visualizing linear combinations of vectors

Vectors with dimension 2 × 1 can be represented in a coordinate system as points in the plane. The vectors

$$\mathbf{v}_1 = \begin{bmatrix} 2 \\ -1 \end{bmatrix}, \mathbf{v}_2 = \begin{bmatrix} 4 \\ 3 \end{bmatrix}$$

are graphed in standard form with initial points at the origin, (0,0), and terminal points (2,-1) and (4,3). You see the two vectors and two points in Figure 5-1. (Refer to Chapter 2 for more on graphing vectors.)

Linear combinations of vectors are represented using parallel lines drawn through the multiples of the points representing the vectors. In Figure 5-1, you see the scalar multiples of $\mathbf{v}_1$: -3$\mathbf{v}_1$, -2$\mathbf{v}_1$, -$\mathbf{v}_1$, 2$\mathbf{v}_1$, and 3$\mathbf{v}_1$, and multiples of $\mathbf{v}_2$: -$\mathbf{v}_2$ and 2$\mathbf{v}_2$. You also see the points representing the linear combinations: -3$\mathbf{v}_1$-$\mathbf{v}_2$, -3$\mathbf{v}_1$+$\mathbf{v}_2$, 2$\mathbf{v}_1$-$\mathbf{v}_2$, and 2$\mathbf{v}_1$+$\mathbf{v}_2$.

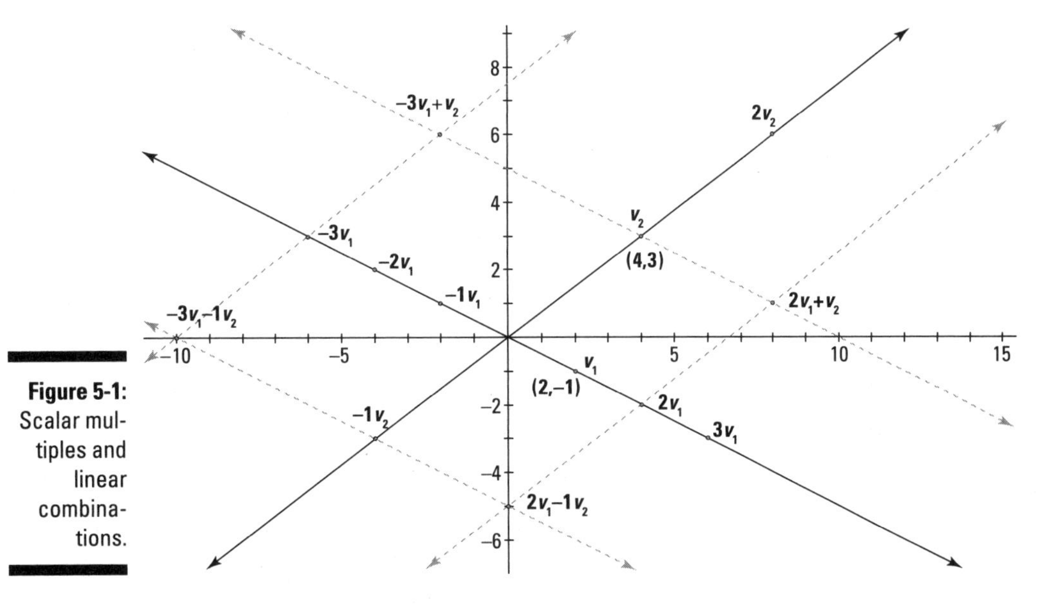

Figure 5-1: Scalar multiples and linear combinations.

Getting Your Attention with Span

You may already be familiar with life *span*, the span of a bridge, wingspan, and a span of horses. Now I get to introduce you to the span of a set of vectors.

The *span* of a set of vectors consists of another set of vectors, all with a relationship to the original set. The span of a vector set is usually infinitely large, because you're dealing with lots of real numbers. But, on the other hand, a span can be finite if you choose to limit the range of the scalars being used.

The concept or idea of span gives a structure to the various linear combinations of a set of vectors. The vectors in a set pretty much determine how wide or small the scope of the results is.

Describing the span of a set of vectors

Consider a set of vectors $\{\mathbf{v}_1, \mathbf{v}_2, \ldots, \mathbf{v}_k\}$. The set of all *linear combinations* of this set is called its *span*. That set of linear combinations of the vectors is *spanned by* the original set of vectors. Each vector in the span$\{\mathbf{v}_1, \mathbf{v}_2, \ldots, \mathbf{v}_k\}$ is of the form $c_1\mathbf{v}_1 + c_2\mathbf{v}_2 + \ldots + c_k\mathbf{v}_k$ where c_i is a real number scalar.

For example, if you have the set of vectors $\{\mathbf{v}_1, \mathbf{v}_2, \mathbf{v}_3\}$ where

$$\mathbf{v}_1 = \begin{bmatrix} 1 \\ -2 \\ 0 \end{bmatrix}, \mathbf{v}_2 = \begin{bmatrix} 4 \\ 2 \\ -1 \end{bmatrix}, \mathbf{v}_3 = \begin{bmatrix} 3 \\ -1 \\ 2 \end{bmatrix}$$

then span$\{\mathbf{v}_1, \mathbf{v}_2, \mathbf{v}_3\}$ =

$$c_1 = \begin{bmatrix} 1 \\ -2 \\ 0 \end{bmatrix} + c_2 \begin{bmatrix} 4 \\ 2 \\ -1 \end{bmatrix} + c_3 \begin{bmatrix} 3 \\ -1 \\ 2 \end{bmatrix}$$

Replacing the scalars with real numbers, you produce new vectors of the same dimension. With all the real number possibilities for the scalars, the resulting set of vectors is infinite.

Broadening a span as wide as possible

A span of a vector set is the set of all vectors that are produced from linear combinations of the original vector set. The most comprehensive or all-encompassing spans are those that include every possibility for a vector — all arrangements of real numbers for elements.

In R^2, the vector set

$$\left\{\begin{bmatrix}1\\0\end{bmatrix},\begin{bmatrix}0\\1\end{bmatrix}\right\}$$

spans all those 2×1 vectors. In R^3, the vector set

$$\left\{\begin{bmatrix}1\\0\\0\end{bmatrix},\begin{bmatrix}0\\1\\0\end{bmatrix},\begin{bmatrix}0\\0\\1\end{bmatrix}\right\}$$

spans all the 3×1 vectors. So, in general, if you want to span all $n \times 1$ vectors, your vector set could look like the following, where each member of the set is an $n \times 1$ vector:

$$\left\{\begin{bmatrix}1\\0\\0\\0\\\vdots\\0\end{bmatrix},\begin{bmatrix}0\\1\\0\\0\\\vdots\\0\end{bmatrix},\begin{bmatrix}0\\0\\1\\0\\\vdots\\0\end{bmatrix},\ldots,\begin{bmatrix}0\\0\\0\\0\\\vdots\\1\end{bmatrix}\right\}$$

Narrowing the scope of a span

Now consider the vector set

$$\left\{\begin{bmatrix}1\\5\\-1\end{bmatrix},\begin{bmatrix}-3\\-4\\2\end{bmatrix}\right\}$$

and the vector

$$\mathbf{b} = \begin{bmatrix}6\\-3\\d\end{bmatrix}$$

For what value or values of d (if any) is the vector **b** in span$\{\mathbf{v}_1, \mathbf{v}_2\}$?

If **b** is in span$\{\mathbf{v}_1, \mathbf{v}_2\}$, then you'll find a linear combination of the vectors such that

$$c_1\begin{bmatrix}1\\5\\-1\end{bmatrix}+c_2\begin{bmatrix}-3\\-4\\2\end{bmatrix}=\begin{bmatrix}6\\-3\\d\end{bmatrix}$$

Rewriting the vector equation as a system of equations, you get

$$\begin{aligned}c_1-3c_2&=6\\5c_1-4c_2&=-3\\-c_1+2c_2&=d\end{aligned}$$

Adding the first and third equations together, you get $-c_2 = 6 + d$. Adding the second equation to five times the third equation, you get $6c_2 = 5d - 3$. Multiplying $-c_2 = 6 + d$ by -6, you get $6c_2 = -36 - 6d$. You have two equations in which one side of the equation is $6c_2$, so set the other sides equal to one another to get $5d - 3 = -36 - 6d$. Add $6d$ to each side, and add 3 to each side to get $11d = -33$, which means that $d = -3$. So the vector **b** is in the span of the vector set only if the last element is -3.

Showing which vectors belong in a span

A span may be all-encompassing, or it may be very restrictive. To determine whether a vector is part of a span, you need to determine which linear combination of the vectors produces that target vector. If you have many vectors to check, you may choose to determine the format for all the scalars used in the linear combinations.

Determining whether a particular vector belongs in the span

Consider the vector set $\{\mathbf{v}_1, \mathbf{v}_2, \mathbf{v}_3\}$ where

$$\mathbf{v}_1=\begin{bmatrix}1\\-2\\0\end{bmatrix},\ \mathbf{v}_2=\begin{bmatrix}4\\2\\-1\end{bmatrix},\ \mathbf{v}_3=\begin{bmatrix}3\\-1\\2\end{bmatrix}$$

You want to determine if the vector

$$\mathbf{b} = \begin{bmatrix} -11 \\ -13 \\ 6 \end{bmatrix}$$

belongs in span$\{\mathbf{v}_1, \mathbf{v}_2, \mathbf{v}_3\}$. To do so, you need to determine the solution of the vector equation

$$c_1 \begin{bmatrix} 1 \\ -2 \\ 0 \end{bmatrix} + c_2 \begin{bmatrix} 4 \\ 2 \\ -1 \end{bmatrix} + c_3 \begin{bmatrix} 3 \\ -1 \\ 2 \end{bmatrix} = \begin{bmatrix} -11 \\ -13 \\ 6 \end{bmatrix}$$

Using the system of equations

$$\begin{cases} c_1 + 4c_2 + 3c_3 = -11 \\ -2c_1 + 2c_2 - c_3 = -13 \\ -c_2 + 2c_3 = 6 \end{cases}$$

and its corresponding augmented matrix (refer to "Determining whether a vector belongs"), the solution is $c_1 = 2$, $c_2 = -4$, and $c_3 = 1$. So you find that a linear combination of the vectors does produce the target vector. The vector **b** does belong in the span.

Writing a general format for all scalars used in a span

Solving a system of equations to determine how a vector belongs in a span is just fine — unless you have to repeat the process over and over again for a large number of vectors. Another option to avoid repeating the process is to create a general format for all the vectors in a span.

For example, consider the vector set $\{\mathbf{v}_1, \mathbf{v}_2, \mathbf{v}_3\}$ where

$$\mathbf{v}_1 = \begin{bmatrix} 1 \\ 0 \\ 1 \end{bmatrix}, \mathbf{v}_2 = \begin{bmatrix} 1 \\ 2 \\ 0 \end{bmatrix}, \mathbf{v}_3 = \begin{bmatrix} 1 \\ 3 \\ 2 \end{bmatrix}$$

Any vector in span$\{\mathbf{v}_1, \mathbf{v}_2, \mathbf{v}_3\}$ satisfies the equation

$$c_1\mathbf{v}_1 + c_2\mathbf{v}_2 + c_3\mathbf{v}_3 = \begin{bmatrix} x \\ y \\ z \end{bmatrix}$$

where x, y, and z are the elements of a vector in the span.

Using the vectors $\mathbf{v}_1$, $\mathbf{v}_2$, $\mathbf{v}_3$, the corresponding system of equations is

$$\begin{cases} c_1 + c_2 + c_3 = x \\ 2c_2 + 3c_3 = y \\ c_1 + 2c_3 = z \end{cases}$$

To solve the system of equations for x, y, and z, you start by adding -2 times row one to row two to get $-2c_1 + c_3 = -2x + y$. Now add this new equation to twice the third equation.

$$\begin{array}{r} 2c_1 + c_3 = -2x + y \\ 2c_1 + 4c_3 = 2z \\ \hline 5c_3 = -2x + y + 2z \end{array}$$

Divide each side by 5 to get

$$c_3 = \frac{-2x + y + 2z}{5}$$

Substituting this value of c_3 into the original second equation, you get

$$c_2 = \frac{3x + y - 3z}{5}$$

Then, substituting the values of c_3 and c_2 into the original first equation, you solve for c_1 and get

$$\frac{4x - 2y + z}{5}$$

So any vector in the span has the format

$$c_1\mathbf{v}_1 + c_2\mathbf{v}_2 + c_3\mathbf{v}_3 = \frac{4x - 2y + z}{5}\mathbf{v}_1 + \frac{3x + y - 3z}{5}\mathbf{v}_2 + \frac{-2x + y + 2z}{5}\mathbf{v}_3$$

For example, to create the vector

$$\begin{bmatrix} 4 \\ 1 \\ 1 \end{bmatrix}$$

you let $x = 4$, $y = 1$, and $z = 1$ and get

$$\begin{aligned} &\frac{4x - 2y + z}{5}\mathbf{v}_1 + \frac{3x + y - 3z}{5}\mathbf{v}_2 + \frac{-2x + y + 2z}{5}\mathbf{v}_3 \\ &= \frac{4(4) - 2(1) + 1}{5}\begin{bmatrix} 1 \\ 0 \\ 1 \end{bmatrix} + \frac{3(4) + 1 - 3(1)}{5}\begin{bmatrix} 1 \\ 2 \\ 0 \end{bmatrix} + \frac{-2(4) + 1 + 2(1)}{5}\begin{bmatrix} 1 \\ 3 \\ 2 \end{bmatrix} \\ &= \frac{15}{5}\begin{bmatrix} 1 \\ 0 \\ 1 \end{bmatrix} + \frac{10}{5}\begin{bmatrix} 1 \\ 2 \\ 0 \end{bmatrix} + \frac{-5}{5}\begin{bmatrix} 1 \\ 3 \\ 2 \end{bmatrix} = 3\begin{bmatrix} 1 \\ 0 \\ 1 \end{bmatrix} + 2\begin{bmatrix} 1 \\ 2 \\ 0 \end{bmatrix} - 1\begin{bmatrix} 1 \\ 3 \\ 2 \end{bmatrix} = \begin{bmatrix} 3 + 2 - 1 \\ 0 + 4 - 3 \\ 3 + 0 - 2 \end{bmatrix} = \begin{bmatrix} 4 \\ 1 \\ 1 \end{bmatrix} \end{aligned}$$

where $c_1 = 3$, $c_2 = 2$, and $c_3 = -1$.

Spanning R^2 and R^3

The 2×1 vectors, represented by points in the coordinate plane, make up R^2. And the 3×1 vectors, represented by points in space, make up R^3. A common question in applications of linear algebra is whether a particular set of vectors spans R^2 or R^3.

Seeking sets that span

When a set of vectors spans R^2, then you can create any possible point in the coordinate plane using linear combinations of that vector set.

For example, the vectors $\mathbf{v}_1$ and $\mathbf{v}_2$ are in the set

$$\left\{\begin{bmatrix}2\\1\end{bmatrix},\begin{bmatrix}3\\2\end{bmatrix}\right\}$$

and span R^2. I show you that the statement is true by writing the linear combination and solving the corresponding system of linear equations.

$$c_1\begin{bmatrix}2\\1\end{bmatrix}+c_2\begin{bmatrix}3\\2\end{bmatrix}=\begin{bmatrix}x\\y\end{bmatrix}$$

$$\begin{cases}2c_1+3c_2=x\\ c_1+2c_2=y\end{cases}$$

Add −2 times the second equation to the first equation, and you get

$$\begin{aligned}-c_2&=x-2y\\ c_1+2c_2&=y\end{aligned}$$

Now multiply the first equation through by −1, and substitute the value of c_2 into the second equation.

$$\begin{aligned}c_2&=2y-x\\ c_1+2(2y-x)&=y\\ c_1+4y-2x&=y\\ c_1&=2x-3y\end{aligned}$$

You get that $c_1 = 2x - 3y$ and that $c_2 = 2y - x$. So any vector can be written as a linear combination of the two vectors in the set. If the elements in the target vector are x and y, then the linear combination $(2x - 3y)\mathbf{v}_1 + (2y - x)\mathbf{v}_2$ results in that vector.

Here's an example of a vector set that spans R^3:

$$\left\{\begin{bmatrix}1\\0\\0\end{bmatrix},\begin{bmatrix}0\\1\\1\end{bmatrix},\begin{bmatrix}1\\0\\1\end{bmatrix}\right\}$$

Writing the corresponding linear combination and solving for the scalar multiples, you get $c_1 = x + y - z$, $c_2 = y$, and $c_3 = z - y$.

So, for any vector in R^3,

$$c_1\mathbf{v}_1 + c_2\mathbf{v}_2 + c_3\mathbf{v}_3 = \begin{bmatrix} x \\ y \\ z \end{bmatrix}$$

$$(x + y - z)\begin{bmatrix} 1 \\ 0 \\ 0 \end{bmatrix} + y\begin{bmatrix} 0 \\ 1 \\ 1 \end{bmatrix} + (z - y)\begin{bmatrix} 1 \\ 0 \\ 1 \end{bmatrix} = \begin{bmatrix} x \\ y \\ z \end{bmatrix}$$

Ferreting out the non-spanning sets

The set of vectors

$$\left\{\begin{bmatrix} 2 \\ 3 \end{bmatrix}, \begin{bmatrix} 4 \\ 6 \end{bmatrix}\right\}$$

does not span R^2. The non-spanning set may seem more apparent to you because one of the vectors is a multiple of the other. Any linear combination written with the vectors in the set has only the zero vector for a solution. Linear combinations of the vectors just give you multiples of the first vector, so the set can't span R^2.

Consider now, the vector set in R^3 where $\mathbf{v}_1$ and $\mathbf{v}_2$ are

$$\left\{\begin{bmatrix} 1 \\ 1 \\ 3 \end{bmatrix}, \begin{bmatrix} 0 \\ 2 \\ 1 \end{bmatrix}\right\}$$

The corresponding linear combination and system of equations is

$$c_1\begin{bmatrix} 1 \\ 1 \\ 3 \end{bmatrix} + c_2\begin{bmatrix} 0 \\ 2 \\ 1 \end{bmatrix} = \begin{bmatrix} x \\ y \\ z \end{bmatrix}$$

$$\begin{cases} c_1 = x \\ c_1 + 2c_2 = y \\ 3c_1 + c_2 = z \end{cases}$$

Adding –1 times the first equation to the second equation and –3 times the first equation to the third equation, you get

$$\begin{cases} c_1 = x \\ 2c_2 = y - x \\ c_2 = z - 3x \end{cases}$$

The second and third equations are now contradictory. The second says that $c_2 = {}^1/_2\,(y - x)$ and the third equation says that $c_2 = z - 3x$. Except for a few select vectors, the equations don't work.

Chapter 6

Investigating the Matrix Equation $A\mathbf{x} = \mathbf{b}$

In This Chapter

- Using matrix multiplication in multiple ways
- Defining the product of a matrix and a vector
- Solving the matrix-vector equation, when possible
- Formulating formulas for multiple solutions

The matrix-vector equation $A\mathbf{x} = \mathbf{b}$ incorporates properties and techniques from matrix arithmetic and algebra's systems of equations. The term $A\mathbf{x}$ indicates that matrix A and vector $\mathbf{x}$ are being multiplied together. To simplify matters, the convention is to just write the two symbols next to one another without an asterisk (*) to indicate multiplication.

In Chapter 3, you find all sorts of information on matrices — from their characteristics to their operations. In Chapter 4, you see systems of equations and how to utilize some of the properties of matrices in their solutions. And here, in Chapter 6, you find the vectors, taken from Chapter 3, and I introduce you to some of the unique properties associated with vectors — properties that lend themselves to solving the equation and its corresponding system.

Joining forces with systems of equations, the matrix equation $A\mathbf{x} = \mathbf{b}$ takes on a life of its own. In this chapter, everything is all rolled together into one happy, (usually) cohesive family of equations and solutions. You find techniques for determining the solutions (when they exist), and you see how to write infinite solutions in a symbolically usable expression. By recognizing the common characteristics of solutions of $A\mathbf{x} = \mathbf{b}$, you see how to take what can be a complicated and messy problem and put it into something workable.

Working Through Matrix-Vector Products

Matrices have two completely different types of products. When preparing to perform multiplication involving matrices, you first take note of why you want to multiply them and then decide how.

The simplest multiplication involving matrices is scalar multiplication, where each element in the matrix is multiplied by the same multiplier. As simple as scalar multiplication may appear, it still plays an important part in solving matrix equations and equations involving the products of matrices and vectors.

The second type of multiplication involving matrices is the more complicated type. You multiply two matrices together only when they have the correct matrix dimensions and correct multiplication order.

Chapter 3 covers matrix multiplication in great detail, so refer to that material if you're feeling a bit shaky on the topic.

In this chapter, I concentrate on multiplying a matrix times a vector (an $n \times 1$ matrix). Because of the special characteristics of matrix-vector multiplication, I'm able to show you a quicker, easier method for determining the product.

Establishing a link with matrix products

In Chapter 3, you find the techniques needed to multiply two matrices together. When you multiply an $m \times n$ matrix times an $n \times p$ matrix, your result is an $m \times p$ matrix. Refer to multiplication of matrices in Chapter 3 if the previous statement doesn't seem familiar to you.

Revisiting matrix multiplication

When multiplying the matrix A times the vector **x**, you still have the same multiplication rule applying: The number of columns in matrix A must match the number of rows in vector **x**. And, because a vector is always some $n \times 1$ matrix, when you multiply a matrix times a vector, you always get another vector.

The product $A\mathbf{x}$ of an $m \times n$ matrix and an $n \times 1$ vector is an $m \times 1$ vector. For example, consider the matrix B and vector **z** shown here:

$$B = \begin{bmatrix} 2 & 3 & 1 & 0 \\ -1 & 4 & 1 & -3 \end{bmatrix} \text{ and } \mathbf{z} = \begin{bmatrix} 1 \\ -3 \\ 2 \\ 2 \end{bmatrix}$$

The product of the 2×4 matrix B and the 4×1 vector **z** is a 2×1 vector.

$$\mathbf{Bz} = \begin{bmatrix} 2 & 3 & 1 & 0 \\ -1 & 4 & 1 & -3 \end{bmatrix} * \begin{bmatrix} 1 \\ -3 \\ 2 \\ 2 \end{bmatrix} = \begin{bmatrix} 2(1)+3(-3)+1(2)+0(2) \\ -1(1)+4(-3)+1(2)+-3(2) \end{bmatrix} = \begin{bmatrix} -5 \\ -17 \end{bmatrix}$$

Looking at systems

Multiplying a matrix and a vector always results in a vector. The matrix multiplication process is actually more quickly and easily performed by rewriting the multiplication problem as the sum of scalar multiplications — a linear combination, in fact.

If A is an $m \times n$ matrix whose *columns* are designated by $\boldsymbol{a}_1, \boldsymbol{a}_2, \ldots, \boldsymbol{a}_n$ and if vector **x** has n rows, then the multiplication $A\mathbf{x}$ is equivalent to having each of the elements in vector **x** times one of the columns in matrix A:

$$A\mathbf{x} = \begin{bmatrix} \mathbf{a}_1 & \mathbf{a}_2 & \mathbf{a}_3 & \cdots & \mathbf{a}_n \end{bmatrix} * \begin{bmatrix} x_1 \\ x_2 \\ x_3 \\ \vdots \\ x_n \end{bmatrix}$$

$$= x_1\mathbf{a}_1 + x_2\mathbf{a}_2 + x_3\mathbf{a}_3 + \cdots + x_n\mathbf{a}_n$$

$$= x_1 \begin{bmatrix} a_{11} \\ a_{21} \\ a_{31} \\ \vdots \\ a_{m1} \end{bmatrix} + x_2 \begin{bmatrix} a_{12} \\ a_{22} \\ a_{32} \\ \vdots \\ a_{m2} \end{bmatrix} + x_3 \begin{bmatrix} a_{13} \\ a_{23} \\ a_{33} \\ \vdots \\ a_{m3} \end{bmatrix} + \cdots + x_n \begin{bmatrix} a_{1n} \\ a_{2n} \\ a_{3n} \\ \vdots \\ a_{mn} \end{bmatrix}$$

So, revisiting the earlier multiplication problem, and using the new procedure, multiply B**z** by using the columns of B and rows of **z**,

$$\mathbf{Bz} = \begin{bmatrix} 2 & 3 & 1 & 0 \\ -1 & 4 & 1 & -3 \end{bmatrix} * \begin{bmatrix} 1 \\ -3 \\ 2 \\ 2 \end{bmatrix} = 1\begin{bmatrix} 2 \\ -1 \end{bmatrix} - 3\begin{bmatrix} 3 \\ 4 \end{bmatrix} + 2\begin{bmatrix} 1 \\ 1 \end{bmatrix} + 2\begin{bmatrix} 0 \\ -3 \end{bmatrix}$$

$$= \begin{bmatrix} 2 \\ -1 \end{bmatrix} + \begin{bmatrix} -9 \\ -12 \end{bmatrix} + \begin{bmatrix} 2 \\ 2 \end{bmatrix} + \begin{bmatrix} 0 \\ -6 \end{bmatrix} = \begin{bmatrix} -5 \\ -17 \end{bmatrix}$$

The following properties apply to the products of matrices and vectors. If A is an $m \times n$ matrix, **u** and **v** are $n \times 1$ vectors, and c is a scalar, then:

- $cA = Ac$ and $A(c\mathbf{u}) = c(A\mathbf{u})$. This is the commutative property of scalar multiplication.
- $A(\mathbf{u} + \mathbf{v}) = A\mathbf{u} + A\mathbf{v}$ and $c(\mathbf{u} + \mathbf{v}) = c\mathbf{u} + c\mathbf{v}$. This is the distributive property of scalar multiplication.

Tying together systems of equations and the matrix equation

Consider matrix A with dimension $m \times n$ whose columns are $\mathbf{a}_1, \mathbf{a}_2, \ldots, \mathbf{a}_n$ and a vector **x** with n rows. Now throw into the mix another vector, **b**, that has m rows. The multiplication $A\mathbf{x}$ results in an $m \times 1$ matrix. And the equation $A\mathbf{x} = \mathbf{b}$ has the same solution set as the augmented matrix $[\mathbf{a}_1\ \mathbf{a}_2\ \mathbf{a}_3 \ldots \mathbf{a}_n\ \mathbf{b}]$.

The matrix equation is written first with A as a matrix of colums $\mathbf{a}_i$ and then with the elements in vector **x** multiplying the columns of A. When multiplied out, you see the results of the scalar multiplication.

$$A\mathbf{x} = \mathbf{b}$$

$$\begin{bmatrix} \mathbf{a}_1 & \mathbf{a}_2 & \mathbf{a}_3 & \cdots & \mathbf{a}_n \end{bmatrix} * \begin{bmatrix} x_1 \\ x_2 \\ x_3 \\ \vdots \\ x_n \end{bmatrix} = \begin{bmatrix} b_1 \\ b_2 \\ b_3 \\ \vdots \\ b_m \end{bmatrix}$$

$$x_1 \begin{bmatrix} a_{11} \\ a_{21} \\ a_{31} \\ \vdots \\ a_{m1} \end{bmatrix} + x_2 \begin{bmatrix} a_{12} \\ a_{22} \\ a_{32} \\ \vdots \\ a_{m2} \end{bmatrix} + x_3 \begin{bmatrix} a_{13} \\ a_{23} \\ a_{33} \\ \vdots \\ a_{m3} \end{bmatrix} + \cdots + x_n \begin{bmatrix} a_{1n} \\ a_{2n} \\ a_{3n} \\ \vdots \\ a_{mn} \end{bmatrix} = \begin{bmatrix} b_1 \\ b_2 \\ b_3 \\ \vdots \\ b_m \end{bmatrix}$$

$$\begin{bmatrix} a_{11}x_1 + & a_{12}x_2 + & a_{13}x_3 + & \cdots & + a_{1n}x_n \\ a_{21}x_1 + & a_{22}x_2 + & a_{23}x_3 + & \cdots & + a_{2n}x_n \\ a_{31}x_1 + & a_{32}x_2 + & a_{33}x_3 + & \cdots & + a_{3n}x_n \\ \vdots & \vdots & \vdots & \ddots & \vdots \\ a_{m1}x_1 + & a_{m2}x_2 + & a_{m3}x_3 + & \cdots & + a_{mn}x_n \end{bmatrix} = \begin{bmatrix} b_1 \\ b_2 \\ b_3 \\ \vdots \\ b_m \end{bmatrix}$$

The corresponding augmented matrix is used to solve the system of equations represented by the $A\mathbf{x} = \mathbf{b}$ equation:

$$\left[\begin{array}{ccccc|c} a_{11} & a_{12} & a_{13} & \cdots & a_{1n} & b_1 \\ a_{21} & a_{22} & a_{23} & \cdots & a_{2n} & b_2 \\ a_{31} & a_{32} & a_{33} & \cdots & a_{3n} & b_3 \\ \vdots & \vdots & \vdots & \ddots & \vdots & \vdots \\ a_{m1} & a_{m2} & a_{m3} & \cdots & a_{mn} & b_m \end{array}\right]$$

I use the augmented matrix as shown in Chapter 4; you can refer to that chapter for more on how augmented matrices are used to solve systems of equations.

Now, let me use an augmented matrix to solve the equation $A\mathbf{x} = \mathbf{b}$ where:

$$A = \begin{bmatrix} -1 & 4 & 2 \\ 3 & 0 & -2 \\ 5 & 4 & 3 \end{bmatrix}, \mathbf{x} = \begin{bmatrix} x_1 \\ x_2 \\ x_3 \end{bmatrix}, \mathbf{b} = \begin{bmatrix} -30 \\ 90 \\ 45 \end{bmatrix}$$

Writing the augmented matrix:

$$A\mathbf{x} = \mathbf{b}$$

$$\begin{bmatrix} -1 & 4 & 2 \\ 3 & 0 & -2 \\ 5 & 4 & 3 \end{bmatrix} * \begin{bmatrix} x_1 \\ x_2 \\ x_3 \end{bmatrix} = \begin{bmatrix} -30 \\ 90 \\ 45 \end{bmatrix}$$

$$\left[\begin{array}{ccc|c} -1 & 4 & 2 & -30 \\ 3 & 0 & -2 & 90 \\ 5 & 4 & 3 & 45 \end{array}\right]$$

When the augmented matrix is row-reduced, the values of x_1, x_2, and x_3 are 16, 7, and -21, respectively. I show you how I got these numbers in the "Singling out a single solution" section, later in this chapter.

For now, this is what the solution to the equation $A\mathbf{x} = \mathbf{b}$ looks like, with the elements of vector $\mathbf{x}$ filled in.

$$\begin{bmatrix} -1 & 4 & 2 \\ 3 & 0 & -2 \\ 5 & 4 & 3 \end{bmatrix} * \begin{bmatrix} 16 \\ 7 \\ -21 \end{bmatrix} = \begin{bmatrix} -30 \\ 90 \\ 45 \end{bmatrix}$$

Confirming the Existence of a Solution or Solutions

Have you ever looked for something and known that it's somewhere close, but can't find it. You *know* it's there, but you can't even prove that it really exists. Well, I can't help you find your keys or that receipt or that earring, but I can help you find the solution of a matrix-vector equation (or state definitively that it doesn't really exist)!

The equation $A\mathbf{x} = \mathbf{b}$ has a solution only when **b** is a linear combination of the columns of matrix A. That's the special matrix-vector multiplication I show you earlier in "Tying together systems of equations and the matrix equation." Also, **b** is in the span of A when there is a solution to $A\mathbf{x} = \mathbf{b}$. (You don't remember much on the *span* of a set of vectors? You'll find that information in Chapter 5.)

Singling out a single solution

The equation $A\mathbf{x} = \mathbf{b}$ has a single solution when only one vector, **x**, makes the equation into a true statement. To determine whether you have a solution, use the tried-and-true method of creating an augmented matrix corresponding to the columns of A and the vector **b** and then performing row operations. For example, given the matrix A and vector **b**, I find the vector **x** that solves the equation.

$$A = \begin{bmatrix} 1 & -2 & 2 & 1 \\ 1 & -1 & 5 & 0 \\ 2 & -2 & 11 & 2 \\ 0 & 2 & 8 & 1 \end{bmatrix} \text{ and } \mathbf{b} = \begin{bmatrix} 2 \\ 10 \\ 20 \\ 17 \end{bmatrix}$$

You see that the dimension of matrix A is 4×4 and the dimension of vector **b** is 4×1. When setting up the matrix multiplication, you know that in order to have the product of A and vector **x** result in a 4×1 vector, then the dimension of vector **x** must also be 4×1. (Refer to Chapter 3 and multiplication of matrices if you need more information on multiplying and dimensions.)

So the matrix equation for $A\mathbf{x} = \mathbf{b}$ is written

$$\begin{bmatrix} 1 & -2 & 2 & 1 \\ 1 & -1 & 5 & 0 \\ 2 & -2 & 11 & 2 \\ 0 & 2 & 8 & 1 \end{bmatrix} * \begin{bmatrix} x_1 \\ x_2 \\ x_3 \\ x_4 \end{bmatrix} = \begin{bmatrix} 2 \\ 10 \\ 20 \\ 17 \end{bmatrix}$$

and the augmented matrix is

$$\left[\begin{array}{cccc|c} 1 & -2 & 2 & 1 & 2 \\ 1 & -1 & 5 & 0 & 10 \\ 2 & -2 & 11 & 2 & 20 \\ 0 & 2 & 8 & 1 & 17 \end{array}\right]$$

Now, performing row operations, I change the matrix to *reduced echelon form* so I can determine the solution — what each x_i represents.

1. **Multiply the first row by −1 and add it to row two; then multiply the first row by −2 and add it to row three.**

   Does this $-1R_1 + R_2$ business look like some sort of secret code? No, it isn't a secret. You can find the explanation in Chapter 4.

$$\left[\begin{array}{cccc|c} 1 & -2 & 2 & 1 & 2 \\ 1 & -1 & 5 & 0 & 10 \\ 2 & -2 & 11 & 2 & 20 \\ 0 & 2 & 8 & 1 & 17 \end{array}\right] \begin{array}{c} \\ -1R_1 + R_2 \to R_2 \\ -2R_1 + R_3 \to R_3 \\ \\ \end{array} \left[\begin{array}{cccc|c} 1 & -2 & 2 & 1 & 2 \\ 0 & 1 & 3 & -1 & 8 \\ 0 & 2 & 7 & 0 & 16 \\ 0 & 2 & 8 & 1 & 17 \end{array}\right]$$

2. **Multiply row two by 2 and add it to row one; multiply row two by −2 and add it to row three; and, finally, multiply row two by −2 and add it to row four.**

$$\left[\begin{array}{cccc|c} 1 & -2 & 2 & 1 & 2 \\ 0 & 1 & 3 & -1 & 8 \\ 0 & 2 & 7 & 0 & 16 \\ 0 & 2 & 8 & 1 & 17 \end{array}\right] \begin{array}{c} 2R_2 + R_1 \to R_1 \\ -2R_2 + R_3 \to R_3 \\ -2R_2 + R_4 \to R_4 \end{array} \left[\begin{array}{cccc|c} 1 & 0 & 8 & -1 & 18 \\ 0 & 1 & 3 & -1 & 8 \\ 0 & 0 & 1 & 2 & 0 \\ 0 & 0 & 2 & 3 & 1 \end{array}\right]$$

3. **Multiply row three by −8 and add it to row one; multiply row three by −3 and add it to row two; then multiply row three by −2 and add it to row four.**

$$\left[\begin{array}{cccc|c} 1 & 0 & 8 & -1 & 18 \\ 0 & 1 & 3 & -1 & 8 \\ 0 & 0 & 1 & 2 & 0 \\ 0 & 0 & 2 & 3 & 1 \end{array}\right] \begin{array}{c} -8R_3 + R_1 \to R_1 \\ -3R_3 + R_2 \to R_2 \\ -2R_3 + R_4 \to R_4 \end{array} \left[\begin{array}{cccc|c} 1 & 0 & 0 & -17 & 18 \\ 0 & 1 & 0 & -7 & 8 \\ 0 & 0 & 1 & 2 & 0 \\ 0 & 0 & 0 & -1 & 1 \end{array}\right]$$

4. **Multiply row four by –1; multiply the new row four by 17 and add it to row one; multiply row four by 7 and add it to row two; and multiply row four by –2 and add it to row three.**

$$\left[\begin{array}{cccc|c} 1 & 0 & 0 & -17 & 18 \\ 0 & 1 & 0 & -7 & 8 \\ 0 & 0 & 1 & 2 & 0 \\ 0 & 0 & 0 & -1 & 1 \end{array}\right] \xrightarrow{-1R_4 \to R_4} \left[\begin{array}{cccc|c} 1 & 0 & 0 & -17 & 18 \\ 0 & 1 & 0 & -7 & 8 \\ 0 & 0 & 1 & 2 & 0 \\ 0 & 0 & 0 & 1 & -1 \end{array}\right]$$

$$\begin{array}{c} 17R_4 + R_1 \to R_1 \\ 7R_4 + R_2 \to R_2 \\ -2R_4 + R_3 \to R_3 \end{array} \left[\begin{array}{cccc|c} 1 & 0 & 0 & 0 & 1 \\ 0 & 1 & 0 & 0 & 1 \\ 0 & 0 & 1 & 0 & 2 \\ 0 & 0 & 0 & -1 & 1 \end{array}\right]$$

The numbers in the last column correspond to the single solution for the vector **x.** In the vector **x,** the elements are $x_1 = 1, x_2 = 1, x_3 = 2, x_4 = -1$. The equation $A\mathbf{x} = \mathbf{b}$ has this one — and only one — solution.

Making way for more than one solution

Many matrix-vector equations have more than one solution. When determining the solutions, you may try to list all the possible solutions or you may just list a rule for quickly determining some of the solutions in the set. When using the matrix-vector equation in an application, you usually have certain *constraints,* which limit the scope of the numbers you're allowed to use for an answer. For example, if your application has to do with the length of a building, you wouldn't choose any of the solutions that make that length a negative number.

Solving for the solutions of a specific vector

When vector **b** has specific numerical values, you don't always know, upfront, if you'll find one solution, no solution, or many solutions. You use the augmented matrix and reduce that matrix to echelon form.

For example, consider the matrix A and vector **b** in the equation $A\mathbf{x} = \mathbf{b}$.

$$A = \begin{bmatrix} 1 & 3 & 1 \\ 3 & -2 & -8 \\ 4 & 5 & -3 \end{bmatrix}, \quad \mathbf{b} = \begin{bmatrix} 6 \\ 7 \\ 17 \end{bmatrix}$$

Writing the corresponding augmented matrix, you get

$$\left[\begin{array}{ccc|c} 1 & 3 & 1 & 6 \\ 3 & -2 & -8 & 7 \\ 4 & 5 & -3 & 17 \end{array}\right]$$

Now, doing row reductions, perfom these steps:

1. **Multiply row one by –3 and add it to row two; multiply row one by –4 and add it to row three.**

$$\left[\begin{array}{ccc|c} 1 & 3 & 1 & 6 \\ 3 & -2 & -8 & 7 \\ 4 & 5 & -3 & 17 \end{array}\right] \begin{array}{c} -3R_1 + R_2 \to R_2 \\ -4R_1 + R_3 \to R_3 \end{array} \left[\begin{array}{ccc|c} 1 & 3 & 1 & 6 \\ 0 & -11 & -11 & -11 \\ 0 & -7 & -7 & -7 \end{array}\right]$$

2. **Divide each element in row two by –11, and divide each element in row three by –7.**

$$\begin{array}{c} -\frac{1}{11}R_2 \to R_2 \\ -\frac{1}{7}R_3 \to R_3 \end{array} \left[\begin{array}{ccc|c} 1 & 3 & 1 & 6 \\ 0 & 1 & 1 & 1 \\ 0 & 1 & 1 & 1 \end{array}\right]$$

3. **Multiply row two by –1 and add it to row three.**

$$\left[\begin{array}{ccc|c} 1 & 3 & 1 & 6 \\ 0 & 1 & 1 & 1 \\ 0 & 1 & 1 & 1 \end{array}\right] -1R_2 + R_3 \to R_3 \left[\begin{array}{ccc|c} 1 & 3 & 1 & 6 \\ 0 & 1 & 1 & 1 \\ 0 & 0 & 0 & 0 \end{array}\right]$$

   The last row in the augmented matrix contains all 0s. Having a row of 0s indicates that you have more than one solution for the equation.

4. **Take the reduced form of the matrix and go back to a system of equations format, with the reduced matrix multiplying vector** x **and setting the product equal to the last column.**

$$\begin{bmatrix} 1 & 3 & 1 \\ 0 & 1 & 1 \\ 0 & 0 & 0 \end{bmatrix} * \begin{bmatrix} x_1 \\ x_2 \\ x_3 \end{bmatrix} = \begin{bmatrix} 6 \\ 1 \\ 0 \end{bmatrix}$$

Carl Friedrich Gauss

One of the most prolific mathematicians of all time was German mathematician Carl Friedrich Gauss. Gauss lived from 1777 until 1855. He made contributions not only to mathematics, but also to astronomy, electrostatics, and optics. Gauss was a child prodigy — to the delight of his parents and, often, to the dismay of his teachers. Many legends are attributed to Gauss's precociousness.

One of the more well-known stories of his childhood has to do with a tired, harried teacher who wanted to have a few minutes of peace and quiet. The teacher gave Gauss's class the assignment of finding the sum of the numbers from 1 through 100, thinking that it would take even the brightest child at least half an hour to complete the task. To the teacher's chagrin, little Carl Friedrich came up with the answer in minutes. Gauss had the answer and had stumbled on the basics to the formula for finding the sum of *n* integers.

Gauss's father wanted Carl Friedrich to become a mason and wasn't all that supportive of Gauss's advanced schooling. But, with the help of his mother and a family friend, Gauss was able to attend the university.

Gauss is credited with developing the technique *Gaussian elimination,* a method or procedure for using elementary row operations on augmented matrices to solve systems of equations.

After Gauss died, his brain was preserved and studied. His brain was found to weigh 1,492 grams (an average male brain weighs between 1,300 and 1,400 grams), and the cerebral area measured 219,588 square millimeters (as compared to about 150,000 square millimeters for an average male). The brain also contained highly developed convolutions. All these measurements were said to confirm Gauss's genius.

5. Do the matrix multiplication and write the corresponding system of equations.

$$\begin{cases} x_1 + 3x_2 + x_3 = 6 \\ x_2 + x_3 = 1 \\ 0 = 0 \end{cases}$$

6. Solve the second equation for x_2, and you get that $x_2 = 1 - x_3$. Replace the x_3 with a parameter, *k*, and you now have $x_2 = 1 - k$. Now solve for x_1 in the first equation, substituting in the equivalences of x_2 and x_3.

$$\begin{aligned} x_1 + 3x_2 + x_3 &= 6 \\ x_1 + 3(1-k) + k &= 6 \\ x_1 + 3 - 3k + k &= 6 \\ x_1 &= 3 + 2k \end{aligned}$$

The vector **b** is now written:

$$\mathbf{b} = \begin{bmatrix} 3+2k \\ 1-k \\ k \end{bmatrix}$$

where k is any real number.

For example, if you let $k = 2$, you have

$$\mathbf{b} = \begin{bmatrix} 3+2(2) \\ 1-2 \\ 2 \end{bmatrix} = \begin{bmatrix} 7 \\ -1 \\ 2 \end{bmatrix}$$

Going more general with infinite solutions

When your matrix-vector equation $A\mathbf{x} = \mathbf{b}$ involves the vector **b** with specific elements, you may have more than one solution. What I show you here is the situation where matrix A has specific values, and vector **b** varies. The vector **b** is different every time you fill in random numbers. Then, after you choose numbers to fill in for the elements in **b,** the solutions in vector **x** correspond to those numbers.

For example, consider matrix A and some random matrix **b.**

$$A = \begin{bmatrix} -1 & 4 & 2 \\ 3 & 0 & -2 \\ 5 & 4 & 3 \end{bmatrix} \text{ and } \mathbf{b} = \begin{bmatrix} b_1 \\ b_2 \\ b_3 \end{bmatrix}$$

The augmented matrix for $A\mathbf{x} = \mathbf{b}$ is

$$\left[\begin{array}{ccc|c} -1 & 4 & 2 & b_1 \\ 3 & 0 & -2 & b_2 \\ 5 & 4 & 3 & b_3 \end{array}\right]$$

Performing row operations to produce an echelon form, follow these steps:

1. **Multiply row one by –1; then multiply row one by –3 and add it to row two; finally, multiply row one by –5 and add it to row 3.**

$$\left[\begin{array}{ccc|c} -1 & 4 & 2 & b_1 \\ 3 & 0 & -2 & b_2 \\ 5 & 4 & 3 & b_3 \end{array}\right] -1R_1 \rightarrow R_1 \left[\begin{array}{ccc|c} 1 & -4 & -2 & -b_1 \\ 3 & 0 & -2 & b_2 \\ 5 & 4 & 3 & b_3 \end{array}\right]$$

$$\begin{array}{l} -3R_1 + R_2 \rightarrow R_2 \\ -5R_1 + R_3 \rightarrow R_3 \end{array} \left[\begin{array}{ccc|c} 1 & -4 & -2 & -b_1 \\ 0 & 12 & 4 & 3b_1 + b_2 \\ 0 & 24 & 13 & 5b_1 + b_3 \end{array}\right]$$

2. Multiply row two by –2 and add it to row three.

$$-2R_2 + R_3 \rightarrow R_3 \left[\begin{array}{ccc|c} 1 & -4 & -2 & -b_1 \\ 0 & 12 & 4 & 3b_1 + b_2 \\ 0 & 0 & 5 & -1b_1 - 2b_2 + b_3 \end{array}\right]$$

3. Divide each term in row two by 12, and divide each term in row three by 5.

$$\begin{array}{l} \frac{1}{12}R_2 \rightarrow R_2 \\ \frac{1}{5}R_3 \rightarrow R_3 \end{array} \left[\begin{array}{ccc|c} 1 & -4 & -2 & -b_1 \\ 0 & 1 & \frac{1}{3} & \frac{1}{4}b_1 + \frac{1}{12}b_2 \\ 0 & 0 & 1 & -\frac{1}{5}b_1 - \frac{2}{5}b_2 + \frac{1}{5}b_3 \end{array}\right]$$

4. Multiply row two by 4 and add it to row one.

$$4R_2 + R_1 \rightarrow R_1 \left[\begin{array}{ccc|c} 1 & 0 & -\frac{2}{3} & \frac{1}{3}b_2 \\ 0 & 1 & \frac{1}{3} & \frac{1}{4}b_1 + \frac{1}{12}b_2 \\ 0 & 0 & 1 & -\frac{1}{5}b_1 - \frac{2}{5}b_2 + \frac{1}{5}b_3 \end{array}\right]$$

5. Multiply row three by $^2/_3$ and add it to row one; then multiply row three by $-^1/_3$ and add it to row two.

$$\begin{array}{l} \frac{2}{3}R_3 + R_1 \rightarrow R_1 \\ -\frac{1}{3}R_3 + R_2 \rightarrow R_2 \end{array} \left[\begin{array}{ccc|c} 1 & 0 & 0 & -\frac{2}{15}b_1 + \frac{1}{15}b_2 + \frac{2}{15}b_3 \\ 0 & 1 & 0 & \frac{19}{60}b_1 + \frac{13}{60}b_2 - \frac{1}{15}b_3 \\ 0 & 0 & 1 & -\frac{1}{5}b_1 - \frac{2}{5}b_2 + \frac{1}{5}b_3 \end{array}\right]$$

The solution for vector **x** is completely determined by the numbers you choose for vector **b.** Once you pick some values for vector **b,** you also have the solution vector **x.** You find the elements of **x** letting

$$x_1 = -\frac{2}{15}b_1 + \frac{1}{15}b_2 + \frac{2}{15}b_3,\ x_2 = \frac{19}{60}b_1 + \frac{13}{60}b_2 - \frac{1}{15}b_3,\ x_3 = -\frac{1}{5}b_1 - \frac{2}{5}b_2 + \frac{1}{5}b_3$$

To demonstrate how this works, I'll first pick some convenient values for vector **b.** By *convenient,* I mean that the elements will change the fractions in the formulas to integers. So, if

$$\mathbf{b} = \begin{bmatrix} 120 \\ -60 \\ 75 \end{bmatrix}$$

then

$$\begin{aligned} x_1 &= -\frac{2}{15}(120) + \frac{1}{15}(-60) + \frac{2}{15}(75) = -16 - 4 + 10 = -10 \\ x_2 &= \frac{19}{60}(120) + \frac{13}{60}(-60) - \frac{1}{15}(75) = 38 - 13 - 5 = 20 \\ x_3 &= -\frac{1}{5}(120) - \frac{2}{5}(-60) + \frac{1}{5}(75) = -24 + 24 + 15 = 15 \end{aligned}$$

and the matrix-vector equation reads

$$A\mathbf{x} = \mathbf{b}$$

$$\begin{bmatrix} -1 & 4 & 2 \\ 3 & 0 & -2 \\ 5 & 4 & 3 \end{bmatrix} \begin{bmatrix} -10 \\ 20 \\ 15 \end{bmatrix} = \begin{bmatrix} 120 \\ -60 \\ 75 \end{bmatrix}$$

But, the beauty of this matrix-vector equation is that any real numbers can be used for elements in vector **b.** You can even use such "inconvenient" numbers as $b_1 = 1$, $b_2 = 2$, and $b_3 = -3$ to get

$$\begin{aligned} x_1 &= -\frac{2}{15}(1) + \frac{1}{15}(2) + \frac{2}{15}(-3) = -\frac{2}{15} + \frac{2}{15} - \frac{6}{15} = -\frac{6}{15} = -\frac{2}{5} \\ x_2 &= \frac{19}{60}(1) + \frac{13}{60}(2) - \frac{1}{15}(-3) = \frac{19}{60} + \frac{26}{60} + \frac{3}{15} = \frac{57}{60} = \frac{19}{20} \\ x_3 &= -\frac{1}{5}(1) - \frac{2}{5}(2) + \frac{1}{5}(-3) = -\frac{1}{5} - \frac{4}{5} - \frac{3}{5} = -\frac{8}{5} \end{aligned}$$

giving you the equation

$$A\mathbf{x} = \mathbf{b}$$

$$\begin{bmatrix} -1 & 4 & 2 \\ 3 & 0 & -2 \\ 5 & 4 & 3 \end{bmatrix} \begin{bmatrix} -\frac{2}{5} \\ \frac{19}{20} \\ -\frac{8}{5} \end{bmatrix} = \begin{bmatrix} 1 \\ 2 \\ -3 \end{bmatrix}$$

Okay, I call this "inconvenient," meaning that the solutions don't come out to be nice integers. You may not feel quite as "inconvenienced," so you can call it whatever you want.

Getting more selective with a few solutions

Now, if you're willing and able to be flexible with the entries in vector **b,** you can create a specialized solution to the matrix-vector equation $A\mathbf{x} = \mathbf{b}$.

Let matrix A and vector **b** have the following values:

$$A = \begin{bmatrix} -1 & 2 & 4 \\ 5 & -3 & 3 \\ 3 & 1 & 11 \end{bmatrix} \text{ and } \mathbf{b} = \begin{bmatrix} b_1 \\ b_2 \\ b_3 \end{bmatrix}$$

Write the augmented matrix and go through the row reductions:

$$\left[\begin{array}{ccc|c} -1 & 2 & 4 & b_1 \\ 5 & -3 & 3 & b_2 \\ 3 & 1 & 11 & b_3 \end{array}\right] -1R_1 \to R_1 \left[\begin{array}{ccc|c} 1 & -2 & -4 & -b_1 \\ 5 & -3 & 3 & b_2 \\ 3 & 1 & 11 & b_3 \end{array}\right]$$

$$\begin{array}{c} -5R_1 + R_2 \to R_2 \\ -3R_1 + R_3 \to R_3 \end{array} \left[\begin{array}{ccc|c} 1 & -2 & -4 & -b_1 \\ 0 & 7 & 23 & 5b_1 + b_2 \\ 0 & 7 & 23 & 3b_1 + b_3 \end{array}\right]$$

$$-1R_2 + R_3 \to R_3 \left[\begin{array}{ccc|c} 1 & -2 & -4 & -b_1 \\ 0 & 7 & 23 & 5b_1 + b_2 \\ 0 & 0 & 0 & -2b_1 - b_2 + b_3 \end{array}\right]$$

The corresponding equation for the last row of the matrix reads $0x_1 + 0x_2 + 0x_3 = -2b_1 - b_2 + b_3$. The equation is not equal to 0 for very many choices of b_1, b_2, and b_3. For example, if you choose to let vector **b** have values $b_1 = 1$,

$b_2 = -1$, and $b_3 = 1$, then you can substitute in those values and finish working on the augmented matrix. Note that I randomly chose numbers that made the value of $-2b_1 - b_2 + b_3$ come out equal to 0. The three entries in the last column become $-b_1 = -1$, $5b_1 + b_2 = 5(1) + (-1) = 4$, and $-2b_1 - b_2 + b_3 = -2(1) - (-1) + 1 = 0$. Using the numbers for the last column and the original matrix A, I now perform row operations on the augmented matrix.

$$\left[\begin{array}{ccc|c} 1 & -2 & -4 & -1 \\ 0 & 7 & 23 & 4 \\ 0 & 0 & 0 & 0 \end{array}\right] \frac{1}{7}R_2 \to R_2 \left[\begin{array}{ccc|c} 1 & -2 & -4 & -1 \\ 0 & 1 & \frac{23}{7} & \frac{4}{7} \\ 0 & 0 & 0 & 0 \end{array}\right]$$

$$2R_2 + R_1 \to R_1 \left[\begin{array}{ccc|c} 1 & 0 & \frac{18}{7} & \frac{1}{7} \\ 0 & 1 & \frac{23}{7} & \frac{4}{7} \\ 0 & 0 & 0 & 0 \end{array}\right]$$

The equations corresponding to the top two lines in the matrix read $x_1 + {}^{18}/_{7}\, x_3 = {}^{1}/_{7}$, and $x_2 + {}^{23}/_{7}\, x_3 = {}^{4}/_{7}$. Choosing k as a parameter to replace x_3, then the vector $\mathbf{x}$ for this particular situation is

$$\begin{bmatrix} \frac{1}{7} - \frac{18}{7}k \\ \frac{4}{7} - \frac{23}{7}k \\ k \end{bmatrix}$$

Now, letting $k = 2$ (I just chose the 2 at random, trying to end up with integers for the elements), you get

$$\begin{bmatrix} -5 \\ -6 \\ 2 \end{bmatrix}$$

and the matrix-vector equation reads

$$\begin{bmatrix} -1 & 2 & 4 \\ 5 & -3 & 3 \\ 3 & 1 & 11 \end{bmatrix}\begin{bmatrix} -5 \\ -6 \\ 2 \end{bmatrix} = \begin{bmatrix} 1 \\ -1 \\ 1 \end{bmatrix}$$

Even though many, many solutions are possible for this matrix equation, remember that the solutions are very carefully constructed after choosing numbers to represent the elements in **b.**

Getting nowhere because there's no solution

Sometimes, a perfectly nice, harmless-looking equation doesn't behave at all. You look for a solution vector, and you find that there's none to be had. Solving a matrix-vector equation seems like a simple enough situation: Just find some elements of a vector to multiply times a matrix so that you have a true statement. You're not fussy about your answer. You'd be satisfied with negative numbers or fractions or both. But you don't always get your wish.

Trying to find numbers to make it work

For example, let matrix A and vector **b** assume the following roles:

$$A = \begin{bmatrix} 2 & -1 & 1 \\ 1 & 1 & -1 \\ 3 & -1 & 1 \end{bmatrix} \text{ and } \mathbf{b} = \begin{bmatrix} 1 \\ 2 \\ 0 \end{bmatrix}$$

Then use the method of an augmented matrix and row operations:

$$\left[\begin{array}{ccc|c} 2 & -1 & 1 & 1 \\ 1 & 1 & -1 & 2 \\ 3 & -1 & 1 & 0 \end{array}\right] R_1 \leftrightarrow R_2 \left[\begin{array}{ccc|c} 1 & 1 & -1 & 2 \\ 2 & -1 & 1 & 1 \\ 3 & -1 & 1 & 0 \end{array}\right]$$

$$\begin{array}{l} -2R_1 + R_2 \rightarrow R_2 \\ -3R_1 + R_3 \rightarrow R_3 \end{array} \left[\begin{array}{ccc|c} 1 & 1 & -1 & 2 \\ 0 & -3 & 3 & -3 \\ 0 & -4 & 4 & -6 \end{array}\right]$$

$$\begin{array}{l} -\frac{1}{3}R_2 \rightarrow R_2 \\ -\frac{1}{4}R_3 \rightarrow R_3 \end{array} \left[\begin{array}{ccc|c} 1 & 1 & -1 & 2 \\ 0 & 1 & -1 & 1 \\ 0 & 1 & -1 & \frac{3}{2} \end{array}\right]$$

$$-1R_2 + R_3 \rightarrow R_3 \left[\begin{array}{ccc|c} 1 & 1 & -1 & 2 \\ 0 & 1 & -1 & 1 \\ 0 & 0 & 0 & \frac{1}{2} \end{array}\right]$$

Everything was going along just wonderfully until the last step in the row reductions. The last row in the matrix has three 0s and then a fraction. Translating this into an equation, you have $0x_1 + 0x_2 + 0x_3 = 1/2$. The statement is impossible. When you multiply by 0, you always get a 0, not some other number. So the matrix-vector equation has no solution.

Expanding your search in hopes of a solution

Consider the matrix C and vector **d**.

$$C = \begin{bmatrix} 1 & a \\ -1 & 3 \\ b & 7 \end{bmatrix} \text{ and } \mathbf{d} = \begin{bmatrix} 2 \\ 13 \\ 17 \end{bmatrix}$$

Determining if the unknowns a and b have any values that make the equation $C\mathbf{x} = \mathbf{d}$ a true statement, you use the same method of writing an augmented matrix. Writing the augmented matrix and using row operations,

$$\left[\begin{array}{cc|c} 1 & a & 2 \\ -1 & 3 & 13 \\ b & 7 & 17 \end{array}\right] \begin{array}{c} R_1 + R_2 \to R_2 \\ -bR_1 + R_3 \to R_3 \end{array} \left[\begin{array}{cc|c} 1 & a & 2 \\ 0 & a+3 & 15 \\ 0 & -ab+7 & -2b+17 \end{array}\right]$$

$$\frac{1}{a+3}R_2 \to R_2 \left[\begin{array}{cc|c} 1 & a & 2 \\ 0 & 1 & \dfrac{15}{a+3} \\ 0 & -ab+7 & -2b+17 \end{array}\right]$$

$$\begin{array}{c} -aR_2 + R_1 \to R_1 \\ (ab-7)R_2 + R_3 \to R_3 \end{array} \left[\begin{array}{cc|c} 1 & 0 & \dfrac{-13a+6}{a+3} \\ 0 & 1 & \dfrac{15}{a+3} \\ 0 & 0 & \dfrac{13ab+17a-6b-54}{a+3} \end{array}\right]$$

the last row has zeros and the fractional expression. The value of

$$\frac{13ab+17a-6b-54}{a+3}$$

must be equal to 0. So, for instance, if you let $a = 2$ and $b = 1$ (aiming for a result of 0), then the elements in the last column are

$$\frac{-13a+6}{a+3} = \frac{-13(2)+6}{2+3} = -4$$
$$\frac{15}{a+3} = \frac{15}{2+3} = 3$$
$$\frac{13ab+17a-6b-54}{a+3} = \frac{13(2)(1)+17(2)-6(1)-54}{2+3} = 0$$

So the equation C**x** = **d** can be written

$$C\mathbf{x} = \mathbf{d}$$
$$\begin{bmatrix} 1 & 2 \\ -1 & 3 \\ 1 & 7 \end{bmatrix} \begin{bmatrix} -4 \\ 3 \end{bmatrix} = \begin{bmatrix} 2 \\ 13 \\ 17 \end{bmatrix}$$

As long as a and b are suitably chosen so as to create a 0 in the last element of the third row, then you'll have a solution. The only time a solution is *not* possible is if you choose $a = -3$. The sum in the denominator must not be equal to 0, so a cannot be -3.

Chapter 7

Homing In on Homogeneous Systems and Linear Independence

In This Chapter

- Working through homogeneous systems of equations
- Solving for nontrivial solutions of homogeneous systems
- Investigating linear independence
- Breaching the subject of basis
- Extending basis to polynomials and matrices
- Delving into dimension based on basis

Solving matrix-vector equations takes on a new flavor in this chapter. You find 0s set equal to sums of linear terms instead of nonzero numbers. Homogeneous systems of equations are set equal to 0, and then you try to find nonzero solutions — an interesting challenge.

Also in this chapter, I tie linear independence with span to produce a whole new topic called *basis*. Old, familiar techniques are used to investigate the new ideas and rules.

Seeking Solutions of Homogeneous Systems

A system of linear equations is said to be *homogeneous* if it is of the form:

$$\begin{cases} a_{11}x_1 + a_{12}x_2 + \cdots + a_{1n}x_n = 0 \\ a_{21}x_1 + a_{22}x_2 + \cdots + a_{2n}x_n = 0 \\ \vdots \\ a_{m1}x_1 + a_{m2}x_2 + \cdots + a_{mn}x_n = 0 \end{cases}$$

where each x_i is a variable and each a_{ij} is a real-number coefficient.

What's special about a homogeneous system of linear equations is that each sum of coefficient-variable multiples is set equal to 0. If you create an $m \times n$ matrix of the coefficients of the variables, x_i, and call it A, then you can write the homogeneous system of equations as a corresponding matrix-vector equation $A\mathbf{x} = \mathbf{0}$. The **x** represents the *variable vector,* and the **0** represents the $m \times 1$ *zero vector.* For more information on matrix-vector equations in the form $A\mathbf{x} = \mathbf{b}$, you should refer to Chapter 6.

Unlike a general system of linear equations, a homogeneous system of linear equations *always* has at least one solution — a solution is guaranteed. The guaranteed solution occurs when you let each variable be equal to 0. The solution where everything is 0 is called the *trivial solution.* The trivial solution has each element in the vector **x** equal to 0. So this isn't a very exciting solution; you might say that there's nothing to it. (Sorry. Couldn't help myself.) What's more interesting about a homogeneous system of equations are the *nontrivial* solutions — if there are any.

Determining the difference between trivial and nontrivial solutions

A system of linear equations, in general, may have one solution, many solutions, or no solution at all. For a system of linear equations to have exactly one solution, the number of variables cannot exceed the number of equations. But just following the guideline of having fewer variables than equations doesn't guarantee a single solution (or any solution at all) — it's just a requirement for the single solution to happen.

A system of linear equations may have more than one solution. Many solutions occur when the number of equations is less than the number of variables. To identify the solutions, you assign one of the variables to be a *parameter* (some real number) and determine the values of the other variables based on formulas developed from the relationships established in the equations.

In the case of a homogeneous system of equations, you always have at least one solution. The guaranteed solution is the trivial solution in which every variable is equal to 0. If a homogeneous system has a nontrivial solution, then it must meet a particular requirement involving the number of equations and number of variables in the system.

If a homogeneous system of linear equations has fewer equations than it has unknowns, then it has a nontrivial solution. Further, a homogeneous system of linear equations has a nontrivial solution if and only if the system has at least one free variable. A *free variable* (also called a parameter) is a variable that can be randomly assigned some numerical value. The other variables are then related to the free variable through some algebraic rules.

Trivializing the situation with a trivial system

The homogeneous system of equations shown next has only a trivial solution — no nontrivial solutions.

$$\begin{cases} 3x + 4y = 0 \\ 4x - 2y = 0 \end{cases}$$

If you multiply the second equation times 2 and add it to the first equation, you get $11x = 0$, which gives you $x = 0$. Substituting $x = 0$ back into either equation, you get $y = 0$. When the variables are equal to 0, the solution is considered trivial.

It isn't always quite this easy to be able to tell if a system has only a trivial solution. In the "Formulating the form for a solution" section, later, I show you how to go through the steps to determine whether you have more than just a trivial solution.

Taking the trivial and adding some more solutions

The next system of equations has nontrivial solutions. Even though, at first glance, you see three equations and three unknowns and each equation is set equal to 0, the system *does* meet the requirement that there be fewer equations than unknowns. The requirement is met because one of the equations was actually created by adding a multiple of one of the other equations to the third equation. In the next section, I show you how to determine when one equation is a linear combination of two others — resulting in one less equation than in the original system.

So, back to a system of equations:

$$\begin{cases} x_1 + 3x_2 - 5x_3 = 0 \\ 3x_1 - x_2 + 5x_3 = 0 \\ 3x_1 + 2x_2 - x_3 = 0 \end{cases}$$

The variable x_3 is a *free variable.* And, letting $x_3 = k$, $x_1 = -k$, and $x_2 = 2k$, you can solve for all the solutions. How did I know this? Just trust me for now — I just want to show you how the nontrivial solutions work. (I show you how to find these values in the next section.)

For example, to find one of the nontrivial solutions using the parameter and formulas determined from the equations, you could pick $k = 3$, then $x_3 = 3$, $x_1 = -3$, and $x_2 = 6$. Substituting these values into the original system of equations, you get

$$\begin{cases} -3+3(6)-5(3) = -3+18-15 = 0 \\ 3(-3)-6+5(3) = -9-6+15 = 0 \\ 3(-3)+2(6)-3 = -9+12-3 = 0 \end{cases}$$

You have an infinite number of solutions possible. Choose any new value for k, and you get a new set of numbers. You could even let $k = 0$.

Formulating the form for a solution

Now I get down to business and show you how to make the determinations regarding trivial and nontrivial solutions. You can tell whether a system of homogeneous equations has only a trivial solution or if it indeed has nontrivial solutions. You can accomplish the task by observation (in the case of small, simple systems) or by changing the system to an echelon form for more complicated systems.

Traveling the road to the trivial

The following system of linear equations has only a trivial solution. Maybe you're a bit suspicious of there being only the one solution, because you have as many equations as there are variables. But the variable-equal-to-equation situation isn't enough to definitely make that assessment. You need to show that none of the equations can be eliminated because it's the linear combination of the others.

So, to determine if your guess is right, you take the system of equations and write its corresponding augmented matrix:

$$\begin{cases} x_1 - 2x_2 + 1x_3 = 0 \\ \quad 5x_2 - 3x_3 = 0 \\ x_1 - 2x_2 - x_3 = 0 \end{cases} \qquad \left[\begin{array}{ccc|c} 1 & -2 & 1 & 0 \\ 0 & 5 & -3 & 0 \\ 1 & -2 & -1 & 0 \end{array}\right]$$

Now, performing the row operation of multiplying the first row by -1 and adding it to row three to create a new row three, you get a matrix where the last row is written as $2x_3 = 0$ for its corresponding equation.

$$\left[\begin{array}{ccc|c} 1 & -2 & 1 & 0 \\ 0 & 5 & -3 & 0 \\ 1 & -2 & -1 & 0 \end{array}\right] -1R_1 + R_3 \rightarrow R_3 \left[\begin{array}{ccc|c} 1 & -2 & 1 & 0 \\ 0 & 5 & -3 & 0 \\ 0 & 0 & -2 & 0 \end{array}\right]$$

Dividing by 2, you get $x_3 = 0$. Substituting that 0 back into the second original equation, you get that x_2 also is equal to 0. And back-substituting into the first or third equation gives you 0 for x_1, too. The clue to the trivial solution in this case was the fact that the last row of the matrix had all but one 0. When you have all 0s in a row, you usually have a nontrivial solution (if the elimination of the row creates fewer equations than variables).

Joining in the journey to the nontrivial

When a homogeneous system of linear equations has nontrivial solutions, you usually have an infinite number of choices of numbers to satisfy the system. Of course, you can't just pick any numbers, willy-nilly. You make a choice for the first number, and then the others fall in line behind that choice. You find the secondary numbers using rules involving algebraic expressions.

For example, consider the next system of four linear equations with four unknowns. You can't have nontrivial solutions unless you have fewer independent equations (where none is a linear combination of the others) than unknowns, so you turn to an augmented matrix, row operations, and changing the matrix to echelon form. First, the system of equations:

$$\begin{cases} x_1 + 2x_2 + 3x_3 + 2x_4 = 0 \\ x_1 + 3x_2 + 5x_3 + 5x_4 = 0 \\ 2x_1 + 4x_2 + 7x_3 + x_4 = 0 \\ -x_1 - 2x_2 - 6x_3 + 7x_4 = 0 \end{cases}$$

And the corresponding augmented matrix:

$$\left[\begin{array}{cccc|c} 1 & 2 & 3 & 2 & 0 \\ 1 & 3 & 5 & 5 & 0 \\ 2 & 4 & 7 & 1 & 0 \\ -1 & -2 & -6 & 7 & 0 \end{array}\right]$$

Now, performing row operations, I change the matrix to reduced row echelon form. (Need a refresher on the row operations notation? Turn to Chapter 3.) Notice that, in the third step, the last row changes to all 0s, meaning that it was a linear combination of the other equations. The number of equations reduces to three, so nontrivial solutions are to be found.

$$\left[\begin{array}{cccc|c} 1 & 2 & 3 & 2 & 0 \\ 1 & 3 & 5 & 5 & 0 \\ 2 & 4 & 7 & 1 & 0 \\ -1 & -2 & -6 & 7 & 0 \end{array}\right] \begin{array}{l} -1R_1 + R_2 \to R_2 \\ -2R_1 + R_3 \to R_3 \\ 1R_1 + R_4 \to R_4 \end{array} \left[\begin{array}{cccc|c} 1 & 2 & 3 & 2 & 0 \\ 0 & 1 & 2 & 3 & 0 \\ 0 & 0 & 1 & -3 & 0 \\ 0 & 0 & -3 & 9 & 0 \end{array}\right]$$

$$-2R_2 + R_1 \to R_1 \left[\begin{array}{cccc|c} 1 & 0 & -1 & -4 & 0 \\ 0 & 1 & 2 & 3 & 0 \\ 0 & 0 & 1 & -3 & 0 \\ 0 & 0 & -3 & 9 & 0 \end{array}\right]$$

$$\begin{array}{l} 1R_3 + R_1 \to R_1 \\ -2R_3 + R_2 \to R_2 \\ 3R_3 + R_4 \to R_4 \end{array} \left[\begin{array}{cccc|c} 1 & 0 & 0 & -7 & 0 \\ 0 & 1 & 0 & 9 & 0 \\ 0 & 0 & 1 & -3 & 0 \\ 0 & 0 & 0 & 0 & 0 \end{array}\right]$$

Writing the corresponding equations for the reduced echelon form, the top row corresponds to $x_1 - 7x_4 = 0$, the second row corresponds to $x_2 + 9x_4 = 0$, and the third row corresponds to $x_3 - 3x_4 = 0$. Letting $x_4 = k$, the other variables are some multiple of k: $x_1 = 7k$, $x_2 = -9k$, and $x_3 = 3k$. So, if I let $k = -2$, for instance, then $x_1 = -14$, $x_2 = 18$, $x_3 = -6$, and $x_4 = -2$. Here's how the particular solution works in the system of equations.

$$\begin{cases} x_1 + 2x_2 + 3x_3 + 2x_4 = -14 + 2(18) + 3(-6) + 2(-2) = -14 + 36 - 18 - 4 = 0 \\ x_1 + 3x_2 + 5x_3 + 5x_4 = -14 + 3(18) + 5(-6) + 5(-2) = -14 + 54 - 30 - 10 = 0 \\ 2x_1 + 4x_2 + 7x_3 + x_4 = 2(-14) + 4(18) + 7(-6) - 2 = -28 + 72 - 42 - 2 = 0 \\ -x_1 - 2x_2 - 6x_3 + 7x_4 = -(-14) - 2(18) - 6(-6) + 7(-2) = 14 - 36 + 36 - 14 = 0 \end{cases}$$

Delving Into Linear Independence

Independence means different things to different people. My mother didn't want to give up her car and lose her independence. Every year, the United States celebrates Independence Day. But if you're a mathematician, you have a completely different take on independence. The word *independence,* in math-speak, often has to do with a set of vectors and the relationship between the vectors in that set.

A collection of vectors is either *linearly independent* or *linearly dependent.*

The vectors $\{\mathbf{v}_1, \mathbf{v}_2, \ldots, \mathbf{v}_n\}$ are *linearly independent* if the equation involving linear combinations, $a_1\mathbf{v}_1 + a_2\mathbf{v}_2 + \ldots + a_n\mathbf{v}_n = \mathbf{0}$, is true only when the scalars (a_i) are *all* equal to 0. The vectors are *linearly dependent* if the equation has a solution when at least one of the scalars is not equal to 0.

The description of *linear independence* is another way of talking about homogeneous systems of linear equations. Instead of discussing the algebraic equations and the corresponding augmented matrix, the discussion now focuses on vectors and vector equations. True, you still use an augmented matrix in your investigations, but the matrix is now created from the vectors. So, looking from the perspective of the homogeneous system of equations, you have linear independence if there's only the trivial solution, and linear dependence if scalars, not all 0, exist to give a result of 0 in the vector equation.

Testing for dependence or independence

For example, consider the following set of vectors and test whether the vectors in the set are linearly independent or linearly dependent.

$$\left\{ \begin{bmatrix} 1 \\ -2 \\ 0 \end{bmatrix}, \begin{bmatrix} 4 \\ 0 \\ 8 \end{bmatrix}, \begin{bmatrix} 3 \\ -1 \\ 5 \end{bmatrix} \right\}$$

To test under what circumstances the equation $a_1\mathbf{v}_1 + a_2\mathbf{v}_2 + a_3\mathbf{v}_3 = \mathbf{0}$, write out the equation with the vectors.

$$a_1 \begin{bmatrix} 1 \\ -2 \\ 0 \end{bmatrix} + a_2 \begin{bmatrix} 4 \\ 0 \\ 8 \end{bmatrix} + a_3 \begin{bmatrix} 3 \\ -1 \\ 5 \end{bmatrix} = \begin{bmatrix} 0 \\ 0 \\ 0 \end{bmatrix}$$

Then write an augmented matrix with the vectors as columns, and perform row reductions.

$$\left[\begin{array}{ccc|c} 1 & 4 & 3 & 0 \\ -2 & 0 & -1 & 0 \\ 0 & 8 & 5 & 0 \end{array}\right] 2R_1 + R_2 \rightarrow R_2 \left[\begin{array}{ccc|c} 1 & 4 & 3 & 0 \\ 0 & 8 & 5 & 0 \\ 0 & 8 & 5 & 0 \end{array}\right]$$

$$-1R_2 + R_3 \rightarrow R_3 \left[\begin{array}{ccc|c} 1 & 4 & 3 & 0 \\ 0 & 8 & 5 & 0 \\ 0 & 0 & 0 & 0 \end{array}\right]$$

$$\frac{1}{8}R_2 \rightarrow R_2 \left[\begin{array}{ccc|c} 1 & 4 & 3 & 0 \\ 0 & 1 & \frac{5}{8} & 0 \\ 0 & 0 & 0 & 0 \end{array}\right]$$

$$\frac{1}{8}R_2 \rightarrow R_2 \left[\begin{array}{ccc|c} 1 & 4 & 3 & 0 \\ 0 & 1 & \frac{5}{8} & 0 \\ 0 & 0 & 0 & 0 \end{array}\right]$$

$$-4R_2 + R_1 \rightarrow R_1 \left[\begin{array}{ccc|c} 1 & 0 & \frac{1}{2} & 0 \\ 0 & 1 & \frac{5}{8} & 0 \\ 0 & 0 & 0 & 0 \end{array}\right]$$

The bottom row has all 0s, so you know that you have nontrivial solutions for the system of equations. (If needed, refer to "Joining in the journey to the nontrivial" for information on this revelation.) The set of vectors is linearly dependent. But you have many instances in which a linear combination of the vectors equals 0 when the vectors themselves are not 0.

I demonstrate the fact that you have more than just the trivial solution by going to the reduced form of the matrix, rewriting the vector equation.

$$a_1 \begin{bmatrix} 1 \\ 0 \\ 0 \end{bmatrix} + a_2 \begin{bmatrix} 0 \\ 1 \\ 0 \end{bmatrix} + a_3 \begin{bmatrix} \frac{1}{2} \\ \frac{5}{8} \\ 0 \end{bmatrix} = \begin{bmatrix} 0 \\ 0 \\ 0 \end{bmatrix}$$

$a_1 + {}^1/_2\, a_3 = 0$ and $a_2 + {}^5/_8\, a_3 = 0$

Letting $a_3 = k$, I also have that $a_2 = -^5/_8 k$ and $a_1 = -^1/_2 k$. So the original vector equation reads

$$-\frac{1}{2}k\begin{bmatrix}1\\-2\\0\end{bmatrix}-\frac{5}{8}k\begin{bmatrix}4\\0\\8\end{bmatrix}+k\begin{bmatrix}3\\-1\\5\end{bmatrix}=\begin{bmatrix}0\\0\\0\end{bmatrix}$$

For example, if you let $k = 8$, you'd have

$$-\frac{1}{2}(8)\begin{bmatrix}1\\-2\\0\end{bmatrix}-\frac{5}{8}(8)\begin{bmatrix}4\\0\\8\end{bmatrix}+8\begin{bmatrix}3\\-1\\5\end{bmatrix}=-4\begin{bmatrix}1\\-2\\0\end{bmatrix}-5\begin{bmatrix}4\\0\\8\end{bmatrix}+8\begin{bmatrix}3\\-1\\5\end{bmatrix}$$

$$=\begin{bmatrix}-4\\8\\0\end{bmatrix}+\begin{bmatrix}-20\\0\\-40\end{bmatrix}+\begin{bmatrix}24\\-8\\40\end{bmatrix}=\begin{bmatrix}-4-20+24\\8+0-8\\0-40+40\end{bmatrix}=\begin{bmatrix}0\\0\\0\end{bmatrix}$$

One requirement for linear dependence is that you have more than just the trivial solution for the related system of equations. The nontrivial solutions occur when you have fewer equations than variables. So, in this next example, I've tried to create the situation of more equations than variables. So, starting with just three vectors, I have more equations than variables.

Here's the vector set I'm considering:

$$\left\{\begin{bmatrix}1\\2\\1\\4\end{bmatrix},\begin{bmatrix}0\\1\\1\\1\end{bmatrix},\begin{bmatrix}2\\0\\1\\7\end{bmatrix}\right\}$$

To check for linear independence, I write the vector equation and corresponding augmented vector.

$$a_1\begin{bmatrix}1\\2\\1\\4\end{bmatrix}+a_2\begin{bmatrix}0\\1\\1\\1\end{bmatrix}+a_3\begin{bmatrix}2\\0\\1\\7\end{bmatrix}=\begin{bmatrix}0\\0\\0\\0\end{bmatrix}\qquad\left[\begin{array}{ccc|c}1&0&2&0\\2&1&0&0\\1&1&1&0\\4&1&7&0\end{array}\right]$$

Performing row operations to reduce the matrix to echelon form, I get:

$$\left[\begin{array}{ccc|c} 1 & 0 & 2 & 0 \\ 2 & 1 & 0 & 0 \\ 1 & 1 & 1 & 0 \\ 4 & 1 & 7 & 0 \end{array}\right] \begin{array}{c} -2R_1 + R_2 \to R_2 \\ -1R_1 + R_3 \to R_3 \\ -4R_1 + R_4 \to R_4 \end{array} \left[\begin{array}{ccc|c} 1 & 0 & 2 & 0 \\ 0 & 1 & -4 & 0 \\ 0 & 1 & -1 & 0 \\ 0 & 1 & -1 & 0 \end{array}\right]$$

$$\begin{array}{c} -1R_2 + R_3 \to R_3 \\ -1R_2 + R_4 \to R_4 \end{array} \to \left[\begin{array}{ccc|c} 1 & 0 & 2 & 0 \\ 0 & 1 & -4 & 0 \\ 0 & 0 & 3 & 0 \\ 0 & 0 & 0 & 0 \end{array}\right]$$

$$\begin{array}{c} -1R_2 + R_3 \to R_3 \\ -1R_2 + R_4 \to R_4 \end{array} \to \left[\begin{array}{ccc|c} 1 & 0 & 2 & 0 \\ 0 & 1 & -4 & 0 \\ 0 & 0 & 3 & 0 \\ 0 & 0 & 0 & 0 \end{array}\right]$$

$$\frac{1}{3}R_3 \to R_3 \left[\begin{array}{ccc|c} 1 & 0 & 2 & 0 \\ 0 & 1 & -4 & 0 \\ 0 & 0 & 1 & 0 \\ 0 & 0 & 0 & 0 \end{array}\right]$$

$$\begin{array}{c} -2R_3 + R_1 \to R_1 \\ 4R_3 + R_2 \to R_2 \end{array} \to \left[\begin{array}{ccc|c} 1 & 0 & 0 & 0 \\ 0 & 1 & 0 & 0 \\ 0 & 0 & 1 & 0 \\ 0 & 0 & 0 & 0 \end{array}\right]$$

The only solution occurs when all the scalars are 0: $a_1 = a_2 = a_3 = 0$, so the vectors are linearly independent.

Characterizing linearly independent vector sets

You always have the tried-and-true method of examining the relationships of vectors in a set using an augmented matrix. When in doubt, go to the matrix form. But, if you're wise to some more obvious characteristics of the vector set you're considering, you may save yourself some time by ferreting out the independent from the dependent sets without going to all the work. You'd like to make a decision with just a simple inspection of the set of vectors, if possible. When applying the following guidelines, you always assume that the dimension of each vector in the set is the same. In other words, you don't have a 2×1 vector and a 3×1 vector in the same set.

Wondering when a set has only one vector

A set, in mathematics, is a collection of objects. A set can have any number of elements or objects, so a set may also contain just one element. And, yes, you can classify a set with one element as independent or dependent.

A set containing only one vector is linearly independent if that one vector is *not* the zero vector.

So, of the four sets shown next, just set D is dependent; the rest are all independent.

$$A = \left\{ \begin{bmatrix} 1 \\ 0 \\ 1 \\ 4 \end{bmatrix} \right\} \quad B = \left\{ \begin{bmatrix} 9 \\ -8 \end{bmatrix} \right\} \quad C = \left\{ \begin{bmatrix} 0 \\ 0 \\ 7 \end{bmatrix} \right\} \quad D = \left\{ \begin{bmatrix} 0 \\ 0 \\ 0 \\ 0 \end{bmatrix} \right\}$$

Doubling your pleasure with two vectors

A set containing two vectors may be linearly independent or dependent. If you're good with your multiplication tables, then you'll be a whiz at determining linear independence of sets of two vectors.

A set containing two vectors is linearly independent as long as one of the vectors is *not* a multiple of the other vector.

Here are two sets containing two vectors. Set E is linearly independent. Set F is linearly dependent, because each element in the second vector is half that of the corresponding element in the first vector.

$$E = \left\{ \begin{bmatrix} 1 \\ 0 \\ -4 \\ 8 \end{bmatrix}, \begin{bmatrix} 2 \\ -2 \\ 1 \\ 3 \end{bmatrix} \right\} \quad F = \left\{ \begin{bmatrix} 8 \\ -6 \\ 0 \end{bmatrix}, \begin{bmatrix} 4 \\ -3 \\ 0 \end{bmatrix} \right\}$$

Three's a crowd and four's even more

A set containing vectors contains all vectors with the same dimension. And that dimension comes into play when making a quick observation about linear dependence or independence.

A set containing $n \times 1$ vectors has linearly dependent vectors if the number of vectors in the set is greater than n.

So, no matter how hopeful you may be that this next set of vectors is linearly independent, it just can't be. The vectors are all 4×1,and I have five vectors in the set.

$$\left\{ \begin{bmatrix} 1 \\ 2 \\ 3 \\ 5 \end{bmatrix}, \begin{bmatrix} 7 \\ 11 \\ 13 \\ 17 \end{bmatrix}, \begin{bmatrix} 19 \\ 23 \\ 29 \\ 31 \end{bmatrix}, \begin{bmatrix} 37 \\ 41 \\ 43 \\ 47 \end{bmatrix}, \begin{bmatrix} 53 \\ 59 \\ 61 \\ 67 \end{bmatrix} \right\}$$

After making up this really awful example of vectors — just arbitrarily putting the number 1 and a bunch of prime numbers in for the elements — I began to wonder if I could demonstrate to you that the set really is linearly dependent, that one of the vectors is a linear combination of the others. So, without showing you all the gory details (feel free, of course, to check my work), the reduced echelon form of the augmented matrix I used is

$$\begin{bmatrix} 1 & 0 & 0 & 0 & -\frac{2}{5} \\ 0 & 1 & 0 & 0 & \frac{47}{150} \\ 0 & 0 & 1 & 0 & -\frac{1}{5} \\ 0 & 0 & 0 & 1 & \frac{223}{150} \end{bmatrix} \rightarrow \left\{ \begin{bmatrix} 1 \\ 0 \\ 0 \\ 0 \end{bmatrix}, \begin{bmatrix} 0 \\ 1 \\ 0 \\ 0 \end{bmatrix}, \begin{bmatrix} 0 \\ 0 \\ 1 \\ 0 \end{bmatrix}, \begin{bmatrix} 0 \\ 0 \\ 0 \\ 1 \end{bmatrix}, \begin{bmatrix} \frac{2}{5} \\ -\frac{47}{150} \\ \frac{1}{5} \\ -\frac{223}{150} \end{bmatrix} \right\}$$

The last vector is a linear combination of the first four vectors. Using the values in the last column of the echelon form as multipliers, you see the linear combination that creates the final vector.

$$-\frac{2}{5}\begin{bmatrix} 1 \\ 2 \\ 3 \\ 5 \end{bmatrix} + \frac{47}{150}\begin{bmatrix} 7 \\ 11 \\ 13 \\ 17 \end{bmatrix} - \frac{1}{5}\begin{bmatrix} 19 \\ 23 \\ 29 \\ 31 \end{bmatrix} + \frac{223}{150}\begin{bmatrix} 37 \\ 41 \\ 43 \\ 47 \end{bmatrix} = \begin{bmatrix} 53 \\ 59 \\ 61 \\ 67 \end{bmatrix}$$

The numbers aren't pretty, but, then, I never intended to demonstrate the property with this set of vectors. It's nice to know that the principle holds up — pretty or not.

Zeroing in on linear dependence

Having the zero vector as the only vector in a set is clearly grounds for your having linear dependence, but what about introducing the zero vector into an otherwise perfectly nice set of nonzero vectors?

A set containing the zero vector is always linearly dependent.

The best and quickest way of convincing you that the zero vector is the spoiler is to say that the zero vector is a multiple of any other vector in the set — it's another vector times the scalar 0.

Reducing the number of vectors in a set

If you already have a linearly independent set of vectors, what happens if you remove one of the vectors from the set? Is the new, reduced set also linearly independent, or have you upset the plan?

If vectors are linearly independent, then removing an arbitrary vector, $\mathbf{v}_i$, does not affect the linear independence.

Unifying the situation with unit vectors

A *unit vector* has one element that's a 1, while the rest of the elements are 0s. For example, the five unit vectors with dimension 5 × 1 are

$$\begin{bmatrix}1\\0\\0\\0\\0\end{bmatrix},\begin{bmatrix}0\\1\\0\\0\\0\end{bmatrix},\begin{bmatrix}0\\0\\1\\0\\0\end{bmatrix},\begin{bmatrix}0\\0\\0\\1\\0\end{bmatrix},\begin{bmatrix}0\\0\\0\\0\\1\end{bmatrix}$$

Why are unit vectors special when considering linear independence?

If the set of vectors $\{\mathbf{v}_1, \mathbf{v}_2, \ldots, \mathbf{v}_n\}$ contains all distinct (different) unit vectors, then that set has linear independence.

Connecting Everything to Basis

In Chapter 13, you find material on a mathematical structure called a *vector space.* What's a vector space, you say? One smart-alecky answer is that a vector space is a place where vectors orbit — and that answer isn't all that far from the truth. In this chapter, I just deal with the vectors that belong in a vector space. In Chapter 13, you find the other important processes and properties needed to establish a vector space.

Getting to first base with the basis of a vector space

Earlier in this chapter, in "Three's a crowd and four's even more," I tell you that a set of vectors can't be linearly independent if you have more vectors than you have rows in each vector. When you have a set of linearly independent vectors, you sort of have a core group of vectors from which other vectors are derived using linear combinations. But, when looking at a set of vectors, which are the core vectors and which are the ones created from the core? Is there a rhyme or reason? The answers to these questions have to do with basis.

A set of vectors $\{\mathbf{v}_1, \mathbf{v}_2, \ldots, \mathbf{v}_n\}$ is said to form a *basis* for a vector space if both of the following are true:

- The vectors $\mathbf{v}_1, \mathbf{v}_2, \ldots, \mathbf{v}_n$ span the vector space.
- The vectors $\mathbf{v}_1, \mathbf{v}_2, \ldots, \mathbf{v}_n$ are linearly independent.

You find information on linear independence earlier in this chapter. You find a complete discussion of the *span* of a set of vectors in Chapter 5. But, to put *span* in just a few words for now: A vector **v** is in the span of a set of vectors, S, if **v** is the result of a linear combination of the vectors in S. So the concept *basis* puts together two other properties or concepts: span and linear independence.

Broadening your horizons with a natural basis

The broadest example of a basis for a set involves unit vectors. The 3×1 unit vectors form a basis for all the possible vectors in R^3.

$$\begin{bmatrix}1\\0\\0\end{bmatrix}, \begin{bmatrix}0\\1\\0\end{bmatrix}, \begin{bmatrix}0\\0\\1\end{bmatrix}$$

The unit vectors are linearly independent, and you can find a linear combination made up of the unit vectors to create any vector in R^3. For example, if you wanted to demonstrate the linear combination of the unit vectors that produce the special vector with the date commemorating the date that Apollo 11 landed on the moon, then you'd use the following:

$$7\begin{bmatrix}1\\0\\0\end{bmatrix} + 20\begin{bmatrix}0\\1\\0\end{bmatrix} + 1969\begin{bmatrix}0\\0\\1\end{bmatrix} = \begin{bmatrix}7\\20\\1969\end{bmatrix}$$

So the 3×1 unit vectors are a *basis* for all the possible 3×1 vectors. When the unit vectors are used as a basis for a vector space, you refer to this as the *natural basis* or *standard basis.*

Alternating your perspective with an alternate basis

The simplest, most *natural* basis for all the 3×1 vectors is the set of the three unit vectors. But the unit vectors are not the only possible set of vectors used to create all other 3×1 vectors. For example, the set of vectors shown next is a basis for all the 3×1 vectors (they span R^3):

$$\begin{bmatrix} 1 \\ 1 \\ 1 \end{bmatrix}, \begin{bmatrix} 0 \\ 1 \\ 0 \end{bmatrix}, \begin{bmatrix} 1 \\ 2 \\ 3 \end{bmatrix}$$

Using the set shown here, you can write any possible 3×1 vector as a linear combination of the vectors in the set. In fact, I have a formula for determining which scalar multipliers in the linear combination correspond to a random vector. If you have some vector

$$\begin{bmatrix} x \\ y \\ z \end{bmatrix}$$

use the linear combination

$$a_1 \begin{bmatrix} 1 \\ 1 \\ 1 \end{bmatrix} + a_2 \begin{bmatrix} 0 \\ 1 \\ 0 \end{bmatrix} + a_3 \begin{bmatrix} 1 \\ 2 \\ 3 \end{bmatrix} = \begin{bmatrix} x \\ y \\ z \end{bmatrix}$$

to create the vector where

$$a_1 = \frac{3x - z}{2}, \; a_2 = \frac{2y - x - z}{2}, \; a_3 = \frac{z - x}{2}$$

So, to create the vector representing the landing on the moon by the Apollo 11 crew, use the setup for the linear combination, substitute the target numbers into the formulas for the scalars, and check by multiplying and adding.

$$\begin{bmatrix} 7 \\ 20 \\ 1969 \end{bmatrix} = a_1 \begin{bmatrix} 1 \\ 1 \\ 1 \end{bmatrix} + a_2 \begin{bmatrix} 0 \\ 1 \\ 0 \end{bmatrix} + a_3 \begin{bmatrix} 1 \\ 2 \\ 3 \end{bmatrix}$$

$$a_1 = \frac{3x - z}{2} = \frac{3(7) - 1969}{2} = \frac{-1948}{2} = -974$$

$$a_2 = \frac{2y - x - z}{2} = \frac{2(20) - 7 - 1969}{2} = \frac{-1936}{2} = -968$$

$$a_3 = \frac{z - x}{2} = \frac{1969 - 7}{2} = \frac{1962}{2} = 981$$

$$-974 \begin{bmatrix} 1 \\ 1 \\ 1 \end{bmatrix} - 968 \begin{bmatrix} 0 \\ 1 \\ 0 \end{bmatrix} + 981 \begin{bmatrix} 1 \\ 2 \\ 3 \end{bmatrix} = \begin{bmatrix} -974 - 0 + 981 \\ -974 - 968 + 1962 \\ -974 - 0 + 2943 \end{bmatrix} = \begin{bmatrix} 7 \\ 20 \\ 1969 \end{bmatrix}$$

I'm sure that you're wondering where in the world I got the lovely formulas for the scalars in the linear combination. I show you the technique in the "Determining basis by spanning out in a search for span" section. For now, I just wanted to demonstrate that it's possible to have more than one basis for a particular vector space.

Charting out the course for determining a basis

A set of vectors B is a *basis* for another set of vectors V if the vectors in B have linear independence and if the vectors in B *span* V. So, if you're considering some vectors as a basis, you have two things to check: linear independence and span. I cover linear independence earlier in this chapter and span in Chapter 5, so I now concentrate on just distinguishing the bases that work (the plural of *basis* is *bases*) from those that don't qualify.

Determining basis by spanning out in a search for span

A set of vectors may be rather diminutive or immensely large. When you're trying to create all the vectors of a particular size, then you identify all those vectors using the notation R^n, where the n refers to the number of rows in the vectors. I show you some more manageable vector sets, for starters, and expand my horizons to the infinitely large.

For example, consider the eight vectors shown here and how I find a basis the vectors.

$$\begin{bmatrix}1\\2\\1\end{bmatrix}, \begin{bmatrix}-1\\-1\\1\end{bmatrix}, \begin{bmatrix}1\\4\\5\end{bmatrix}, \begin{bmatrix}3\\4\\-1\end{bmatrix}, \begin{bmatrix}0\\1\\2\end{bmatrix}, \begin{bmatrix}4\\9\\6\end{bmatrix}, \begin{bmatrix}-2\\-3\\0\end{bmatrix}, \begin{bmatrix}1\\0\\-3\end{bmatrix}$$

The vectors are clearly not linearly independent, because you see more vectors than there are rows. (I show you this earlier in "Three's a crowd and four's even more.")

I want to find a basis for the span of the eight vectors. The set of unit vectors is a natural, but can I get away with fewer than three vectors for the particular basis? I have a procedure to determine just what might constitute a basis for a span.

If you have vectors $\mathbf{v}_1, \mathbf{v}_2, \ldots, \mathbf{v}_n$, to find vectors which form a basis for the span of the given vectors, you:

1. **Form the linear combination $a_1\mathrm{v}_1 + a_2\mathrm{v}_2 + \ldots + a_n\mathrm{v}_n = 0$, where each a_i is a real number.**
2. **Construct the corresponding augmented matrix.**
3. **Transform the matrix to reduced row echelon form.**
4. **Identify the vectors in the original matrix corresponding to columns in the reduced matrix that contain leading 1s (the first nonzero element in the row is a 1).**

So, using the eight vectors as a demonstration, I write the augmented matrix and perform row reductions.

$$\left[\begin{array}{cccccccc|c}1 & -1 & 1 & 3 & 0 & 4 & -2 & 1 & 0\\2 & -1 & 4 & 4 & 1 & 9 & -3 & 0 & 0\\1 & 1 & 5 & -1 & 2 & 6 & 0 & -3 & 0\end{array}\right]$$

$$\begin{array}{l}-2R_1 + R_2 \rightarrow R_2\\-1R_1 + R_3 \rightarrow R_3\end{array}\left[\begin{array}{cccccccc|c}1 & -1 & 1 & 3 & 0 & 4 & -2 & 1 & 0\\0 & 1 & 2 & -2 & 1 & 1 & 1 & -2 & 0\\0 & 2 & 4 & -4 & 2 & 2 & 2 & -4 & 0\end{array}\right]$$

$$\begin{array}{l}R_2 + R_1 \rightarrow R_1\\-2R_2 + R_3 \rightarrow R_3\end{array}\left[\begin{array}{cccccccc|c}1 & 0 & 3 & 1 & 1 & 5 & -1 & -1 & 0\\0 & 1 & 2 & -2 & 1 & 1 & 1 & -2 & 0\\0 & 0 & 0 & 0 & 0 & 0 & 0 & 0 & 0\end{array}\right]$$

The first two columns in the reduced echelon form have the leading 1s. The first two columns correspond to the first two vectors in the vector set. So a basis for the eight vectors is a set containing the first two vectors:

$$\begin{bmatrix}1\\2\\1\end{bmatrix},\begin{bmatrix}-1\\-1\\1\end{bmatrix}$$

But, wait a minute! What if you had written the vectors in a different order? What does that do to the basis? Consider the same vectors in a different order.

$$\begin{bmatrix}3\\4\\-1\end{bmatrix},\begin{bmatrix}0\\1\\2\end{bmatrix},\begin{bmatrix}4\\9\\6\end{bmatrix},\begin{bmatrix}1\\2\\1\end{bmatrix},\begin{bmatrix}1\\4\\5\end{bmatrix},\begin{bmatrix}-2\\-3\\0\end{bmatrix},\begin{bmatrix}-1\\-1\\1\end{bmatrix},\begin{bmatrix}1\\0\\-3\end{bmatrix}$$

Now I write the augmented matrix and perform row operations.

$$\left[\begin{array}{cccccccc|c}3&0&4&1&1&-2&-1&1&0\\4&1&9&2&4&-3&-1&0&0\\-1&2&6&1&5&0&1&-3&0\end{array}\right]$$

$$-1R_3 \leftrightarrow R_1 \left[\begin{array}{cccccccc|c}1&-2&-6&-1&-5&0&-1&3&0\\4&1&9&2&4&-3&-1&0&0\\3&0&4&1&1&-2&-1&1&0\end{array}\right]$$

$$\begin{array}{r}-4R_1+R_2\rightarrow R_2\\-3R_1+R_3\rightarrow R_3\end{array}\left[\begin{array}{cccccccc|c}1&-2&-6&-1&-5&0&-1&3&0\\0&9&33&6&24&-3&3&-12&0\\0&6&22&4&16&-2&2&-8&0\end{array}\right]$$

$$\frac{1}{9}R_2\rightarrow R_2\left[\begin{array}{cccccccc|c}1&-2&-6&-1&-5&0&-1&3&0\\0&1&\frac{11}{3}&\frac{2}{3}&\frac{8}{3}&-\frac{1}{3}&\frac{1}{3}&-\frac{4}{3}&0\\0&6&22&4&16&-2&2&-8&0\end{array}\right]$$

$$\begin{array}{r}2R_2+R_1\rightarrow R_1\\-6R_2+R_3\rightarrow R_3\end{array}\left[\begin{array}{cccccccc|c}1&0&\frac{4}{3}&\frac{1}{3}&\frac{1}{3}&-\frac{2}{3}&-\frac{1}{3}&\frac{1}{3}&0\\0&1&\frac{11}{3}&\frac{2}{3}&\frac{8}{3}&-\frac{1}{3}&\frac{1}{3}&-\frac{4}{3}&0\\0&0&0&0&0&0&0&0&0\end{array}\right]$$

Again, the first two columns of the reduced matrix have leading 1s. So the first two vectors in the new listing can also be a basis for the span of vectors.

Assigning a basis to a list with a rule

Rather than list all the vectors in a particular set — if that's at all possible — you often can describe the vectors in a set with a rule. For example, you may have a set of 3×1 vectors in which the first and second elements are randomly chosen and the third element is twice the first element minus three times the second element. You write the rule

$$\left\{ \begin{bmatrix} x_1 \\ x_2 \\ 2x_1 - 3x_2 \end{bmatrix} \right\}$$

Now you want the basis for some vectors, even though you don't have a complete listing. Instead, write an expression in which the rule shown in the vector is the sum of two scalar multiples.

$$\begin{bmatrix} x_1 \\ x_2 \\ 2x_1 - 3x_2 \end{bmatrix} = x_1 \begin{bmatrix} 1 \\ 0 \\ 2 \end{bmatrix} + x_2 \begin{bmatrix} 0 \\ 1 \\ -3 \end{bmatrix}$$

The two vectors in the linear combination are linearly independent, because neither is a multiple of the other. The vectors with the specifications can all be written as linear combinations of the two vectors. So the basis of the vectors is written with the rule

$$\begin{bmatrix} 1 \\ 0 \\ 2 \end{bmatrix}, \begin{bmatrix} 0 \\ 1 \\ -3 \end{bmatrix}$$

Extending basis to matrices and polynomials

The basis of a set of $n \times 1$ vectors meets requirements of linear independence and span. The basis must span the set — you must be able to construct linear combinations that result in all the vectors — and the vectors in the basis must be linearly independent. In this section, I explain how basis applies to matrices, in general, and even to polynomials.

Moving matrices into the arena

When dealing with span and linear independence in $n \times 1$ vectors, the vectors under consideration must all have the same number of rows. Likewise, matrices in a basis must all have the same dimension.

For example, consider the 3×2 matrices shown here:

$$\begin{bmatrix} 1 & 0 \\ 0 & 0 \\ 0 & 1 \end{bmatrix}, \begin{bmatrix} 0 & 1 \\ 0 & 0 \\ 1 & 0 \end{bmatrix}, \begin{bmatrix} 0 & 0 \\ 0 & 1 \\ 0 & 0 \end{bmatrix}, \begin{bmatrix} 0 & 0 \\ 1 & 0 \\ 0 & 0 \end{bmatrix}$$

The span of the matrices consists of all the linear combinations of the form

$$a_1 \begin{bmatrix} 1 & 0 \\ 0 & 0 \\ 0 & 1 \end{bmatrix} + a_2 \begin{bmatrix} 0 & 1 \\ 0 & 0 \\ 1 & 0 \end{bmatrix} + a_3 \begin{bmatrix} 0 & 0 \\ 0 & 1 \\ 0 & 1 \end{bmatrix} + a_4 \begin{bmatrix} 0 & 0 \\ 1 & 0 \\ 0 & 0 \end{bmatrix} = \begin{bmatrix} a_1 & a_2 \\ a_4 & a_3 \\ a_2 & a_1 + a_3 \end{bmatrix}$$

So all 3×2 matrices resulting from the linear combination of the scalars and matrices is in the span of the matrices. To determine if the four matrices form a basis for all the matrices in the span, you need to know if the matrices are linearly independent. In the case of matrices, you want to determine if there's more than just the trivial solution when setting the linear combination of matrices equal to the zero matrix.

$$a_1 \begin{bmatrix} 1 & 0 \\ 0 & 0 \\ 0 & 1 \end{bmatrix} + a_2 \begin{bmatrix} 0 & 1 \\ 0 & 0 \\ 1 & 0 \end{bmatrix} + a_3 \begin{bmatrix} 0 & 0 \\ 0 & 1 \\ 0 & 1 \end{bmatrix} + a_4 \begin{bmatrix} 0 & 0 \\ 1 & 0 \\ 0 & 0 \end{bmatrix} = \begin{bmatrix} 0 & 0 \\ 0 & 0 \\ 0 & 0 \end{bmatrix}$$

$$\begin{bmatrix} a_1 & a_2 \\ a_4 & a_3 \\ a_2 & a_1 + a_3 \end{bmatrix} = \begin{bmatrix} 0 & 0 \\ 0 & 0 \\ 0 & 0 \end{bmatrix}$$

The linear equations arising out of the matrix equation are $a_1 = 0$, $a_2 = 0$, $a_3 = 0$, $a_4 = 0$, and $a_1 + a_3 = 0$. The only solution to this system of equations is the trivial solution, in which each $a_i = 0$. So the matrices are linearly independent, and the set S is the basis of the set of all matrices fitting the prescribed format.

Pulling in polynomials

When dealing with $n \times 1$ vectors, you find R^2, R^3, R^4, and so on to represent all the vectors possible where $n = 2$, $n = 3$, and $n = 4$, respectively. Now I introduce P^2, P^3, P^4, and so on to represent second-degree polynomials, third-degree polynomials, fourth-degree polynomials, and so on, respectively.

A polynomial is the sum of variables and their coefficients where the variables are raised to whole-number powers. The general format for a polynomial is $a_n x^n + a_n\text{-}1x^{n-1} + \ldots + a_2x^2 + a_1x^1 + a_0$; the degree of the polynomial is whichever is the highest power of any variable.

The terms in the set $\{x^2, x, 1\}$ are a span of P^2, because all second-degree polynomials are the result of some linear combination $a_2x^2 + a_1x + a_0(1)$. The elements in the set are linearly independent, because $a_2x^2 + a_1x + a_0(1) = 0$ is true only in the case of the trivial solution where each coefficient is 0. So the elements in the set form a basis for P^2.

Now consider a certain basis for P^3. Let set $Q = \{x^3 + 3, 2x^2 - x, x + 1, 2\}$ and determine if Q spans P^3 and if the elements in Q are linearly independent.

Writing the linear combination $a_1(x^3 + 3) + a_2(2x^2 - x) + a_3(x + 1) + 2a_4$, the expression simplifies to $a_1x^3 + 2a_2x^2 + (-a_2 + a_3)x + (3a_1 + a_3 + 2a_4)$. Now, letting a general third-degree polynomial be represented by $ax^3 + bx^2 + cx + d$, I set up a system of equations matching the scalar multiples with the coefficients so that I can solve for each a_i in terms of the coefficients in the polynomial.

$$\begin{cases} a_1 = a \\ 2a_2 = b \\ -a_2 + a_3 = c \\ 3a_1 + a_3 + 2a_4 = d \end{cases}$$

Solving for the values of each a_i,

$$a_1 = a,\ a_2 = \frac{b}{2},\ a_3 = \frac{b + 2c}{2},\ a_4 = \frac{-6a - b - 2c + 2d}{4}$$

The set Q spans a third-degree polynomial when the multipliers assume the values determined by the coefficients and constants in the polynomials.

For example, to determine the multipliers of the terms in Q needed to write the polynomial $x^3 + 4x^2 - 5x + 3$, you let $a = 1$, $b = 4$, $c = -5$, and $d = 3$. The linear combination becomes

$$1(x^3 + 3) + \frac{4}{2}(2x^2 - x) + \frac{4 + 2(-5)}{2}(x + 1) + \frac{-6(1) - 4 - 2(-5) + 2(3)}{4}(2)$$
$$= 1(x^3 + 3) + 2(2x^2 - x) + (-3)(x + 1) + \left(\frac{3}{2}\right)(2)$$
$$= x^3 + 3 + 4x^2 - 2x - 3x - 3 + 3$$
$$= x^3 + 4x^2 - 5x + 3$$

Legendre

Adrien-Marie Legendre was a French mathematician, living in the late 1700s through the early 1800s. He made significant contributions to several areas of mathematics, statistics, and physics. Getting his feet wet with the trajectories of cannonballs, Legendre then moved on to various challenges in mathematics. When you look up at the moon, you may even recognize the Legendre crater, named for this mathematician.

Legendre made some good starts, producing mathematics that were later completed or proven by others. But, on his own, he is credited with developing the *least squares method* used in fitting lines and curves to data sets. Also, Legendre worked with polynomials, beginning with discoveries involving the roots of polynomials and culminating with establishing structures called *Legendre polynomials,* which are found in applications of mathematics to physics and engineering.

Legendre took a hit starting at the fall of the Bastille in 1789 and continuing during the French Revolution — losing all his money. He continued to work in mathematics and stayed clear of any political activism during the revolution, producing significant contributions.

And, finally, to check for linear independence, determine if you have more than just the trivial solution. Solve the system of equations involving the multipliers in which the linear combination is set equal to 0.

$$\begin{cases} a_1 = 0 \\ 2a_2 = 0 \\ -a_2 + a_3 = 0 \\ 3a_1 + a_3 + 2a_4 = 0 \end{cases}$$

The multipliers a_1 and a_2 are equal to 0. Substituting 0 for a_2 in the third equation, you get $a_3 = 0$. And substituting 0 for a_1 and a_3 in the fourth equation, you have $a_4 = 0$. Only the trivial solution exists, so the elements are linearly independent, and Q is a basis for P^3.

Finding the dimension based on basis

The dimension of a matrix or vector is tied to the number of rows and columns in that matrix or vector. The dimension of a vector *space* (see Chapter 13), is the number of vectors in the basis of the vector space. I discuss dimension here, to tie the concept to the overall picture of basis in this chapter.

In "Determining basis by spanning out in a search for span," earlier in this chapter, I show you that it's possible to have more than one basis for a particular span — there's more than one way to create a set of vectors from a smaller set. But no matter how many bases a span may have, it's always the same number of vectors in the basis. Look at the following two sets of vectors, F and T:

$$F = \left\{ \begin{bmatrix} 1 \\ 0 \\ 0 \\ 0 \end{bmatrix}, \begin{bmatrix} 1 \\ 1 \\ 1 \\ 1 \end{bmatrix}, \begin{bmatrix} 0 \\ 1 \\ 0 \\ 1 \end{bmatrix}, \begin{bmatrix} 0 \\ 1 \\ 1 \\ 0 \end{bmatrix} \right\} \qquad T = \left\{ \begin{bmatrix} 1 \\ 4 \end{bmatrix}, \begin{bmatrix} 2 \\ -3 \end{bmatrix} \right\}$$

So if set F is a basis for R^4 and T is a basis for R^2, then the span of F has dimension 4, because it contains four vectors, and T has dimension 2, because this basis contains two vectors.

Consider the next situation, where you're given a set of vectors, V, that spans another set of vectors, A. You want to find the basis for V, which will help in determining more about set A.

$$V = \left\{ \begin{bmatrix} 1 \\ -2 \\ 5 \end{bmatrix}, \begin{bmatrix} -2 \\ 4 \\ -10 \end{bmatrix}, \begin{bmatrix} 3 \\ 1 \\ 1 \end{bmatrix}, \begin{bmatrix} 1 \\ 5 \\ -9 \end{bmatrix}, \begin{bmatrix} 5 \\ -3 \\ 11 \end{bmatrix} \right\}$$

To find the basis, write the vectors in V as an augmented matrix, and go through row reductions.

$$V = \begin{bmatrix} 1 & -2 & 3 & 1 & 5 \\ -2 & 4 & 1 & 5 & -3 \\ 5 & -10 & 1 & -9 & 11 \end{bmatrix} \rightarrow \begin{bmatrix} 1 & -2 & 3 & 1 & 5 \\ 0 & 0 & 7 & 7 & 7 \\ 0 & 0 & -14 & -14 & -14 \end{bmatrix}$$

$$\rightarrow \begin{bmatrix} 1 & -2 & 3 & 1 & 5 \\ 0 & 0 & 1 & 1 & 1 \\ 0 & 0 & 0 & 0 & 0 \end{bmatrix} \rightarrow \begin{bmatrix} 1 & -2 & 0 & -2 & 2 \\ 0 & 0 & 1 & 1 & 1 \\ 0 & 0 & 0 & 0 & 0 \end{bmatrix}$$

In reduced echelon form, the matrix has leading 1s in the first and third columns, corresponding to the first and third vectors in V. So the basis is

$$\left\{ \begin{bmatrix} 1 \\ -2 \\ 5 \end{bmatrix}, \begin{bmatrix} 3 \\ 1 \\ 1 \end{bmatrix} \right\}$$

and the dimension is 2.

The set V contains five vectors, but what can be said about the set of vectors — all the vectors — that are in the span of the basis? Perhaps you want to be able to extend your set of vectors to more than five. Write the linear combinations of the vectors in the basis and the corresponding equations.

$$a_1\begin{bmatrix}1\\-2\\5\end{bmatrix}+a_2\begin{bmatrix}3\\1\\1\end{bmatrix}=\begin{bmatrix}x\\y\\z\end{bmatrix}$$

$$\begin{cases}a_1+3a_2=x\\-2a_1+a_2=y\\5a_1+a_2=z\end{cases}$$

Solving for a_1 and a_2, you get

$$a_1=\frac{x-3y}{7},\ a_2=\frac{2x+y}{7},\text{ or } a_3=\frac{-5x+z}{-14}$$

Because the choices for x, y, and z have no limitations, you can create any 3×1 vector. You have a basis for R^3.

Chapter 8

Making Changes with Linear Transformations

In This Chapter

- Describing and investigating linear transformations
- Perusing the properties associated with linear transformations
- Interpreting linear transformations as rotations, reflections, and translations
- Saluting the kernel and scoping out the range of a linear transformation

If you're old enough (or young enough at heart), you remember the Transformers toys — and all the glitz that went with that fad. Little did you know (if you were a Transformers fan) that you were being set up for a fairly serious mathematical subject. Some transformations found in geometry are performed by moving objects around systematically and not changing their shape or size. Other transformations make more dramatic changes to geometric figures. Linear transformations even incorporate some of the geometric transformational processes. But linear transformations in this chapter have some restrictions, as well as opening up many more mathematical possibilities.

In this chapter, I describe what linear transformations are. Then I take you through some examples of linear transformations. I show you many of the operational properties that accompany linear transformations. And, finally, I describe the kernel and range of a linear transformation — two concepts that are very different but tied together by the transformation operator.

Formulating Linear Transformations

Linear transformations are very specific types of processes in mathematics. They often involve mathematical structures such as vectors and matrices; the transformations also incorporate mathematical operations. Some of the operations used by linear transformations are your everyday addition and multiplication; others are specific to the type of mathematical structure that the operation is being performed upon.

In this section, I lay out the groundwork of what constitutes a linear transformation. In later sections, I show you examples of different types of linear transformations — even drawing pictures where appropriate.

Delineating linear transformation lingo

In mathematics, a *transformation* is an operation, function, or mapping in which one set of elements is transformed into another set of elements. For example, the function $f(x) = 2x$ takes any number in the set of whole numbers and transforms it to an even number (if it wasn't even already). The absolute value function, $f(x) = |x|$ takes any number and transforms it to a non-negative number, if the number wasn't non-negative already. And the trigonometric functions are truly amazing. The trig functions, such as $\sin x$, take angle measures (in degrees or radians) and transform them into real numbers.

A *linear transformation* is a particular type of transformation in which one set of vectors is transformed into another vector set using some *linear operator*. Also part of the definition of a linear transformation is that the sets and operator are bound by the following two properties:

- Performing the transformation on the sum of any two vectors in the set has the same result as performing the transformation on the vectors independently and then adding the resulting vectors together.
- Performing the transformation on a scalar multiple of a vector has the same result as performing the transformation on the vector and then multiplying by that scalar.

Examining how a transformation transforms

It's one thing to describe in words how a linear transformation changes a set of vectors into another set, but it's usually more helpful to illustrate how a linear transformation works with several examples.

Consider the following linear transformation, which I choose to name with the capital letter T. The description of T is

$T: R^3 \to R^2$ where

$$T\left(\begin{bmatrix} x \\ y \\ z \end{bmatrix}\right) \to \begin{bmatrix} x+y \\ y-z \end{bmatrix}$$

which is read: "Transformation T takes a 3×1 vector and transforms it into a 2×1 vector. The elements in the 2×1 vector are a sum and difference of some elements in the 3×1 vector."

After you've defined what a particular transformation does, you indicate that you want to perform that transformation on a vector **v** by writing T(**v**).

So, if you want T(**v**) when **v** is the following vector, you get

$$\mathbf{v} = \begin{bmatrix} 4 \\ 2 \\ -3 \end{bmatrix} \quad T(\mathbf{v}) = T\left(\begin{bmatrix} 4 \\ 2 \\ -3 \end{bmatrix}\right) = \begin{bmatrix} 4+2 \\ 2-(-3) \end{bmatrix} = \begin{bmatrix} 6 \\ 5 \end{bmatrix}$$

The linear operator (the rule describing the transformation) of a linear transformation might also involve the multiplication of a matrix times the vector being operated upon. For example, if you have a matrix A, then the notation: T(**x**) = A**x** reads: "The transformation T on vector **x** is performed by multiplying matrix A times vector **x**." When a linear transformation is defined using matrix multiplication, the input vectors all have a prescribed dimension so that the matrix can multiply them. (The need for the correct dimension is covered fully in Chapter 3.)

For example, consider the transformation involving the matrix A, shown next. The transformation T(**x**) = A**x** (also written A * **x**). Matrix A can multiply any 3×1 vector, and the result is a 3×1 vector.

$$A = \begin{bmatrix} 2 & 0 & 1 \\ 3 & 1 & -1 \\ 4 & 3 & 0 \end{bmatrix}$$

Performing the linear transformation T on the vector **x** as shown below,

$$\mathbf{x} = \begin{bmatrix} 5 \\ 1 \\ -2 \end{bmatrix} \quad T(\mathbf{x}) = A^*\mathbf{x} = \begin{bmatrix} 2 & 0 & 1 \\ 3 & 1 & -1 \\ 4 & 3 & 0 \end{bmatrix} * \begin{bmatrix} 5 \\ 1 \\ -2 \end{bmatrix}$$

$$= \begin{bmatrix} 10+0-2 \\ 15+1+2 \\ 20+3+0 \end{bmatrix} = \begin{bmatrix} 8 \\ 18 \\ 23 \end{bmatrix}$$

Another example of a matrix being involved in the transformation operator is one where the matrix, the vectors being operated on, and the resulting vectors all have different dimensions. For example, look at the linear transformation W in which W(**v**) = B**v.** Note that the matrix B has dimension 2×3, the vector **v** is 3×1, and the resulting vectors are 2×1.

$$B = \begin{bmatrix} 2 & 3 & 0 \\ -1 & -2 & 3 \end{bmatrix} \quad \mathbf{v} = \begin{bmatrix} 5 \\ 1 \\ -2 \end{bmatrix}$$

$$W(\mathbf{v}) = B^*\mathbf{v} = \begin{bmatrix} 2 & 3 & 0 \\ -1 & -2 & 3 \end{bmatrix} * \begin{bmatrix} 5 \\ 1 \\ -2 \end{bmatrix} = \begin{bmatrix} 10+3+0 \\ -5-2-6 \end{bmatrix} = \begin{bmatrix} 13 \\ -13 \end{bmatrix}$$

Again, for how and why this change in dimension occurs after matrix multiplication, refer to the material in Chapter 3.

Completing the picture with linear transformation requirements

The two properties required to make a transformation perform as a *linear transformation* involve vector addition and scalar multiplication. Both properties require that you get the same result when performing the transformation on a sum or product *after* the operation as you do if you perform the transformation and *then* the operation.

If T is a linear transformation and **u** and **v** are vectors and *c* is a scalar, then:

- T(**u** + **v**) = T(**u**) + T(**v**)
- T(*c***v**) = *c*T(**v**)

For example, consider the linear transformation T that transforms 2×1 vectors into 3×1 vectors.

$$T\left(\begin{bmatrix} x \\ y \end{bmatrix}\right) \rightarrow \begin{bmatrix} x-y \\ x+y \\ 3x \end{bmatrix}$$

I demonstrate the first of the two requirements — the additive requirement — needed for transformation T to be a linear transformation, using random vectors **u** and **v.** First, the two vectors are added and the transformation performed on the result. Then I show how to perform the transformation on the two original vectors and add the transformed results.

$$\mathbf{u}=\begin{bmatrix}6\\-5\end{bmatrix},\ \mathbf{v}=\begin{bmatrix}-4\\-1\end{bmatrix}\quad \mathrm{T}\left(\begin{bmatrix}x\\y\end{bmatrix}\right)\rightarrow\begin{bmatrix}x-y\\x+y\\3x\end{bmatrix}$$

$$\mathrm{T}(\mathbf{u}+\mathbf{v})=\mathrm{T}\left(\begin{bmatrix}6\\-5\end{bmatrix}+\begin{bmatrix}-4\\-1\end{bmatrix}\right)=\mathrm{T}\left(\begin{bmatrix}2\\-6\end{bmatrix}\right)=\begin{bmatrix}2-(-6)\\2+(-6)\\3(2)\end{bmatrix}=\begin{bmatrix}8\\-4\\6\end{bmatrix}$$

$$\begin{aligned}\mathrm{T}(\mathbf{u})+\mathrm{T}(\mathbf{v})&=\mathrm{T}\left(\begin{bmatrix}6\\-5\end{bmatrix}\right)+\mathrm{T}\left(\begin{bmatrix}-4\\-1\end{bmatrix}\right)\\&=\begin{bmatrix}6-(-5)\\6+(-5)\\3(6)\end{bmatrix}+\begin{bmatrix}-4-(-1)\\-4+(-1)\\3(-4)\end{bmatrix}=\begin{bmatrix}11\\1\\18\end{bmatrix}+\begin{bmatrix}-3\\-5\\-12\end{bmatrix}=\begin{bmatrix}8\\-4\\6\end{bmatrix}\end{aligned}$$

The end results are the same, as they should be.

Now, using vector **u** and a scalar $c = 5$, I show the requirement involving scalar multiplication.

$$\mathrm{T}(c\mathbf{u})=\mathrm{T}\left(5\begin{bmatrix}6\\-5\end{bmatrix}\right)=\mathrm{T}\left(\begin{bmatrix}30\\-25\end{bmatrix}\right)=\begin{bmatrix}30-(-25)\\30+(-25)\\3(30)\end{bmatrix}=\begin{bmatrix}55\\5\\90\end{bmatrix}$$

$$c\mathrm{T}(\mathbf{u})=5\mathrm{T}\left(\begin{bmatrix}6\\-5\end{bmatrix}\right)=5\begin{bmatrix}6-(-5)\\6+(-5)\\3(6)\end{bmatrix}=5\begin{bmatrix}11\\1\\18\end{bmatrix}=\begin{bmatrix}55\\5\\90\end{bmatrix}$$

Again, the results are the same.

Recognizing when a transformation is a linear transformation

In the section "Completing the picture with linear transformation requirements," I demonstrate for you how the rules for a linear transformation work. I start with a linear transformation, so, of course, the rules work. But how do you know that a transformation is really a linear transformation? If you have a candidate and want to determine if it's a linear transformation, you might pick random vectors and try using the rules, but you may be very lucky and happen to choose the only vectors that *do* work — accidentally avoiding all those that don't work with the addition and multiplication rules. Instead of trying to come up with all possible vectors (which is usually impractical, if not impossible), you apply the transformation rule to some general vector and determine if the rules hold.

Establishing a process for determining whether you have a linear transformation

For example, using the same transformation T, as described in the previous section, and two general vectors **u** and **v**, I first prove that the transformation of a vector sum is the same as the sum of the transformations on the vector.

T and the vectors **u** and **v**:

$$\mathrm{T}\left(\begin{bmatrix} x \\ y \end{bmatrix}\right) \rightarrow \begin{bmatrix} x-y \\ x+y \\ 3x \end{bmatrix}, \ \mathbf{u}=\begin{bmatrix} x_1 \\ x_2 \end{bmatrix}, \ \mathbf{v}=\begin{bmatrix} v_1 \\ v_2 \end{bmatrix}$$

I now determine if the rule involving addition and the transformation holds for all candidate vectors.

$$\mathrm{T}(\mathbf{u}+\mathbf{v})=\mathrm{T}\left(\begin{bmatrix} u_1 \\ u_2 \end{bmatrix}+\begin{bmatrix} v_1 \\ v_2 \end{bmatrix}\right)=\mathrm{T}\left(\begin{bmatrix} u_1+v_1 \\ u_2+v_2 \end{bmatrix}\right)$$

$$=\begin{bmatrix} u_1+v_1-(u_2+v_2) \\ u_1+v_1+(u_2+v_2) \\ 3(u_1+v_1) \end{bmatrix}=\begin{bmatrix} u_1+v_1-u_2-v_2 \\ u_1+v_1+u_2+v_2 \\ 3u_1+3v_1 \end{bmatrix}$$

$$\mathrm{T}(\mathbf{u})+\mathrm{T}(\mathbf{v})=\mathrm{T}\left(\begin{bmatrix} u_1 \\ u_2 \end{bmatrix}\right)+\mathrm{T}\left(\begin{bmatrix} v_1 \\ v_2 \end{bmatrix}\right)$$

$$=\begin{bmatrix} u_1-u_2 \\ u_1+u_2 \\ 3u_1 \end{bmatrix}+\begin{bmatrix} v_1-v_2 \\ v_1+v_2 \\ 3v_1 \end{bmatrix}=\begin{bmatrix} u_1+v_1-u_2-v_2 \\ u_1+v_1+u_2+v_2 \\ 3u_1+3v_1 \end{bmatrix}$$

The sums (and differences) in the two end results are the same.

Now, to show that the transformation of the scalar multiple is equal to the scalar times the transformation:

$$T(c\mathbf{u}) = T\left(c\begin{bmatrix} u_1 \\ u_2 \end{bmatrix}\right) = T\left(\begin{bmatrix} cu_1 \\ cu_2 \end{bmatrix}\right) = \begin{bmatrix} cu_1 - cu_2 \\ cu_1 + cu_2 \\ 3cu_1 \end{bmatrix}$$

$$cT(\mathbf{u}) = cT\left(\begin{bmatrix} u_1 \\ u_2 \end{bmatrix}\right) = c\begin{bmatrix} u_1 - u_2 \\ u_1 + u_2 \\ 3u_1 \end{bmatrix} = \begin{bmatrix} c(u_1 - u_2) \\ c(u_1 + u_2) \\ c(3u_1) \end{bmatrix} = \begin{bmatrix} cu_1 - cu_2 \\ cu_1 + cu_2 \\ 3cu_1 \end{bmatrix}$$

Voilá! Confirmation is achieved.

Sorting out the transformations that aren't linear

A transformation is not termed a linear transformation unless the two operational qualifications are met. For example, the transformation W, shown next, fails when the addition property is applied.

$$W\left(\begin{bmatrix} x \\ y \end{bmatrix}\right) \to \begin{bmatrix} x + y \\ y + 1 \end{bmatrix} \quad \mathbf{u} = \begin{bmatrix} u_1 \\ u_2 \end{bmatrix}, \mathbf{v} = \begin{bmatrix} v_1 \\ v_2 \end{bmatrix}$$

$$W(\mathbf{u} + \mathbf{v}) = W\left(\begin{bmatrix} u_1 \\ u_2 \end{bmatrix} + \begin{bmatrix} v_1 \\ v_2 \end{bmatrix}\right) = W\left(\begin{bmatrix} u_1 + v_1 \\ u_2 + v_2 \end{bmatrix}\right) = \begin{bmatrix} u_1 + v_1 + u_2 + v_2 \\ u_2 + v_2 + 1 \end{bmatrix}$$

$$W(\mathbf{u}) + W(\mathbf{v}) = W\left(\begin{bmatrix} u_1 \\ u_2 \end{bmatrix}\right) + W\left(\begin{bmatrix} v_1 \\ v_2 \end{bmatrix}\right)$$

$$= \begin{bmatrix} u_1 + u_2 \\ u_2 + 1 \end{bmatrix} + \begin{bmatrix} v_1 + v_2 \\ v_2 + 1 \end{bmatrix} = \begin{bmatrix} u_1 + v_1 + u_2 + v_2 \\ u_2 + v_2 + 2 \end{bmatrix}$$

The final vectors don't match.

And the next transformation, S, doesn't meet the scalar multiplication property.

$$S\left(\begin{bmatrix} x \\ y \\ z \end{bmatrix}\right) \rightarrow \begin{bmatrix} 1 \\ z \\ x \end{bmatrix}, \quad \mathbf{u} = \begin{bmatrix} u_1 \\ u_2 \\ u_3 \end{bmatrix}$$

$$S(c\mathbf{u}) = S\left(c\begin{bmatrix} u_1 \\ u_2 \\ u_3 \end{bmatrix}\right) = S\left(\begin{bmatrix} cu_1 \\ cu_2 \\ cu_3 \end{bmatrix}\right) = \begin{bmatrix} 1 \\ cu_3 \\ cu_1 \end{bmatrix}$$

$$cS(\mathbf{u}) = cS\left(\begin{bmatrix} u_1 \\ u_2 \\ u_3 \end{bmatrix}\right) = c\begin{bmatrix} 1 \\ u_3 \\ u_1 \end{bmatrix} = \begin{bmatrix} c \\ cu_3 \\ cu_1 \end{bmatrix}$$

The top elements in the two end results don't match; the vectors are different.

Proposing Properties of Linear Transformations

The definition of a linear transformation involves two operations and their properties. Many other properties of algebra also apply to linear operations when one or more vectors or transformations are involved. Some algebraic properties that you find associated with linear transformations are those of commutativity, associativity, distribution, and working with 0.

Summarizing the summing properties

In algebra, the *associative property* of *addition* establishes that, when adding the three terms x, y, and z, you get the same result by adding the sum of x and y to z as you do if you add x to the sum of y and z. In algebraic symbols, the associative property of addition is $(x + y) + z = x + (y + z)$.

The associative property of addition applies to linear transformations when you perform more than one transformation on a vector.

Given the vector **v** and the linear transformations T_1, T_2, and T_3:

$$[T_1(\mathbf{v}) + T_2(\mathbf{v})] + T_3(\mathbf{v}) = T_1(\mathbf{v}) + [T_2(\mathbf{v}) + T_3(\mathbf{v})]$$

or, stated more simply:

$$[T_1 + T_2] + T_3 = T_1 + [T_2 + T_3]$$

For example, let T_1, T_2, and T_3 be linear transformations defined as shown:

$$T_1\left(\begin{bmatrix} x \\ y \end{bmatrix}\right) \to \begin{bmatrix} x \\ 2x \\ -y \end{bmatrix},\ T_2\left(\begin{bmatrix} x \\ y \end{bmatrix}\right) \to \begin{bmatrix} y \\ 2y \\ x+y \end{bmatrix},\ T_3\left(\begin{bmatrix} x \\ y \end{bmatrix}\right) \to \begin{bmatrix} 2x \\ x \\ 3y \end{bmatrix}$$

Comparing the two different groupings defined by the associative property, and applying the transformations, you end up with the same result.

$$\left[T_1\left(\begin{bmatrix} v_1 \\ v_2 \end{bmatrix}\right) + T_2\left(\begin{bmatrix} v_1 \\ v_2 \end{bmatrix}\right)\right] + T_3\left(\begin{bmatrix} v_1 \\ v_2 \end{bmatrix}\right) =$$

$$\left[\begin{bmatrix} v_1 \\ 2v_1 \\ -v_2 \end{bmatrix} + \begin{bmatrix} v_2 \\ 2v_2 \\ v_1+v_2 \end{bmatrix}\right] + \begin{bmatrix} 2v_1 \\ v_1 \\ 3v_2 \end{bmatrix} = \begin{bmatrix} v_1+v_2 \\ 2v_1+2v_2 \\ v_1 \end{bmatrix} + \begin{bmatrix} 2v_1 \\ v_1 \\ 3v_2 \end{bmatrix} = \begin{bmatrix} 3v_1+v_2 \\ 3v_1+2v_2 \\ v_1+3v_2 \end{bmatrix}$$

$$T_1\left(\begin{bmatrix} v_1 \\ v_2 \end{bmatrix}\right) + \left[T_2\left(\begin{bmatrix} v_1 \\ v_2 \end{bmatrix}\right) + T_3\left(\begin{bmatrix} v_1 \\ v_2 \end{bmatrix}\right)\right] =$$

$$\begin{bmatrix} v_1 \\ 2v_1 \\ -v_2 \end{bmatrix} + \left[\begin{bmatrix} v_2 \\ 2v_2 \\ v_1+v_2 \end{bmatrix} + \begin{bmatrix} 2v_1 \\ v_1 \\ 3v_2 \end{bmatrix}\right] = \begin{bmatrix} v_1 \\ 2v_1 \\ -v_2 \end{bmatrix} + \begin{bmatrix} 2v_1+v_2 \\ v_1+2v_2 \\ v_1+4v_2 \end{bmatrix} = \begin{bmatrix} 3v_1+v_2 \\ 3v_1+2v_2 \\ v_1+3v_2 \end{bmatrix}$$

The commutative property is important in algebra and other mathematical areas because you have more flexibility when the order doesn't matter. The addition of real numbers is commutative, because you can add two numbers in either order and get the same answer: $x + y = y + x$. Subtraction is *not* commutative, because you get a different answer if you subtract 10 - 7 than if you subtract 7 - 10.

The commutative property of addition applies to linear transformations when you perform more than one transformation on a vector.

Given the vector **v** and the linear transformations T_1 and T_2:

$$T_1(\mathbf{v}) + T_2(\mathbf{v}) = T_2(\mathbf{v}) + T_1(\mathbf{v})$$

or, more simply:

$$T_1 + T_2 = T_2 + T_1$$

So, using the two transformations T_1 and T_2 given next, you see that changing the order does not change the final result.

$$T_1\left(\begin{bmatrix} x \\ y \end{bmatrix}\right) \to \begin{bmatrix} x \\ 2x \\ x+3y \end{bmatrix},\quad T_2\left(\begin{bmatrix} x \\ y \end{bmatrix}\right) \to \begin{bmatrix} y \\ -2y \\ x+y \end{bmatrix}$$

$$T_1\left(\begin{bmatrix} v_1 \\ v_2 \end{bmatrix}\right) + T_2\left(\begin{bmatrix} v_1 \\ v_2 \end{bmatrix}\right) = \begin{bmatrix} v_1 \\ 2v_1 \\ v_1+3v_2 \end{bmatrix} + \begin{bmatrix} v_2 \\ -2v_2 \\ v_1+v_2 \end{bmatrix} = \begin{bmatrix} v_1+v_2 \\ 2v_1-2v_2 \\ 2v_1+4v_2 \end{bmatrix}$$

$$T_2\left(\begin{bmatrix} v_1 \\ v_2 \end{bmatrix}\right) + T_1\left(\begin{bmatrix} v_1 \\ v_2 \end{bmatrix}\right) = \begin{bmatrix} v_2 \\ -2v_2 \\ v_1+v_2 \end{bmatrix} + \begin{bmatrix} v_1 \\ 2v_1 \\ v_1+3v_2 \end{bmatrix} = \begin{bmatrix} v_1+v_2 \\ 2v_1-2v_2 \\ 2v_1+4v_2 \end{bmatrix}$$

Because addition of vectors and matrices must follow the rules involving proper dimensions, this commutative property only applies when it makes any sense. See Chapter 3 for more on dimension and the addition of matrices.

Introducing transformation composition and some properties

Transformation composition is actually more like an *embedded* operation. When you perform the composition transformation T_1 followed by transformation T_2, you first perform transformation T_2 on the vector **v** and then perform transformation T_1 on the result.

Transformation composition is defined as $(T_1T_2)(\mathbf{v}) = T_1(T_2(\mathbf{v}))$.

Being able to perform transformation composition is dependent upon the vectors and matrices being of the correct dimension. When the transformation T_1 has a rule acting on 3×1 vectors, then the transformation T_2 must have an output or result of 3×1 vectors.

For example, consider the following two transformations, T_1 and T_2, and the operation of transformation composition on the selected vector:

$$T_1\left(\begin{bmatrix} x \\ y \\ z \end{bmatrix}\right) \to \begin{bmatrix} x-y \\ 2z \\ x-3y \end{bmatrix},\ T_2\left(\begin{bmatrix} x \\ y \end{bmatrix}\right) \to \begin{bmatrix} y \\ -2y \\ x+3y \end{bmatrix}$$

$$T_1\left(T_2\left(\begin{bmatrix} 4 \\ -3 \end{bmatrix}\right)\right) = T_1\left(\begin{bmatrix} -3 \\ 6 \\ -5 \end{bmatrix}\right) = \begin{bmatrix} -9 \\ -10 \\ -21 \end{bmatrix}$$

Performing T_2, first, on the 2×1 vector results in a 3×1 vector. T_1 is then performed on that result.

And, in general, the composition of these two transformations is

$$T_1\left(T_2\left(\begin{bmatrix} v_1 \\ v_2 \end{bmatrix}\right)\right) = T_1\left(\begin{bmatrix} v_2 \\ -2v_2 \\ v_1 + 3v_2 \end{bmatrix}\right) = \begin{bmatrix} v_2 + 2v_2 \\ 2(v_1 + 3v_2) \\ v_2 - 3(-2v_2) \end{bmatrix} = \begin{bmatrix} 3v_2 \\ 2v_1 + 6v_2 \\ 7v_2 \end{bmatrix}$$

Associating with vectors and the associative property of composition

The associative property has to do with the grouping of the transformations, not the order. In general, you can't change the order of performing transformations and get the same result. But, with careful arrangements of transformations and dimension, you do get to see the associative property in action when composing transformations. The associative property states that when you perform transformation composition on the first of two transformations and then the result on a third, you get the same result as performing the composition on the last two transformations and then performing the transformation described by the first on the result of those second two. Whew! It's more understandable written symbolically:

Given the vector **v** and the linear transformations T_1, T_2, and T_3:

$$([T_1T_2]T_3)(\mathbf{v}) = (T_1[T_2T_3])(\mathbf{v})$$

For example, let T_1, T_2, and T_3 be linear transformations defined as shown:

$$T_1(\mathbf{x}) = A_1\mathbf{x} \text{ where } A_1 = \begin{bmatrix} 1 & 0 \\ 0 & 2 \end{bmatrix}$$

$$T_2(\mathbf{x}) = A_2\mathbf{x} \text{ where } A_2 = \begin{bmatrix} 1 & 1 \\ 0 & 1 \end{bmatrix}$$

$$T_3(\mathbf{x}) = A_3\mathbf{x} \text{ where } A_3 = \begin{bmatrix} 1 & -1 \\ 3 & 1 \end{bmatrix}$$

The results of performing the transformations are shown here.

$$\left(\left[T_1T_2\right]T_3\right)(\mathbf{x}) =$$

$$\left(\left[T_1\left(\begin{bmatrix}1 & 1\\0 & 1\end{bmatrix}\right)\right]T_3\right)(\mathbf{x})$$

$$= \left(\begin{bmatrix}1 & 1\\0 & 2\end{bmatrix}T_3\right)(\mathbf{x})$$

$$= \begin{bmatrix}4 & 0\\6 & 2\end{bmatrix}(\mathbf{x})$$

$$\left(T_1\left[T_2T_3\right]\right)(\mathbf{x}) =$$

$$= \left(T_1\left[T_2\left(\begin{bmatrix}1 & -1\\3 & 1\end{bmatrix}\right)\right]\right)(\mathbf{x})$$

$$= \left(T_1\left(\begin{bmatrix}4 & 0\\3 & 1\end{bmatrix}\right)\right)(\mathbf{x})$$

$$= \begin{bmatrix}4 & 0\\6 & 2\end{bmatrix}(\mathbf{x})$$

Because multiplication of vectors and matrices must follow the rules involving proper dimensions, this associative property only applies when it makes any sense.

Scaling down the process with scalar multiplication

Performing scalar multiplication on matrices or vectors amounts to multiplying each element in the matrix or vector by a constant number. I cover scalar multiplication thoroughly in Chapter 3, if you want just a bit more information on what's involved. Because transformations performed on vectors result in other vectors, the properties of scalar multiplication do hold in transformation multiplication. In fact, the scalar can be introduced at any one of three places in the multiplication.

Given the vector **v** and the linear transformations T_1 and T_2:

$$c([T_1T_2])(\mathbf{v}) = ([cT_1]T_2)(\mathbf{v}) = (T_1[cT_2])(\mathbf{v})$$

or,

$$c[T_1T_2] = [cT_1]T_2 = T_1[cT_2]$$

No matter where you introduce the scalar multiple, the result is the same.

Performing identity checks with identity transformations

Addition and multiplication of real numbers include different *identity* elements for each operation. The identity for addition is 0, and the identity for multiplication is 1. When you add any number to 0 (or add 0 to the number), you don't change that number's identity. The same goes for multiplication. And, because this chapter deals in linear transformations, I get to introduce identity *transformations*.

Adding to the process with the additive identity transformation

The additive identity transformation, T_0, takes a vector and transforms that vector into a zero vector.

The linear transformation $T_0(\mathbf{v}) = \mathbf{0}$.

So, if you have a 2 × 1 vector, the additive identity transformation changes it into the 2 × 1 zero vector. Here's the additive identity transformation at work:

$$T_0\left(\begin{bmatrix}2\\3\end{bmatrix}\right)=\begin{bmatrix}0\\0\end{bmatrix},\ T_0\left(\begin{bmatrix}-5\\0\\4\end{bmatrix}\right)=\begin{bmatrix}0\\0\\0\end{bmatrix},\ T_0\left(\begin{bmatrix}x\\y\\z\\w\end{bmatrix}\right)=\begin{bmatrix}0\\0\\0\\0\end{bmatrix}$$

Now I show how the additive identity transformation behaves when combined with other transformations.

When combining the additive identity transformation T_0 with another linear transformation T you get:

$$T_0(\mathbf{v}) + T(\mathbf{v}) = T(\mathbf{v}) + T_0(\mathbf{v}) = T(\mathbf{v})$$

or

$$T_0 + T = T + T_0 = T$$

For example, consider the following transformation, T, and vector, **v:**

$$T\left(\begin{bmatrix} x \\ y \\ z \end{bmatrix}\right) \rightarrow \begin{bmatrix} 2x \\ z \\ -3y \end{bmatrix}, \quad \mathbf{v} = \begin{bmatrix} 4 \\ 5 \\ 6 \end{bmatrix}$$

$$T_0\left(\begin{bmatrix} 4 \\ 5 \\ 6 \end{bmatrix}\right) + T\left(\begin{bmatrix} 4 \\ 5 \\ 6 \end{bmatrix}\right) = T\left(\begin{bmatrix} 4 \\ 5 \\ 6 \end{bmatrix}\right) + T_0\left(\begin{bmatrix} 4 \\ 5 \\ 6 \end{bmatrix}\right)$$

$$\begin{bmatrix} 0 \\ 0 \\ 0 \end{bmatrix} + \begin{bmatrix} 8 \\ 6 \\ -15 \end{bmatrix} = \begin{bmatrix} 8 \\ 6 \\ -15 \end{bmatrix} + \begin{bmatrix} 0 \\ 0 \\ 0 \end{bmatrix} = \begin{bmatrix} 8 \\ 6 \\ -15 \end{bmatrix}$$

Making a move with the multiplicative identity transformation

In general, the commutative property of multiplication does *not* apply to linear transformations when you perform more than one transformation on a vector. The exception comes into play when dealing with the multiplicative identity linear transformation.

The linear transformation I is the multiplicative identity linear transformation when, given the vector **v**,

$$I(\mathbf{v}) = \mathbf{v}$$

Illustrating how the identity transformation works is quite simple. You perform the transformation on any vector and don't change the vector at all.

$$I\left(\begin{bmatrix} x \\ y \\ z \end{bmatrix}\right) = \begin{bmatrix} x \\ y \\ z \end{bmatrix}, \quad I\left(\begin{bmatrix} 3 \\ 4 \end{bmatrix}\right) = \begin{bmatrix} 3 \\ 4 \end{bmatrix}, \quad I\left(\begin{bmatrix} 1 \\ 0 \\ -1 \\ 1 \end{bmatrix}\right) = \begin{bmatrix} 1 \\ 0 \\ -1 \\ 1 \end{bmatrix}$$

What I show you here are just three different identity transformations. Each is different, but each has the same property of preserving the vector.

Furthermore, when multiplying the multiplicative identity I and some other linear transformation T, you have commutativity:

$$(I*T)(\mathbf{v}) = (T*I)(\mathbf{v}) = T(\mathbf{v})$$

or

$$IT = TI = T$$

Again, the dimensions of the vectors have to be such that the multiplication makes sense.

Delving into the distributive property

The distributive property in algebra involves spreading out, or *distributing,* the product of some factor over the sum of two or more terms. For example, distributing the number 2 over three variable terms: $2(x + 2y + 3z) = 2x + 4y + 6z$. Linear transformations also have distributive laws.

Given the vector **v** and the linear transformations T_1, T_2, and T_3:

$$(T_1 * [T_2 + T_3])(\mathbf{v}) = (T_1 * T_2)(\mathbf{v}) + (T_1 * T_3)(\mathbf{v})$$

and

$$([T_1 + T_2] * T_3)(\mathbf{v}) = (T_1 * T_3)(\mathbf{v}) + (T_2 * T_3)(\mathbf{v})$$

or, stated more simply:

$$T_1[T_2 + T_3] = T_1T_2 + T_1T_3$$

and

$$[T_1 + T_2]T_3 = T_1T_3 + T_2T_3$$

Dimension qualifications must prevail, of course.

And finally, tying distribution, scalar multiplication, and one of the main properties of linear transformations:

$$T(c_1\mathbf{v}_1 + c_2\mathbf{v}_2 + c_3\mathbf{v}_3 + \ldots + c_k\mathbf{v}_k) = c_1T\mathbf{v}_1 + c_2T\mathbf{v}_2 + c_3T\mathbf{v}_3 + \ldots + c_kT\mathbf{v}_k$$

Writing the Matrix of a Linear Transformation

One way of describing a linear transformation is to give rules involving the elements of the input vector and then the output vector. The preceding sections of this chapter use various rules to show what the linear transformation does. Another way of describing a linear transformation is to use a matrix multiplier instead of a rule, making computations quicker and easier.

For example, consider the following linear transformation T and its rule:

$$T\left(\begin{bmatrix} x \\ y \end{bmatrix}\right) \rightarrow \begin{bmatrix} -x \\ y \end{bmatrix}$$

The rule is simple enough. The transformation takes a 2×1 vector and changes the first element to its opposite, otherwise leaving the vector alone.

The corresponding matrix multiplication for this particular rule is as follows, where A is a 2×2 matrix:

$$A = \begin{bmatrix} -1 & 0 \\ 0 & 1 \end{bmatrix}$$

$$T(\mathbf{v}) = A * \begin{bmatrix} x \\ y \end{bmatrix} = \begin{bmatrix} -1 & 0 \\ 0 & 1 \end{bmatrix} * \begin{bmatrix} x \\ y \end{bmatrix} = \begin{bmatrix} -x+0 \\ 0+y \end{bmatrix} = \begin{bmatrix} -x \\ y \end{bmatrix}$$

So, in this case $T(\mathbf{v}) = A*\mathbf{v}$.

Manufacturing a matrix to replace a rule

In the previous section, it appeared that I magically introduced a matrix A to use for a linear transformation. But I really didn't just dream up the matrix or have it drop out of thin air for me; I solved for that matrix using scalar multiples of two vectors. Beginning with the rule for transformation T, I wrote the sum of the two elements in the original vector multiplied times vectors containing the coefficients of elements in the resulting vector.

$$\begin{bmatrix} -x \\ y \end{bmatrix} = \begin{bmatrix} -1x+0y \\ 0x+y \end{bmatrix} = x\begin{bmatrix} -1 \\ 0 \end{bmatrix} + y\begin{bmatrix} 0 \\ 1 \end{bmatrix}$$

Then I took the two vectors of coefficients and created a matrix whose columns are the vectors.

$$\begin{bmatrix} -1 \\ 0 \end{bmatrix} \text{ and } \begin{bmatrix} 0 \\ 1 \end{bmatrix} \rightarrow \begin{bmatrix} -1 & 0 \\ 0 & 1 \end{bmatrix}$$

Okay, that example was easy because I started out with the answer. Now let me show you *starting from scratch*. Consider next the rule given for transformation S:

$$S\left(\begin{bmatrix} x \\ y \\ z \end{bmatrix}\right) \rightarrow \begin{bmatrix} y-2z \\ -2x+y-3z \\ x+3y \end{bmatrix}$$

Writing the rule as the sum of the three elements times three vectors of coefficients:

$$\begin{bmatrix} y-2z \\ -2x+y-3z \\ x+3y \end{bmatrix} = \begin{bmatrix} 0+1y-2z \\ -2x+1y-3z \\ 1x+3y+0z \end{bmatrix} \rightarrow x\begin{bmatrix} 0 \\ -2 \\ 1 \end{bmatrix} + y\begin{bmatrix} 1 \\ 1 \\ 3 \end{bmatrix} + z\begin{bmatrix} -2 \\ -3 \\ 0 \end{bmatrix}$$

And then, writing the transformation matrix, A, using the vectors as columns:

$$A = \begin{bmatrix} 0 & 1 & -2 \\ -2 & 1 & -3 \\ 1 & 3 & 0 \end{bmatrix}$$

The transformation S is equal to the product of matrix A times a 3×1 vector. Now I show you the matrix in action with an example.

$$S\left(\begin{bmatrix} 2 \\ 4 \\ 5 \end{bmatrix}\right) = A^* \begin{bmatrix} 2 \\ 4 \\ 5 \end{bmatrix} = \begin{bmatrix} 0 & 1 & -2 \\ -2 & 1 & -3 \\ 1 & 3 & 0 \end{bmatrix} * \begin{bmatrix} 2 \\ 4 \\ 5 \end{bmatrix} = \begin{bmatrix} 0+4-10 \\ -4+4-15 \\ 2+12+0 \end{bmatrix} = \begin{bmatrix} -6 \\ -15 \\ 14 \end{bmatrix}$$

Visualizing transformations involving rotations and reflections

A very nice way to illustrate the effect of a linear transformation is to draw a picture. And some of the nicest pictures to consider are those in two-space that involve rotations about the origin, reflections about a line, translations along a line, and size changes. In this section, I show you how rotations and reflections are written as rules and matrix multiplications.

Rotating about the origin

Rotations about the origin are typically in a counterclockwise direction. In Figure 8-1, for example, I show the point (6,4) being rotated 90 degrees to the left — or counterclockwise. The distance from the original point (6,4) to the origin stays the same, and the angle measured from the beginning point to the ending point (with the vertex at the origin) is 90 degrees.

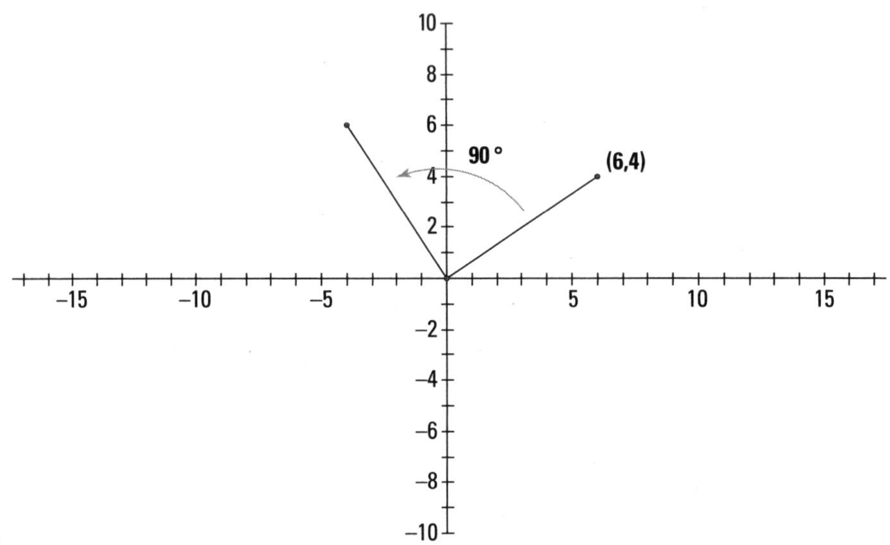

Figure 8-1: The point rotates to the left.

"All well and good," you say, "but what are the coordinates of the point resulting from the rotation?" To determine the coordinates, first I show you the general format for a matrix used in any counterclockwise rotations about the origin.

When rotating the point (x,y) θ degrees about the origin in a counterclockwise direction, let matrix A multiply the vector **v**, containing the coordinates of the point, resulting in the coordinates of the ending point.

$$A = \begin{bmatrix} \cos\theta & -\sin\theta \\ \sin\theta & \cos\theta \end{bmatrix}$$

$$A^*\mathbf{v} = \begin{bmatrix} \cos\theta & -\sin\theta \\ \sin\theta & \cos\theta \end{bmatrix} * \begin{bmatrix} x \\ y \end{bmatrix} = \begin{bmatrix} x\cos\theta - y\sin\theta \\ x\sin\theta + y\cos\theta \end{bmatrix}$$

You'll find the values of the sine and cosine of the basic angles in this book's Cheat Sheet.

Rotating the point (6,4) 90 degrees to a point in the second quadrant, you multiply matrix A times the vector formed from the coordinates. The sine of 90 degrees is 1, and the cosine of 90 degrees is 0.

$$\begin{aligned} A^*\mathbf{v} &= \begin{bmatrix} \cos 90° & -\sin 90° \\ \sin 90° & \cos 90° \end{bmatrix} * \begin{bmatrix} 6 \\ 4 \end{bmatrix} \\ &= \begin{bmatrix} 0 & -1 \\ 1 & 0 \end{bmatrix} * \begin{bmatrix} 6 \\ 4 \end{bmatrix} = \begin{bmatrix} 0 \cdot 6 - 1 \cdot 4 \\ 1 \cdot 6 + 0 \cdot 4 \end{bmatrix} = \begin{bmatrix} -4 \\ 6 \end{bmatrix} \end{aligned}$$

The coordinates of the point after the rotation are (–4,6). If you want to write the transformation as a rule rather than a matrix multiplication, you have

$$T\left(\begin{bmatrix} x \\ y \end{bmatrix}\right) \rightarrow \begin{bmatrix} -y \\ x \end{bmatrix}$$

The matrix A also corresponds to the sum of the two vectors:

$$\begin{bmatrix} -y \\ x \end{bmatrix} = x\begin{bmatrix} 0 \\ 1 \end{bmatrix} + y\begin{bmatrix} -1 \\ 0 \end{bmatrix} = \begin{bmatrix} 0 & -1 \\ 1 & 0 \end{bmatrix} * \begin{bmatrix} x \\ y \end{bmatrix} = A^* \begin{bmatrix} x \\ y \end{bmatrix}$$

Now, using a not-quite-so-nice example, I rotate the point (–2,4) about the origin with an angle of 30 degrees. I say that the example isn't quite as nice, because the value of the sine of 30 degrees is the fraction $^1/_2$, and the cosine of 30 degrees has a radical in it. Fractions and radicals aren't quite as nice as 1s, 0s, and -1s in a matrix.

$$\begin{aligned} \begin{bmatrix} \cos 30° & -\sin 30° \\ \sin 30° & \cos 30° \end{bmatrix} * \begin{bmatrix} -2 \\ 4 \end{bmatrix} &= \begin{bmatrix} \frac{\sqrt{3}}{2} & -\frac{1}{2} \\ \frac{1}{2} & \frac{\sqrt{3}}{2} \end{bmatrix} * \begin{bmatrix} -2 \\ 4 \end{bmatrix} \\ &= \begin{bmatrix} -\sqrt{3} - 2 \\ -1 + 2\sqrt{3} \end{bmatrix} \approx \begin{bmatrix} -3.73 \\ 2.46 \end{bmatrix} \end{aligned}$$

Rounding to the nearer hundredth, the coordinates of the point resulting from the rotation are (–3.73,2.46).

Reflecting across an axis or line

When a point is reflected over a line, the image point is the same distance from the line as the original point, and the segment drawn between the original point and its image is perpendicular to the line of reflection. In Figure 8-2, I show you four different reflections. The left side shows the point (-2,4) reflected over the x-axis and y-axis. The second sketch shows you that same point, (-2,4) reflected over the line $y = x$ and the line $y = -x$.

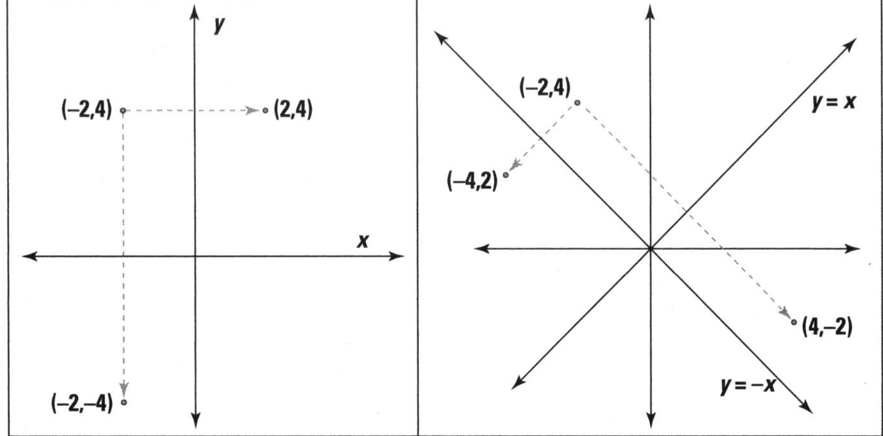

Figure 8-2: A point reflected across four different lines.

The linear transformations resulting in reflections of a point over an axis or one of the lines shown in the figure are written as matrix multiplications and the rules are given here:

- Reflection over the x-axis:

$$A^*\begin{bmatrix} x \\ y \end{bmatrix} = \begin{bmatrix} 1 & 0 \\ 0 & -1 \end{bmatrix}\begin{bmatrix} x \\ y \end{bmatrix} = \begin{bmatrix} x \\ -y \end{bmatrix}$$

- Reflection over the y-axis:

$$A^*\begin{bmatrix} x \\ y \end{bmatrix} = \begin{bmatrix} -1 & 0 \\ 0 & 1 \end{bmatrix}\begin{bmatrix} x \\ y \end{bmatrix} = \begin{bmatrix} -x \\ y \end{bmatrix}$$

- Reflection over the line $y = x$:

$$A^*\begin{bmatrix} x \\ y \end{bmatrix} = \begin{bmatrix} 0 & 1 \\ 1 & 0 \end{bmatrix}\begin{bmatrix} x \\ y \end{bmatrix} = \begin{bmatrix} y \\ x \end{bmatrix}$$

- Reflection over the line $y = -x$:

$$A^*\begin{bmatrix} x \\ y \end{bmatrix} = \begin{bmatrix} 0 & -1 \\ -1 & 0 \end{bmatrix}\begin{bmatrix} x \\ y \end{bmatrix} = \begin{bmatrix} -y \\ -x \end{bmatrix}$$

Insert your favorite coordinates for x and y and reflect away!

Translating, dilating, and contracting

The title of this section sounds like you've arrived at the delivery room in a foreign country. Sorry, no bundles of joy being produced here, but the results are mathematically pleasing!

Mathematically speaking, a *translation* is a slide. In two-space, a translation occurs when a point slides from one position to another along some straight line.

When the point (x,y) moves x_0 distance parallel to the x-axis and moves y_0 distance parallel to the y-axis, then the linear transformation T and matrix multiplication A result in the image point:

$$T\left(\begin{bmatrix} x \\ y \end{bmatrix}\right) = \begin{bmatrix} x + x_0 \\ y + y_0 \end{bmatrix}$$

$$A^*\begin{bmatrix} x \\ y \\ 1 \end{bmatrix} = \begin{bmatrix} 1 & 0 & x_0 \\ 0 & 1 & y_0 \end{bmatrix}\begin{bmatrix} x \\ y \\ 1 \end{bmatrix} = \begin{bmatrix} x + x_0 \\ y + y_0 \end{bmatrix}$$

For example, if you want to translate the point (–2,4) a distance of five units to the right and three units down, then $x_0 = 5$ and $y_0 = -3$. And the transformation performed is

$$A^*\begin{bmatrix} -2 \\ 4 \\ 1 \end{bmatrix} = \begin{bmatrix} 1 & 0 & 5 \\ 0 & 1 & -3 \end{bmatrix}\begin{bmatrix} -2 \\ 4 \\ 1 \end{bmatrix} = \begin{bmatrix} -2+5 \\ 4+(-3) \end{bmatrix} = \begin{bmatrix} 3 \\ 1 \end{bmatrix}$$

I show you the translation of the point (–2,4) five units right and three units down in Figure 8-3.

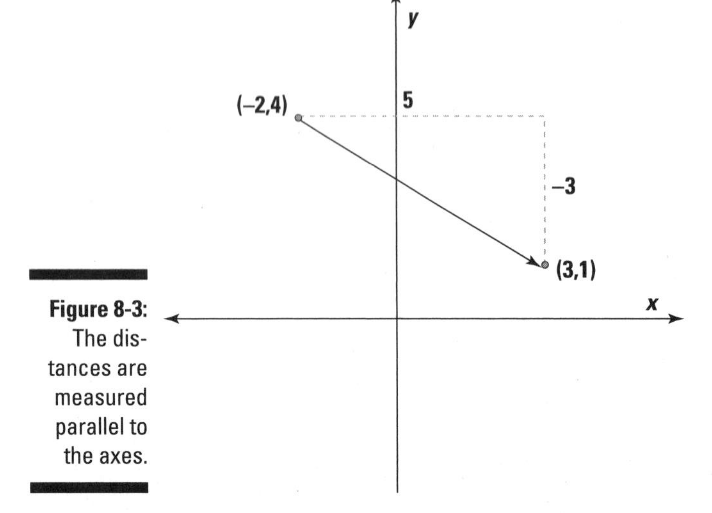

Figure 8-3: The distances are measured parallel to the axes.

A *dilation* increases the magnitude of a vector, and a *contraction* reduces the magnitude of a vector. In Chapter 2, I describe the magnitude of a vector — which is illustrated as its length when working in two-space and three-space. The linear transformation shown next is D, which doubles the magnitude of the vector. I also show what the transformation D does to vector **v.**

$$D = \begin{bmatrix} 2 & 0 \\ 0 & 2 \end{bmatrix}, \mathbf{v} = \begin{bmatrix} 4 \\ -3 \end{bmatrix}$$

$$D^*\mathbf{v} = \begin{bmatrix} 2 & 0 \\ 0 & 2 \end{bmatrix} * \begin{bmatrix} 4 \\ -3 \end{bmatrix} = \begin{bmatrix} 8 \\ -6 \end{bmatrix}$$

Comparing the magnitudes of the beginning and ending vectors:

$$\mathbf{v} = \begin{bmatrix} 4 \\ -3 \end{bmatrix}, |\mathbf{v}| = \sqrt{4^2 + (-3)^2} = \sqrt{16+9} = \sqrt{25} = 5$$

$$\mathbf{v} = \begin{bmatrix} 8 \\ -6 \end{bmatrix}, |\mathbf{v}| = \sqrt{8^2 + (-6)^2} = \sqrt{64+36} = \sqrt{100} = 10$$

As you can see, the magnitude has doubled as a result of the transformation. A contraction works much the same way, except that, in a contraction, the nonzero elements in the transformation matrix are always between 0 and 1 (resulting in a smaller magnitude).

So, in general, when applying a dilation or contraction transformation to a vector, you have the same multiplier along the main diagonal and 0s in all other positions of the matrix. If that multiplier is larger than 1, then you have a dilation. If the multiplier is between 0 and 1, you have a contraction. A multiplier of 1 has no effect; you're multiplying by the multiplicative identity matrix.

Determining the Kernel and Range of a Linear Transformation

Linear transformations performed on vectors result in other vectors. Sometimes the resulting vectors are of the same dimension as those you started with, and sometimes the result has a different dimension. A special result of a linear transformation is that in which every element in the resulting vector is a 0. The kernel of a linear transformation is also referred to as the *null space*. The kernel and range of a linear transformation are related in that they're both results of performing the linear transformation. The kernel is a special portion, and the range is everything that happens.

Keeping up with the kernel

When the linear transformation T from a given set of vectors to another set of vectors results in one or more 0 vectors, then the original vectors, $\mathbf{v}_1, \mathbf{v}_2, \mathbf{v}_3, \ldots$ for which $T(\mathbf{v}) = 0$ are called the *kernel* or *null space* of the set of vectors.

Note that T is not the additive identity vector, T_0, which changes all vectors to zero vectors.

For example, consider the transformation T shown next and its effect on the vector **v**.

$$T\left(\begin{bmatrix} x \\ y \\ z \end{bmatrix}\right) = \begin{bmatrix} x-z \\ x+y \\ y+z \end{bmatrix}, \mathbf{v} = \begin{bmatrix} 1 \\ -1 \\ 1 \end{bmatrix}$$

$$T\left(\begin{bmatrix} 1 \\ -1 \\ 1 \end{bmatrix}\right) = \begin{bmatrix} 1-1 \\ 1+(-1) \\ -1+1 \end{bmatrix} = \begin{bmatrix} 0 \\ 0 \\ 0 \end{bmatrix}$$

The vector resulting from performing T on **v** is a zero vector, so the vector **v** must be in the null space of the set of vectors. The next question is "Are there any other vectors in the null space?"

To determine the vectors in a null space, you determine what the elements of the null-space vectors must be to satisfy the statement T(**v**) = **0** where **0** is a zero vector. Using the transformation T shown in the previous example, each of the elements in the resulting vector must equal 0.

$$T\left(\begin{bmatrix} x \\ y \\ z \end{bmatrix}\right) = \begin{bmatrix} x - z \\ x + y \\ y + z \end{bmatrix} = \begin{bmatrix} 0 \\ 0 \\ 0 \end{bmatrix}$$

To determine what the desired elements (those in the kernel) must be, you solve the system of homogeneous equations. Chapter 7 tells you all about systems of homogeneous equations and how to find the solutions. The system of equations to be solved is

$$\begin{cases} x - z = 0 \\ x + y = 0 \\ y + z = 0 \end{cases}$$

Solving the system, you get that $x = z$ and $y = -z$, so any vector of the form

$$\begin{bmatrix} a \\ -a \\ a \end{bmatrix}$$

(where the first and last elements are the same and the middle element is the opposite of the other two) is in the kernel of the vector set.

Ranging out to find the range

The *range* of a linear transformation T is the set of all vectors that result from performing T(**v**) where **v** is in a particular set of vectors.

The range of a linear transformation may be rather restricted or it may be infinitely large. The range is dependent on the vector set that the transformation is being applied to and the way in which the transformation is defined.

For example, a transformation T changes a 4×1 vector into a 2×1 vector in the following way:

$$T\left(\begin{bmatrix} x \\ y \\ z \\ w \end{bmatrix}\right) \to \begin{bmatrix} x-z \\ y+w \end{bmatrix}$$

The way T is defined, any 2×1 matrix is in the range of the transformation.

The range of a linear transformation may be restricted by some format or rule. For example, the transformation S may be defined by matrix multiplication. Consider the following transformation:

$$S(\mathbf{v}) = A^*\mathbf{v} = \begin{bmatrix} 1 & -2 \\ -1 & 3 \\ 2 & -2 \end{bmatrix} * \begin{bmatrix} x \\ y \end{bmatrix}$$

The range of S is the vector resulting from the following multiplication:

$$\begin{bmatrix} 1 & -2 \\ -1 & 3 \\ 2 & -2 \end{bmatrix} * \begin{bmatrix} x \\ y \end{bmatrix} = \begin{bmatrix} x-2y \\ -x+3y \\ 2x-2y \end{bmatrix}$$

Create an augmented matrix in which the elements of the resulting vector are represented by v_1, v_2, and v_3. Then reduce the matrix to solve for the elements of the resulting vector in terms of the transformation matrix entries. For more on augmented matrices and solving systems using the reduced row echelon form, refer to Chapter 4.

$$\begin{bmatrix} x-2y \\ -x+3y \\ 2x-2y \end{bmatrix} = \begin{bmatrix} v_1 \\ v_2 \\ v_3 \end{bmatrix}$$

$$\left[\begin{array}{cc|c} 1 & -2 & v_1 \\ -1 & 3 & v_2 \\ 2 & -2 & v_3 \end{array}\right] \begin{array}{c} R_1+R_2 \to R_2 \\ -2R_1+R_3 \to R_3 \end{array} \left[\begin{array}{cc|c} 1 & -2 & v_1 \\ 0 & 1 & v_1+v_2 \\ 0 & 2 & v_3-2v_1 \end{array}\right]$$

$$\begin{array}{c} 2R_2+R_1 \to R_1 \\ -2R_2+R_3 \to R_3 \end{array} \left[\begin{array}{cc|c} 1 & 0 & 3v_1+v_2 \\ 0 & 1 & v_1+v_2 \\ 0 & 0 & -4v_1-2v_2+v_3 \end{array}\right]$$

The last row in the reduced echelon form tells you that $v_3 = 4v_1 + 2v_2$. The first entry tells you that $x = 3v_1 + v_2$ and the second entry has $y = v_1 + v_2$. So every vector in the range must have the following form:

$$\begin{bmatrix} v_1 \\ v_2 \\ 4v_1+2v_2 \end{bmatrix}$$

Part III
Evaluating Determinants

"He ran some linear equations, threw in a few theorems, and before I knew it I was buying rustproofing."

In this part . . .

Step right up in line at the DMV! You don't drive? No problem. Instead of heading to the Department of Motor Vehicles, in this part you cruise along with determinants, matrices, and vectors. Learn the rules of the roadway for evaluating determinants, and avoid the speed traps when manipulating matrices corresponding to the determinants. Put on your seat belt and get ready for the ride of your life!

Chapter 9

Keeping Things in Order with Permutations

In This Chapter

- Counting the number of arrangements possible
- Making lists of the different permutations
- Involving inversions in permutations

How many different ways can you arrange the members of your family for the yearly portrait? You can't put Emily next to Tommy? Then you have fewer arrangements than you might if Tommy weren't always pulling Emily's hair.

Don't worry — this chapter isn't about the problems of a photographer. This is just my way of introducing you to the type of situation where you might apply permutations. A *permutation* is an arrangement or ordering of the elements of a set. One permutation of the letters CAT is ACT. In fact, the letters in the word *cat* can be arranged in six different ways: CAT, CTA, ACT, ATC, TAC, and TCA. Not all these arrangements are legitimate English words of course, but these are all the possible orderings.

Figuring out how many permutations are possible is the first step to actually listing those arrangements. If you know how many arrangements are possible, you know when to stop looking for more.

In this chapter, I also introduce you to *inversions* (special circumstances arising in some permutations). All these counting concerns are important when you deal with determinants (see Chapter 10).

Computing and Investigating Permutations

A permutation is an ordering of the elements of a set. In the introduction to this chapter, I show you the six different ways of ordering the letters in the word *cat.* Now look at the set containing the first four letters of the alphabet and how the letters can be ordered. The four letters in the set {a, b, c, d} have 24 permutations: abcd, abdc, acbd, acdb, adbc, adcb, bacd, badc, bcad, bcda, bdac, bdca, cabd, cadb, cbad, cbda, cdab, cdba, dabc, dacb, dbac, dbca, dcab, or dcba. The number of possible permutations is dependent on the number of elements involved.

Counting on finding out how to count

The first task in determining and listing all the permutations of a set of items is to determine how many permutations or orderings there are to be found. The *how many* is the easy part. The harder part is incorporating some sort of systematic method for listing all these permutations — if you really do need a list of them.

Before showing you the formula that counts how many permutations you have of a particular set, let me introduce you to a mathematical operation called *factorial.* The factorial operation, designated by **!**, indicates that you multiply the number in front of the factorial symbol by every positive integer smaller than that number:

$$n! = n \cdot (n-1) \cdot (n-2) \cdots 4 \cdot 3 \cdot 2 \cdot 1$$

You start with the number n, decrease n by 1 and multiply it times the first number, decrease it by another 1 and multiply it times the first two, and keep doing the decreasing and multiplying until you get to the number 1 — always the smallest number in the product. So $4! = 4 \cdot 3 \cdot 2 \cdot 1 = 24$ and $7! = 7 \cdot 6 \cdot 5 \cdot 4 \cdot 3 \cdot 2 \cdot 1 = 5{,}040$. Also, by definition or convention, the value of $0! = 1$. Assigning 0! the value of 1 makes some other mathematical formulas work properly.

Now, armed with what the factorial notation means, here's how to find the number of permutations you have when you choose a number of items to permute:

- If you want the number of permutations of all n items of a list, then the number of permutations is $n!$
- If you want the number of permutations of just r of the items chosen from the entire list of n items, then you use the following formula:

$$_{n}P_{r} = \frac{n!}{(n-r)!}$$

So, if you have ten people in your family and want to line up all ten of them side-by-side for a picture, you have $10! = 10 \cdot 9 \cdot 8 \cdot 7 \cdot 6 \cdot 5 \cdot 4 \cdot 3 \cdot 2 \cdot 1 =$ 3,628,800 different ways to arrange them. Of course, if Emily and Tommy can't stand next to each other, you have to eliminate a few choices (but we don't want to go there right now).

Sometimes, you have a set of items to choose from but won't use every item. For example, if you have seven different CDs and you want to load three of them into your car's CD player, how many different arrangements (permutations or orders) are possible? In this case, using the formula, the n is 7 and the r is 3.

$$_{7}P_{3} = \frac{7!}{(7-3)!} = \frac{7!}{4!} = \frac{7 \cdot 6 \cdot 5 \cdot \cancel{4} \cdot \cancel{3} \cdot \cancel{2} \cdot \cancel{1}}{\cancel{4} \cdot \cancel{3} \cdot \cancel{2} \cdot \cancel{1}} = 7 \cdot 6 \cdot 5 = 210$$

Another way of looking at the result is that you have seven choices for the first CD you load, then six choices for the second CD, and five choices for the third CD. You have 210 different ways to choose three CDs and arrange them in some order.

Making a list and checking it twice

After you've determined how many different permutations or arrangements you can create from a list of items, you have to go about actually listing all those arrangements. The key to making a long list (or even a not-so-long list) is to use some sort of system. Two very effective methods of listing all the different permutations are making a systematic listing like a table and making a tree.

Listing the permutations table-wise

The best way for me to explain making a listing of all the permutations of a set is to just demonstrate how it's done. For example, say I decide that I want to list all the arrangements of the digits 1, 2, 3, and 4 if I take just three of them at a time. The first step is to determine how many arrangements I'm talking about.

Letting $n = 4$ and $r = 3$ in the formula,

$$_4P_3 = \frac{4!}{(4-3)!} = \frac{4!}{1!} = \frac{4 \cdot 3 \cdot 2 \cdot 1}{1} = 24$$

I need a list of 24 different arrangements.

Of the 24 arrangements, 6 must start with 1, 6 start with 2, 6 start with 3, and 6 start with 4. (I have just those four numbers to choose from.) I write the six 1s in a row, leaving room for two more digits after each.

1 1 1 1 1 1

Then, after two of the 1s I put 2s; after the next two 1s I put 3s; and after the last two 1s I put 4s.

12 12 13 13 14 14

After the listing of 12 in the first two positions, my only choices to finish the arrangement are a 3 or a 4. After the 13 listings, I can finish with a 2 or a 4. And after the 14 listings, I can finish with a 2 or a 3.

123 124 132 134 142 143

That gives me the first six arrangements — all the arrangements starting with the digit 1. I do the same thing with the 2s, 3s, and 4s. Write the first digit for each grouping (six of them start with 2, six with 3, six with 4).

2 2 2 2 2 2

3 3 3 3 3 3

4 4 4 4 4 4

The next digit after a 2 can be a 1, a 3, or a 4, and so on.

21 21 23 23 24 24

31 31 32 32 34 34

41 41 42 42 43 43

Finally, finish off with the last two choices for each.

213	214	231	234	241	243
312	314	321	324	341	342
412	413	421	423	431	432

These last 18 arrangements, plus the earlier 6 arrangements, give you the 24 possible arrangements of three of the first four digits.

Branching out with a tree

A very effective and very visual way of listing all the possible arrangements of numbers is with a horizontally moving tree. The only drawback of using the tree method is that, if your set of objects is very big, the tree can get rather unwieldy.

I show you the top half of a tree used to produce the 24 permutations of three of the first four digits in Figure 9-1. Note that the tree is branching from left to right.

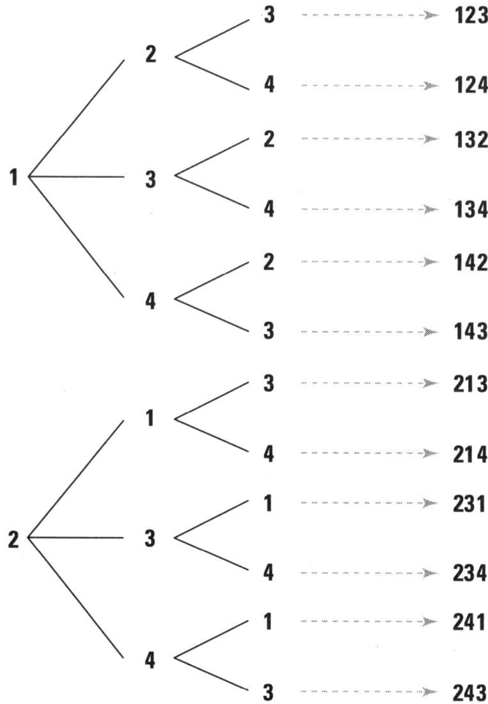

Figure 9-1: Creating a tree to list permutations.

On the left, the first beginning digits (you could call them the trunks from which the branches are built) are 1s, then the next are 2s. If you could see the bottom half of the tree, you'd see the 3s and 4s. After each beginning digit, you have three choices for the second digit, which are shown at the end of the three branches. Each of the three second digits has two more choices, because you aren't repeating any digits. You read the orderings from left to right, listing the digits you find from the leftmost branch to the tip of the last branch.

Bringing permutations into matrices (or matrices into permutations)

Rearranging elements in a list of numbers results in a permutation of the numbers — an ordering. When working with eigenvalues and eigenvectors in Chapter 16, it's sometimes important to be able to recognize a *permutation matrix*.

A permutation matrix is formed by rearranging the columns of an identity matrix.

For example, the 3×3 identity matrix has six different permutation matrices (including the original identity matrix):

$$\begin{bmatrix} 1 & 0 & 0 \\ 0 & 1 & 0 \\ 0 & 0 & 1 \end{bmatrix}, \begin{bmatrix} 1 & 0 & 0 \\ 0 & 0 & 1 \\ 0 & 1 & 0 \end{bmatrix}, \begin{bmatrix} 0 & 1 & 0 \\ 1 & 0 & 0 \\ 0 & 0 & 1 \end{bmatrix}$$

$$\begin{bmatrix} 0 & 0 & 1 \\ 1 & 0 & 0 \\ 0 & 1 & 0 \end{bmatrix}, \begin{bmatrix} 0 & 1 & 0 \\ 0 & 0 & 1 \\ 1 & 0 & 0 \end{bmatrix}, \begin{bmatrix} 0 & 0 & 1 \\ 0 & 1 & 0 \\ 1 & 0 & 0 \end{bmatrix}$$

One way of identifying the different permutation matrices is to use numbers such as 312 or 213 to indicate where the digit 1 is positioned in a particular row. For example, permutation 312 is represented by the fourth matrix, because row 1 has its 1 in the third column, row 2 has its 1 in the first column, and row 3 has its 1 in the second column.

The 3×3 identity matrix shown here has $3! = 6$ permutation matrices. You count the number of permutation matrices associated with each identity matrix by using the permutation formula $n!$. If n is the number of rows and columns in the identity matrix, then $n!$ is the number of permutation matrices possible.

Involving Inversions in the Counting

The permutation of a set of numbers is said to have an *inversion* if a larger digit precedes a smaller one. For example, consider the permutation 5423. The 5 precedes 4, the 5 precedes 2, the 5 precedes 3, the 4 precedes 2, and the 4 precedes 3. You count five inversions in this number. ***Remember:*** I said you count when a larger digit comes before a smaller one — when they seem to be out of the natural order.

Investigating inversions

A permutation of a set of numbers has a countable number of inversions — even if that number is zero. The permutation 5423 has five inversions. Another way of looking at inversions is that they're the minimum number of interchanges of consecutive elements necessary to arrange the numbers in their natural order. This ordering process is basic to many computer programming tasks.

For example, see how I rearrange the permutation 5423 by interchanging two consecutive (adjacent) numbers at a time. My demonstration is not the only way possible. You could do this in one of many more ways, but any choice will take five interchanges:

- Interchange the 4 and the 2: 5243
- Interchange the 5 and the 2: 2543
- Interchange the 4 and the 3: 2534
- Interchange the 5 and the 3: 2354
- Interchange the 5 and the 4: 2345

You think you can order the permutation in fewer steps using consecutive elements? I invite you to try! Now I introduce you to the *inverse* of a permutation.

An *inverse permutation* is a permutation in which each number and the number of the place it occupies are exchanged.

For example, consider the permutation of the first five positive integers: 25413. The inverse permutation associated with this particular arrangement is: 41532. The way the inverse works is that, in the original permutation,

- The 1 was in the fourth position, so I put the 4 in the first position.
- The 2 was in the first position, so I put the 1 in the second position.

- The 3 was in the fifth position, so I put the 5 in the third position.
- The 4 was in the third position, so I put the 3 in the fourth position.
- The 5 was in the second position, so I put the 2 in the fifth position.

Another way to think about the inverse permutation is to look at the matrix permutations involved. The permutation 25413 is represented by the 5×5 matrix A. And the permutation 41532 is represented by matrix B. I declare that A and B are inverses of one another and show it by multiplying the two matrices together — resulting in the 5×5 identity matrix.

$$A = \begin{bmatrix} 0 & 1 & 0 & 0 & 0 \\ 0 & 0 & 0 & 0 & 1 \\ 0 & 0 & 0 & 1 & 0 \\ 1 & 0 & 0 & 0 & 0 \\ 0 & 0 & 1 & 0 & 0 \end{bmatrix}, \; B = \begin{bmatrix} 0 & 0 & 0 & 1 & 0 \\ 1 & 0 & 0 & 0 & 0 \\ 0 & 0 & 0 & 0 & 1 \\ 0 & 0 & 1 & 0 & 0 \\ 0 & 1 & 0 & 0 & 0 \end{bmatrix}$$

$$A*B = \begin{bmatrix} 0 & 1 & 0 & 0 & 0 \\ 0 & 0 & 0 & 0 & 1 \\ 0 & 0 & 0 & 1 & 0 \\ 1 & 0 & 0 & 0 & 0 \\ 0 & 0 & 1 & 0 & 0 \end{bmatrix} * \begin{bmatrix} 0 & 0 & 0 & 1 & 0 \\ 1 & 0 & 0 & 0 & 0 \\ 0 & 0 & 0 & 0 & 1 \\ 0 & 0 & 1 & 0 & 0 \\ 0 & 1 & 0 & 0 & 0 \end{bmatrix}$$

$$= \begin{bmatrix} 1 & 0 & 0 & 0 & 0 \\ 0 & 1 & 0 & 0 & 0 \\ 0 & 0 & 1 & 0 & 0 \\ 0 & 0 & 1 & 0 & 0 \\ 0 & 0 & 0 & 0 & 1 \end{bmatrix} = I$$

And now, I tell you an interesting fact about a permutation and its inverse.

The number of inversions in a permutation is equal to the number of inversions in its inverse.

So, considering the permutation 25413 and its inverse, 41532, I count the number of inversions in each. The inversion 25413 has six inversions: 21, 54, 51, 53, 41, and 43. The inversion 41532 also has six inversions: 41, 43, 42, 53, 52, and 32. The inversions are different, but their count is the same.

Inviting even and odd inversions to the party

A permutation (or ordering) is called *even* if it has an even number of inversions, and it's odd if it has an odd number of inversions. Can't get much more straightforward than that. And with the categorization of permutation inversions as even or odd, you find some interesting properties of the inversions.

If a set has n elements, and $n \geq 2$, then there are $n! \div 2$ even and $n! \div 2$ odd permutations of the set. For example, the set {1, 2, 3} has three elements, so you'll find $3! \div 2 = [3 \cdot (3 - 1) \cdot (3 - 2)] \div 2 = 3$ even permutations and 3 odd permutations. The three even permutations (those with an even number of inversions) are 123 (with zero inversions), 312 (with two inversions), and 231 (with two inversions). The three odd permutations are 132 (with one inversion), 213 (with one inversion), and 321 (with three inversions).

Another interesting property of inversions has to do with making changes to a particular permutation.

If you form a new permutation from a given permutation of a set of numbers by interchanging two elements, then the difference between the number of inversions in the two permutations is always an odd number.

For example, consider the permutation 543672, which has eight inversions: 54, 53, 52, 43, 42, 32, 62, 72. I interchange the 5 and the 2, resulting in 243675, which has just three inversions: 43, 65, 75. The difference between the number of inversions is $8 - 3 = 5$. Go ahead and try it yourself!

Chapter 10

Determining Values of Determinants

In This Chapter

- Determining how to compute determinants
- Linking permutations to determinants
- Using cofactor expansion to evaluate large determinants
- Computing area and volume of figures in two-space and three-space

Determinants are linked with square matrices. You compute determinants of matrices using the elements in the matrices. Determinants are used in many applications that you find in this part. In this chapter, you see how to compute area and volume using determinants.

The values of determinants are computed using one of several methods and can be evaluated using a graphing calculator or computer program. The beauty of being able to use a calculator or computer to compute determinants is especially evident as the determinants get larger — more than three rows and columns. The process of cofactor expansion extends to all sizes of square matrices and lends itself very nicely to a tidy computer program.

Evaluating the Determinants of 2 × 2 Matrices

A *determinant* is a number associated with a square matrix; the determinant is a single number of *worth.* The number you get for a determinant is created using products of elements in the matrix along with sums and differences of those products. The products involve permutations of the elements in the matrix, and the signs (indicating a sum or a difference) of the products are dependent on the number of inversions found in the permutations. If you need a refresher on permutations or inversions, you'll find what you need in Chapter 9.

Involving permutations in determining the determinant

A 2 × 2 matrix has a determinant involving 2! = 2 products of the elements in the matrix. A 3 × 3 matrix has a determinant involving 3! = 6 products of the elements in the matrix. A 4 × 4 matrix has a determinant involving 4! = 24 products of the elements in the matrix. The pattern goes on for as large as the matrix becomes. I ease into the process of computing the determinant by starting with the 2 × 2 matrix.

Tackling the 2 × 2 determinant first

Look at the 2 × 2 matrix.

$$\begin{bmatrix} a_{11} & a_{12} \\ a_{21} & a_{22} \end{bmatrix}$$

The products involved in computing the determinant of a 2 × 2 matrix consist of multiplying the elements in the general form:

$$a_{1j_1}a_{2j_2}$$

The 1 and 2 in the indices indicate the rows in the matrix, and the indices j_1 and j_2 represent the number of the columns being permuted — in this case, just two of them.

When referring to elements in a matrix, you use subscripts (indices) to indicate the row and column where the element belongs. So the element a_{13} refers to the element in the first row and third column. For more on matrices and their elements, refer to Chapter 3.

In a 2 × 2 matrix, you find two columns and need the two permutations of the numbers 1 and 2. The two permutations are 12 and 21. Replacing the *j*'s in

$$a_{1j_1}a_{2j_2}$$

with 12 and then 21, you get the two products $a_{11}a_{22}$ and $a_{12}a_{21}$.

The products $a_{11}a_{22}$ and $a_{12}a_{21}$ are created with the row and column structure of the indices and are also found by cross-multiplying — by multiplying the upper-left element by the lower-right element and multiplying the upper-right element by the lower-left element. The product $a_{11}a_{22}$ has no inversions (even), and the product $a_{12}a_{21}$ has the one inversion (odd). The even and odd designations are used when determining whether -1 or 1 multiplies the products.

When determining the sign of the product of elements in a determinant, if the indices of the elements in a product have an odd number of inversions, then the product is multiplied by a -1. If the product has an even number of inversions of the indices, then the product is multiplied by 1 (stays the same).

Now, after dealing with the preliminaries, I define the computations needed for the determinant. The determinant of the matrix A is designated with vertical lines around the name of the matrix: |A|.

For a 2 × 2 matrix, the determinant is:

$$\begin{vmatrix} a_{11} & a_{12} \\ a_{21} & a_{22} \end{vmatrix} = a_{11}a_{22} - a_{12}a_{21}$$

The product $a_{11}a_{22}$ has an even number of inversions, so it's multiplied by 1; and the product $a_{12}a_{21}$ has an odd number of inversions, so it's multiplied by -1.

Here is the numerical value associated with a particular 2 × 2 matrix:

$$\begin{vmatrix} 4 & -2 \\ 5 & -3 \end{vmatrix} = 4(-3)-(-2)(5) = -12-(-10) = -2$$

Another way of describing the value of the determinant of a 2 × 2 matrix is to say it's the difference between the cross-products.

Triumphing with a 3 × 3 determinant

The products involved in computing the determinant of a 3 × 3 matrix consist of the general elements:

$$a_{1j_1}a_{2j_2}a_{3j_3}$$

Each product involves three elements from the matrix, one from each row. The *j* parts of the indices indicate columns — each a different column in each product. You have a total of 3! = 6 different products, all created by using the permutations of the numbers 1, 2, and 3 for the rows and columns. The six products are $a_{11}a_{22}a_{33}$, $a_{11}a_{23}a_{32}$, $a_{12}a_{22}a_{33}$, $a_{12}a_{23}a_{31}$, $a_{13}a_{21}a_{32}$, and $a_{13}a_{22}a_{31}$. Note that each product has one of each row and one of each column.

The product $a_{11}a_{22}a_{33}$ has no inversions (even) and is positive; the product $a_{11}a_{23}a_{32}$ has one inversion (odd) and is negative; the product $a_{12}a_{22}a_{33}$ has one inversion and is negative; the product $a_{12}a_{23}a_{31}$ has two inversions (even) and is positive; the product $a_{13}a_{21}a_{32}$ has two inversions (even) and is positive; and the product $a_{13}a_{22}a_{31}$ has three inversions (odd) and is negative.

For a 3 × 3 matrix, the determinant is:

$$\begin{vmatrix} a_{11} & a_{12} & a_{13} \\ a_{21} & a_{22} & a_{23} \\ a_{31} & a_{32} & a_{33} \end{vmatrix} = a_{11}a_{22}a_{33} - a_{11}a_{23}a_{32} - a_{12}a_{21}a_{33} + a_{12}a_{23}a_{31} + a_{13}a_{21}a_{32} - a_{13}a_{22}a_{31}$$

"Egad!" you say. No, please don't throw your pencil across the room in despair (like my brother did, way back a bunch of years, when I tried to help him with his algebra). I've shown you the proper definition of a 3 × 3 determinant and why the different signs on the different products work, but now I can show you a very quick and practical way of computing the determinant of a 3 × 3 matrix without having to count inversions and line up signs.

The quick-and-easy, down-and-dirty method involves copying the first and second columns of the determinant outside, to the right of the determinant.

$$\left|\begin{matrix} a_{11} & a_{12} & a_{13} \\ a_{21} & a_{22} & a_{23} \\ a_{31} & a_{32} & a_{33} \end{matrix}\right| \begin{matrix} a_{11} & a_{12} \\ a_{21} & a_{22} \\ a_{31} & a_{32} \end{matrix}$$

Now, the positive (even) products lie along diagonal lines starting at the top of the three columns inside the determinant and moving to the right. The negative (odd) products start at the top of the last column in the determinant and the two columns outside the determinant and move left.

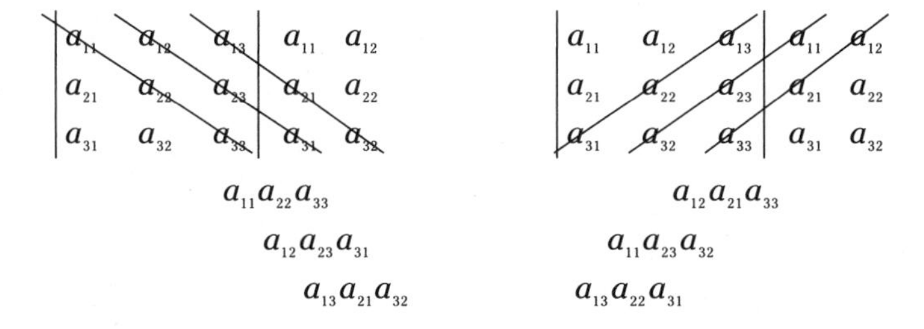

You add the products from the left-to-right slants, and subtract (multiply by -1) the products from the right-to-left slants:

$$\begin{vmatrix} a_{11} & a_{12} & a_{13} \\ a_{21} & a_{22} & a_{23} \\ a_{31} & a_{32} & a_{33} \end{vmatrix} \begin{matrix} a_{11} & a_{12} \\ a_{21} & a_{22} \\ a_{31} & a_{32} \end{matrix}$$
$$= \left(a_{11}a_{22}a_{33} + a_{12}a_{23}a_{31} + a_{13}a_{21}a_{32}\right) - \left(a_{13}a_{22}a_{31} + a_{11}a_{23}a_{32} + a_{12}a_{21}a_{33}\right)$$

Here's an example using actual numbers:

$$\begin{vmatrix} 3 & 0 & -1 \\ 2 & 4 & 1 \\ 7 & 2 & 3 \end{vmatrix} \rightarrow \begin{vmatrix} 3 & 0 & -1 \\ 2 & 4 & 1 \\ 7 & 2 & 3 \end{vmatrix} \begin{matrix} 3 & 0 \\ 2 & 4 \\ 7 & 2 \end{matrix}$$
$$= 3(4)(3) + 0(1)(7) + (-1)(2)(2) - \left[(-1)(4)(7) + 3(1)(2) + 0(2)(3)\right]$$
$$= 36 + 0 - 4 - \left[-28 + 6 + 0\right] = 32 - \left[-22\right] = 54$$

Coping with cofactor expansion

You evaluate determinants of any size square matrix, but, as you see from the simple 3×3 determinant, the number of products to consider grows rapidly — you might even say *permutationally.* (Well, no one would really say that.)

Anyway, the determinant of a 4×4 matrix involves $4! = 24$ products, half of which have even permutations of the indices and half of which have odd permutations. The determinant of a 5×5 matrix involves $5! = 120$ products, and on and on it goes. The best way to handle the larger determinants is to reduce the large problems into several smaller, more manageable problems. The method I'm referring to is *cofactor expansion.* Evaluating determinants using cofactor expansion amounts to choosing a row or column of the target matrix and multiplying either a positive or negative version of each element in the chosen row or column times a minor of the determinant formed by eliminating the row and column of that element. Huh? Say that again. No, I won't do that, but I'll explain with a demonstration of what I mean by the elements and their respective minors.

The *minor* of a determinant is a smaller determinant created by eliminating a specific row and column of the original determinant.

For example, one minor of the 4 × 4 matrix A is formed when you remove the second row and third column of A to form a 3 × 3 matrix C.

$$A = \begin{bmatrix} 1 & 4 & -5 & 6 \\ 3 & 5 & 2 & 0 \\ 1 & 1 & 8 & 6 \\ 4 & 5 & 6 & 7 \end{bmatrix}$$

$$\text{eliminate } \square \rightarrow \begin{bmatrix} 1 & 4 & \boxed{-5} & 6 \\ \boxed{3} & \boxed{5} & \boxed{2} & \boxed{0} \\ 1 & 1 & \boxed{8} & 6 \\ 4 & 5 & \boxed{6} & 7 \end{bmatrix}$$

$$C_{23} = \begin{bmatrix} 1 & 4 & 6 \\ 1 & 1 & 6 \\ 4 & 5 & 7 \end{bmatrix}$$

Since the minor was created by elimination of the second row and third column, you denote the minor with C_{23}. When using *cofactor expansion* to evaluate a determinant, you create several minors of the determinant and multiply them by positive or negative versions of the elements in a row or column. Here's a formal explanation of this expansion:

The value of the determinant of matrix A is:

$$|A| = \sum (-1)^{i+j} a_{ij} |C_{ij}|$$

The formula is read: *The determinant of matrix A is the sum of the products formed by multiplying (1) the number −1 raised to the i + j power, (2) the element a_{ij}, and (3) $|C_{ij}|$ the determinant of the minor associated with the element* a_{ij}. The elements a_{ij} are taken from one selected row or column of the original matrix A.

The number obtained from $(-1)^{i+j} \det A_{ij} = (-1)^{i+j} |A_{ij}|$ is called the *ij*-cofactor of matrix A.

For example, consider the 4 × 4 matrix A shown here:

$$A = \begin{bmatrix} 1 & 2 & -5 & 6 \\ 3 & 5 & 2 & 0 \\ 1 & 1 & 8 & 6 \\ 4 & 5 & 6 & 0 \end{bmatrix}$$

I chose to do cofactor expansion along the first row of the matrix. So the summation of the products along that row becomes:

$$|A| = (-1)^{1+1} a_{11}|C_{11}| + (-1)^{1+2} a_{12}|C_{12}| + (-1)^{1+3} a_{13}|C_{13}| + (-1)^{1+4} a_{14}|C_{14}|$$

You see how the -1 factors are being raised to powers formed from the sum of the numbers in the index of the particular element. The C_{ij} is the minor associated with the particular element.

$$|A| = (-1)^2(1)\begin{vmatrix} 5 & 2 & 0 \\ 1 & 8 & 6 \\ 5 & 6 & 0 \end{vmatrix} + (-1)^3(2)\begin{vmatrix} 3 & 2 & 0 \\ 1 & 8 & 6 \\ 4 & 6 & 0 \end{vmatrix}$$

$$+(-1)^4(-5)\begin{vmatrix} 3 & 5 & 0 \\ 1 & 1 & 6 \\ 4 & 5 & 0 \end{vmatrix} + (-1)^5(6)\begin{vmatrix} 3 & 5 & 2 \\ 1 & 1 & 8 \\ 4 & 5 & 6 \end{vmatrix}$$

Next, the powers of -1 and element multipliers are combined to simplify the statement:

$$= 1\begin{vmatrix} 5 & 2 & 0 \\ 1 & 8 & 6 \\ 5 & 6 & 0 \end{vmatrix} - 2\begin{vmatrix} 3 & 2 & 0 \\ 1 & 8 & 6 \\ 4 & 6 & 0 \end{vmatrix} - 5\begin{vmatrix} 3 & 5 & 0 \\ 1 & 1 & 6 \\ 4 & 5 & 0 \end{vmatrix} - 6\begin{vmatrix} 3 & 5 & 2 \\ 1 & 1 & 8 \\ 4 & 5 & 6 \end{vmatrix}$$

And then the 3×3 determinants are evaluated and all the products formed. The results are then combined and the answer simplified:

$$= 1[0 + 60 + 0 - (0 + 180 + 0)] - 2[0 + 48 + 0 - (0 + 108 + 0)]$$

$$- 5[0 + 120 + 0 - (0 + 90 + 0)] - 6[18 + 160 + 10 - (8 + 120 + 30)]$$

$$= 1[60 - 180] - 2[48 - 108] - 5[120 - 90] - 6[188 - 158]$$

$$= -120 + 120 - 150 - 180 = -330$$

You can use any column or row when performing cofactor expansion to evaluate a determinant. So it makes more sense to pick a column or row with a large number of zeros, if that's possible. In the previous example, if I'd used column four instead of row one, then two of the four products would be 0, because the element multiplier is 0. Here's what the computation looks like when expanding over the fourth column:

$$\begin{aligned}|A| &= (-1)^{4+1} a_{41}|C_{41}| + (-1)^{4+2} a_{42}|C_{42}| \\ &\quad + (-1)^{4+3} a_{43}|C_{43}| + (-1)^{4+4} a_{44}|C_{44}| \\ &= (-1)^5(6)\begin{vmatrix}3 & 5 & 2\\1 & 1 & 8\\4 & 5 & 6\end{vmatrix} + 0 + (-1)^7(6)\begin{vmatrix}1 & 2 & -5\\3 & 5 & 2\\4 & 5 & 6\end{vmatrix} + 0 \\ &= -6\begin{vmatrix}3 & 5 & 2\\1 & 1 & 8\\4 & 5 & 6\end{vmatrix} - 6\begin{vmatrix}1 & 2 & -5\\3 & 5 & 2\\4 & 5 & 6\end{vmatrix} \\ &= -6\left[18+160+10-(8+120+30)\right] - 6\left[30+16-75-(-100+10+36)\right] \\ &= -6\left[188-158\right] - 6\left[-29-(-54)\right] = -180-150 = -330\end{aligned}$$

It's nice to have a smaller number of computations to perform.

Using Determinants with Area and Volume

A determinant is a number associated with a square matrix. It's what you do with a determinant that makes it useful and important. Two very nice applications of the value of a determinant are associated with area and volume. By evaluating a determinant, you can find the area of a triangle, the vertices of which are described in the coordinate plane. And, even better, the value of a determinant is linked to the volume of a parallelepiped.

Think of a parallelepiped as a cardboard box that may have been sat upon by an NFL linebacker. I give a better description later, in the "Paying the piper with volumes of parallel epipeds" section.

Finding the areas of triangles

The traditional method used for finding the area of a triangle is to find the height measured from one of the vertices, perpendicular to the opposite side. Another method involves Heron's formula, which you use if you happen to know the length of each side of the triangle. Now I show you a method for finding the area of a triangle when you know the coordinates of the vertices of the triangle in two-space (on the coordinate axes).

The area of a triangle is found by taking one half the absolute value of the determinant of the matrix formed using the coordinates of the triangle and a column of 1s.

Taking flight with Heron's formula

If you need to compute the area of a triangular figure, you can do so with the measures of the sides of the triangle. Heron's formula for the area of a triangle, the sides of which measure *a, b,* and *c,* is

$$\text{Area} = \sqrt{s(s-a)(s-b)(s-c)}$$

where *s* is the *semiperimeter* (half the perimeter of the triangle).

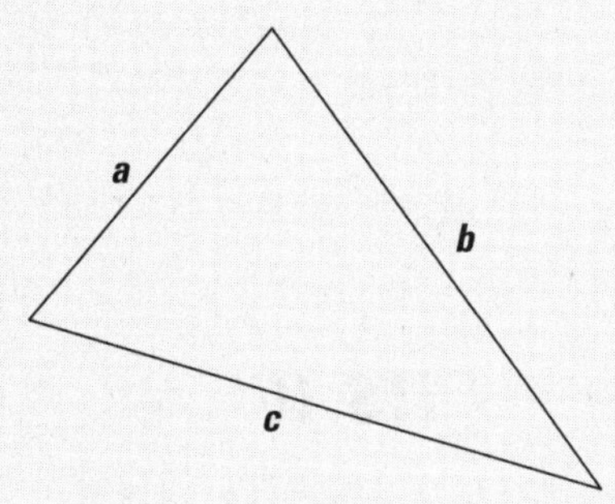

$$\text{perimeter} = p = a + b + c$$

$$\text{semiperimeter} = s = \frac{a+b+c}{2}$$

For example, if the sides of a triangular area measure 5, 7, and 8 inches, then the perimeter is 20 inches, the semiperimeter, *s,* is half that or 10 inches, and the area is

$$\text{Area} = \sqrt{10(10-5)(10-7)(10-8)}$$
$$= \sqrt{10(5)(3)(2)} = \sqrt{300} \approx 17.32 \text{ square inches}$$

The area of a triangle with vertex coordinates (x_1,y_1), (x_2,y_2), and (x_3,y_3) is

$$\text{Area} = \frac{1}{2}\left\| \begin{matrix} x_1 & y_1 & 1 \\ x_2 & y_2 & 1 \\ x_3 & y_3 & 1 \end{matrix} \right\|$$

The absolute value of a number n is shown with vertical segments, $|n|$, and the determinant of a matrix is also designated with the vertical bars. So you see some *nested* bars to show both absolute value and a determinant. That type of notation can get confusing, so sometimes an alternate form, using the letters *det* in front of the matrices' brackets, is used to indicate the determinant of the matrix. So you'd see:

$$\text{Area} = \frac{1}{2}\left|\det\begin{bmatrix} x_1 & y_1 & 1 \\ x_2 & y_2 & 1 \\ x_3 & y_3 & 1 \end{bmatrix}\right|$$

An example of using the determinant to find the area of a triangle is when the vertices of the triangle are (2,4), (9,6), and (7,–3), as shown in Figure 10-1. You could find the area using Heron's formula, once you use the distance formula to find the lengths of the three sides. But the determinant method will be much quicker and easier.

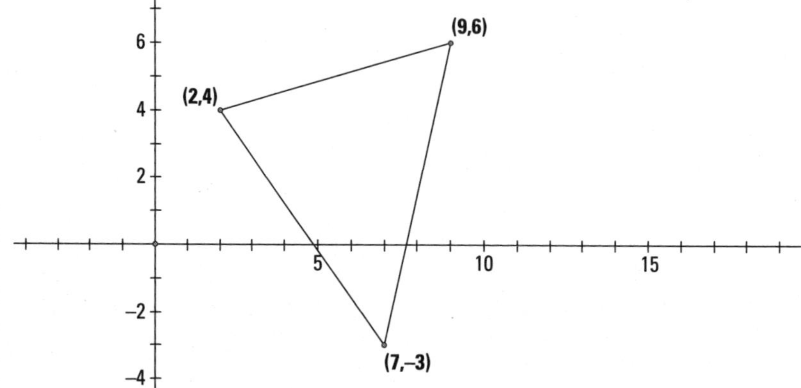

Figure 10-1: The coordinates of the acute triangle.

It doesn't really matter which coordinates you choose to be (x_1,y_1), (x_2,y_2), or (x_3,y_3). And the absolute value part of the formula removes the necessity of having to choose between moving clockwise or counterclockwise around the triangle to select the vertices. So, placing the coordinates as elements in the determinant:

$$\text{Area} = \frac{1}{2}\left|\det\begin{bmatrix} 2 & 4 & 1 \\ 9 & 6 & 1 \\ 7 & -3 & 1 \end{bmatrix}\right|$$

$$= \frac{1}{2}\left|12 + 28 - 27 - (42 - 6 + 36)\right| = \frac{1}{2}\left|-59\right| = 29.5$$

The area of the triangle is 29.5 square units.

Pursuing parallelogram areas

A parallelogram is a four-sided polygon in which both pairs of opposite sides are parallel to one another and equal in length. The traditional method of finding the area of a parallelogram is to multiply the length of one of the sides times the perpendicular distance between that side and its opposite. The biggest stumbling block to finding the area of a parallelogram is in measuring the necessary lengths, especially that perpendicular distance. Plotting a parallelogram on the coordinate plane makes computing the area much easier.

A parallelogram graphed on the coordinate plane has an area that can be computed using a determinant. One of two different situations may occur with the parallelograms graphed in two-space: (1) One of the vertices of the parallelogram is at the origin, or (2) none of the vertices is at the origin. Having a vertex at the origin makes the work easier, but it's not really all that hard to deal with a parallelogram that's *out there.*

Originating parallelograms with the origin

To find the area of a parallelogram that has one of its vertices at the origin, you just need the coordinates of the vertices that are directly connected with the origin by one of the sides of the parallelogram. In Figure 10-2, you see a parallelogram drawn in the first quadrant. The two vertices connected to the origin by one of the sides of the parallelogram are at the points (1,2) and (6,5).

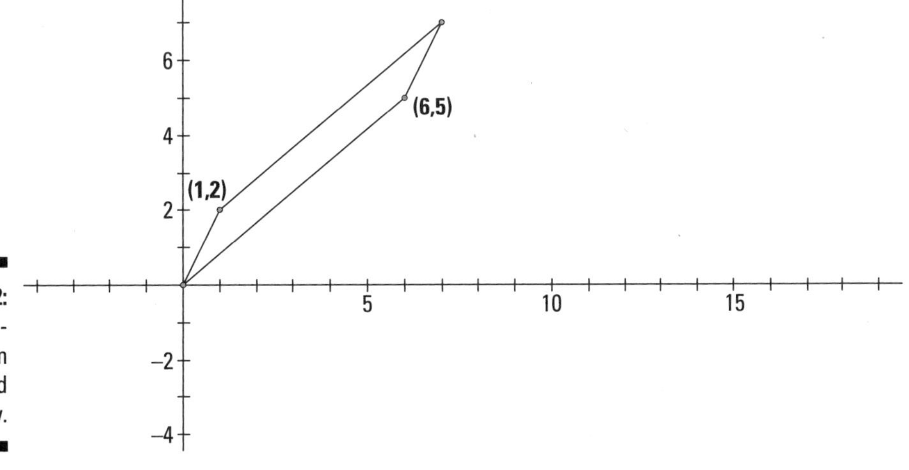

Figure 10-2: The parallelogram is long and narrow.

The two vertices link with the origin and determine the two different side lengths of the parallelogram. The area of the parallelogram is found using a determinant of the coordinates.

The area of a parallelogram with one vertex at the origin and with coordinates of points (x_1,y_1) and (x_2,y_2) connecting a side of the parallelogram with the origin has an area equal to the absolute value of the determinant formed as follows:

$$\begin{vmatrix} x_1 & y_1 \\ x_2 & y_2 \end{vmatrix}$$

So, in the case of the parallelogram in Figure 10-2, the area is found:

$$\text{Area} = \left| \det \begin{bmatrix} 1 & 6 \\ 2 & 5 \end{bmatrix} \right| = |5-12| = |-7| = 7$$

The area of the parallelogram is 7 square units.

Branching out with random parallelograms in two-space

When a parallelogram doesn't have a vertex at the origin, you adjust or slide the parallelogram — maintaining its size and shape — until one of the vertices ends up at the origin. In Figure 10-3, you see a parallelogram with all the vertices in the first quadrant.

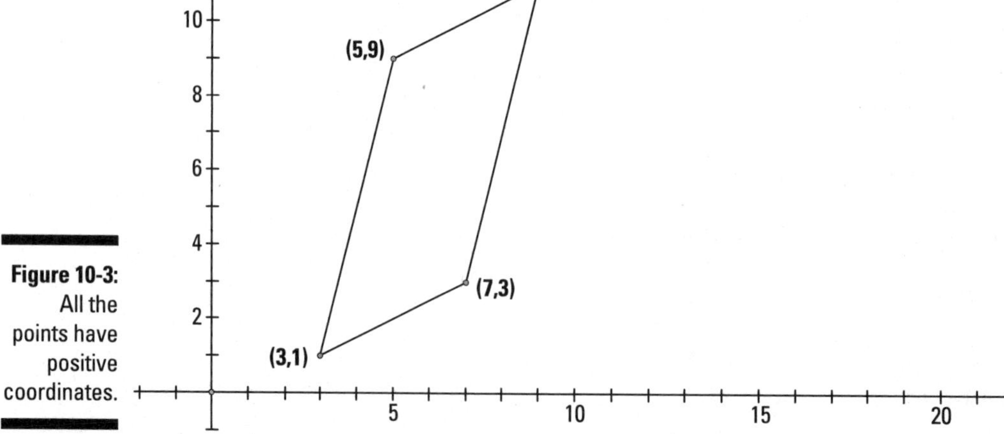

Figure 10-3: All the points have positive coordinates.

You can slide the parallelogram any way you want to put any of the vertices at the origin. But the move that makes most sense and seems to be easiest is to slide the parallelogram until the vertex at the point (3,1) meets up with the origin.

I've been calling the movement of the parallelogram a *slide*. The proper term in transformational geometry is *translation*. You *translate* a point or segment or figure from one position to another without changing the dimensions or shape of the figure. When performing a translation, the coordinates of each point change in the same manner.

When a figure is translated in two-space, each point (x,y) is transformed into its image point (x',y') using the formulas $x' = x + h$ and $y' = y + k$, where h is the horizontal change (and direction) of the translation and k is the vertical change (and direction) of the translation.

If the parallelogram in Figure 10-3 is translated three units to the left and one unit down, then the vertex (3,1) will end up at the origin. Here's what happens to the four vertices using the translation $x' = x - 3$, $y' = y - 1$: $(3,1) \rightarrow (0,0)$, $(7,3) \rightarrow (4,2)$, $(9,11) \rightarrow (6,10)$, and $(5,9) \rightarrow (2,8)$. Figure 10-4 shows you what the new parallelogram looks like.

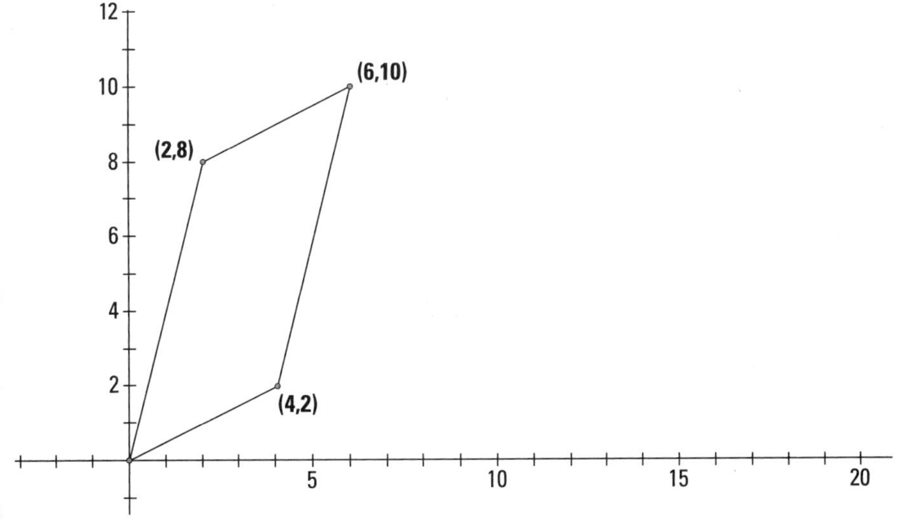

Figure 10-4: The size and shape stay the same with the translation.

Now you can find the area of the parallelogram using the determinant and the vertices at (4,2) and (2,8).

$$\text{Area} = \left|\det\begin{bmatrix} 4 & 2 \\ 2 & 8 \end{bmatrix}\right| = |32 - 4| = 28$$

The examples here have involved parallelograms — where the angles haven't been right angles. Because rectangles, squares, and rhombi (the plural of *rhombus*) are special parallelograms, you can use this determinant method to find the areas of these special quadrilaterals, too.

Paying the piper with volumes of parallelepipeds

A *parallelepiped* is a fancy name for a squished box. Actually, I'm making light of a very precise mathematical structure. Just picture a six-sided figure, like a cardboard box, with opposite sides parallel to one another and opposite sides with the same dimensions. Also, a parallelepiped doesn't need sides to be perpendicular to one another or corners of the sides to be 90 degrees. Every side is a parallelogram. So even a squished box is special, with its rectangular sides. In Figure 10-5, I show you a parallelepiped graphed in one of the eight octants of three-space. The vertices of three of the corners have labels shown as (x,y,z) coordinates.

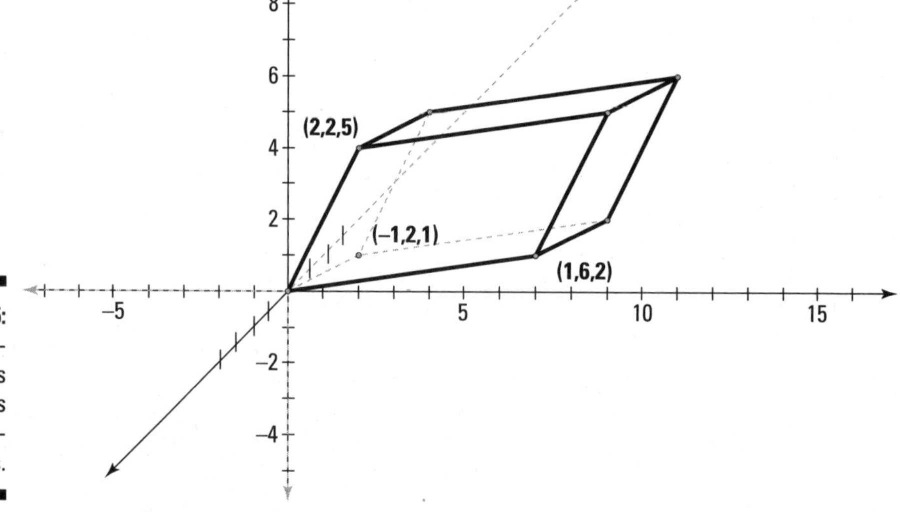

Figure 10-5: A parallelepiped has three pairs of congruent sides.

When a parallelepiped has one of its vertices at the origin, then the volume of that parallelepiped is found with the determinant of a 3 × 3 matrix using the coordinates of the vertices at the other end of the sides of the parallelepiped that all meet at the origin.

The volume of a parallelepiped with one vertex at the origin and with coordinates of points (x_1,y_1,z_1), (x_2,y_2,z_2), and (x_3,y_3,z_3) connecting an edge of the parallelepiped with the origin has a volume equal to the absolute value of the determinant formed as follows:

$$\text{Volume} = \left| \det \begin{bmatrix} x_1 & y_1 & z_1 \\ x_2 & y_2 & z_2 \\ x_3 & y_3 & z_3 \end{bmatrix} \right|$$

So, in the case of the parallelepiped found in Figure 10-5, the volume is 38 cubic units:

$$\text{V} = \left| \det \begin{bmatrix} 1 & -1 & 2 \\ 6 & 2 & 2 \\ 2 & 1 & 5 \end{bmatrix} \right| = |38| = 38$$

If none of the vertices lies at the origin, then you can do a translation similar to that in two-space to get one of the vertices in position at the origin.

Chapter 11

Personalizing the Properties of Determinants

In This Chapter

- Performing operations and manipulations on determinants
- Determining how the values of determinants change
- Putting manipulations together with computations to make life easier
- Taking advantage of upper triangular or lower triangular matrices

Determinants are functions applied to square matrices. When the elements of the square matrix are all numbers, then the determinant results in a single numerical value. In Chapter 10, I show you how to evaluate reasonably sized determinants and how to do cofactor expansion to make the computations easier.

In this chapter, I show you the many different properties of determinants — which actions leave the value of the determinant the same and how the value of the determinant changes when you perform other operations or actions. Many of the properties and processes involving determinants that I show you in this chapter are related to matrix operations and matrices, so a journey back to Chapter 3 isn't a bad idea if you need a bit of a brush-up on those processes.

The properties presented in this chapter have a big payoff when working with determinants. As interesting as the properties are on their own, they truly become important when you want to make adjustments to matrices so that the computation of their determinants is easier. In this chapter, I introduce you to the properties and then show you how very nice those properties are.

Transposing and Inverting Determinants

The transpose of a matrix takes each row of the matrix and changes it into a column and each column of a matrix and makes it into a row. I cover matrix transposition in Chapter 3; but in this chapter you only see square matrices and their transposes. The inverse of a matrix is another matrix that's closely linked to the first. Also, in Chapter 3, you find that the product of a matrix and its inverse is the identity matrix.

I introduce these ideas of transposing matrices and finding the inverses of matrices, because the processes are closely related when transposing determinants and finding determinants of the inverses of matrices.

Determining the determinant of a transpose

You can compute the determinant of a square matrix using one of several methods (as found in Chapter 10) or by using a calculator or computer. And after you've found the determinant of a matrix, you don't have to work much harder to find the determinant of the matrix's transpose — mainly because the determinants are exactly the same.

The determinant of matrix A is equal to the determinant of the transpose of matrix A: det (A) = det (A^T).

For example, look at the following matrix A and its transpose:

$$A = \begin{bmatrix} 1 & 2 & 3 \\ 4 & 5 & 0 \\ 3 & 1 & 2 \end{bmatrix}, A^T = \begin{bmatrix} 1 & 4 & 3 \\ 2 & 5 & 1 \\ 3 & 0 & 2 \end{bmatrix}$$

The quick, easy way of evaluating the two determinants is to write the first three columns to the right of the determinant and multiply along the diagonals (see Chapter 10 for more on this procedure).

$$\det(A) = \left|\begin{array}{ccc|cc} 1 & 2 & 3 & 1 & 2 \\ 4 & 5 & 0 & 4 & 5 \\ 3 & 1 & 2 & 3 & 1 \end{array}\right.$$

$$= 10 + 0 + 12 - (45 + 0 + 16) = 22 - 61 = -39$$

$$\det(A^T) = \left|\begin{array}{ccc|cc} 1 & 4 & 3 & 1 & 4 \\ 2 & 5 & 1 & 2 & 5 \\ 3 & 0 & 2 & 3 & 0 \end{array}\right.$$

$$= 10 + 12 + 0 - (45 + 0 + 16) = 22 - 61 = -39$$

The values of the determinants are the same. If you were to examine the different products created in the determinants of a matrix and its transpose, you'd find the same combinations of elements aligned with one another in both, resulting in the same final answer.

Investigating the determinant of the inverse

When you multiply a matrix times its inverse, you get an identity matrix. The identity matrix is square with a diagonal of 1s and the rest of the elements 0s. A similar type of product occurs when you multiply the determinant of a matrix times the determinant of the inverse of the matrix: You get the number 1, which is the multiplicative identity.

The determinant of matrix A is equal to the reciprocal (multiplicative inverse) of the determinant of the inverse of matrix A:

$$\det(A) = \left[\det(A^{-1})\right]^{-1} = \frac{1}{\det(A^{-1})}$$

Be careful with the –1 superscript. The inverse of matrix B is written B^{-1}; this indicates the inverse matrix, not a reciprocal. The determinant of matrix B, $|B|$ can have a reciprocal (if it isn't zero). You can write that as $|B|^{-1}$

The inverse of a matrix exists only if the determinant is not equal to 0.

The matrix A must be *invertible* — it must have an inverse. For example, the matrix B you see next has an inverse matrix B^{-1}.

$$B = \begin{bmatrix} 1 & 3 & 3 \\ 1 & 6 & 3 \\ 1 & 3 & 4 \end{bmatrix}, B^{-1} = \begin{bmatrix} 5 & -1 & -3 \\ -\frac{1}{3} & \frac{1}{3} & 0 \\ -1 & 0 & 1 \end{bmatrix}$$

The product of B and B^{-1} is the identity matrix.

$$BB^{-1} = \begin{bmatrix} 1 & 3 & 3 \\ 1 & 6 & 3 \\ 1 & 3 & 4 \end{bmatrix} \begin{bmatrix} 5 & -1 & -3 \\ -\frac{1}{3} & \frac{1}{3} & 0 \\ -1 & 0 & 1 \end{bmatrix}$$

$$= \begin{bmatrix} 1(5)+3\left(-\frac{1}{3}\right)+3(-1) & 1(-1)+3\left(\frac{1}{3}\right)+3(0) & 1(-3)+3(0)+3(1) \\ 1(5)+6\left(-\frac{1}{3}\right)+3(-1) & 1(-1)+6\left(\frac{1}{3}\right)+3(0) & 1(-3)+6(0)+3(1) \\ 1(5)+3\left(-\frac{1}{3}\right)+4(-1) & 1(-1)+3\left(\frac{1}{3}\right)+4(0) & 1(-3)+3(0)+4(1) \end{bmatrix}$$

$$= \begin{bmatrix} 1 & 0 & 0 \\ 0 & 1 & 0 \\ 0 & 0 & 1 \end{bmatrix}$$

Now look at the values of the determinants of B and B^{-1}.

$$\det(B) = \begin{vmatrix} 1 & 3 & 3 \\ 1 & 6 & 3 \\ 1 & 3 & 4 \end{vmatrix} = 24+9+9-(18+9+12) = 42-39 = 3$$

$$\det\left(B^{-1}\right) = \begin{vmatrix} 5 & -1 & -3 \\ -\frac{1}{3} & \frac{1}{3} & 0 \\ -1 & 0 & 1 \end{vmatrix} = \frac{5}{3}+0+0-\left(1+0+\frac{1}{3}\right) = \frac{5}{3}-\frac{4}{3} = \frac{1}{3}$$

The determinant of B is the reciprocal of the determinant of B^{-1}. The product of 3 and $1/3$ is 1.

The determinant of an identity matrix comes out to be equal to 1.

Interchanging Rows or Columns

When using matrices to solve systems of equations, you sometimes switch rows of the matrices to make a more convenient arrangement for the solving process. I go into great detail about solving systems of equations in Chapter 4. Switching rows of a matrix around has absolutely no effect on the final answer to a system of equations — the solution is preserved. But switching rows in a determinant *does* have an effect, as does switching columns.

To switch or interchange a row or column of a determinant, you just interchange the corresponding elements. I show you, next, the determinant of matrix A, what the determinant looks like when you interchange the first and third rows, and what the determinant looks like when you interchange the second and third columns.

$$\det(A) = \begin{vmatrix} 1 & 2 & 3 & 4 \\ 5 & 6 & 7 & 8 \\ 8 & 7 & 6 & 5 \\ 4 & 3 & 2 & 1 \end{vmatrix}$$

$$\det(R_1 \leftrightarrow R_3) = \begin{vmatrix} 8 & 7 & 6 & 5 \\ 5 & 6 & 7 & 8 \\ 1 & 2 & 3 & 4 \\ 4 & 3 & 2 & 1 \end{vmatrix} \qquad \det(C_2 \leftrightarrow C_3) = \begin{vmatrix} 1 & 3 & 2 & 4 \\ 5 & 7 & 6 & 8 \\ 8 & 6 & 7 & 5 \\ 4 & 2 & 3 & 1 \end{vmatrix}$$

Interchanging two rows (or columns) of a matrix results in a determinant of the opposite sign. For example, take a look at matrix D and its determinant. I'll interchange columns two and three and name the new matrix D'; then I'll compute the determinant of D'.

$$D = \begin{bmatrix} 1 & 3 & 5 \\ 1 & 6 & 0 \\ -2 & 3 & 4 \end{bmatrix}$$

$$\det(D) = 24 + 0 + 15 - (-60 + 0 + 12) = 39 + 48 = 87$$

$$D(C_2 \leftrightarrow C_3) = D' = \begin{bmatrix} 1 & 5 & 3 \\ 1 & 0 & 6 \\ -2 & 4 & 3 \end{bmatrix}$$

$$\det(D') = 0 - 60 + 12 - (15 + 0 + 24) = -48 - 39 = -87$$

You might ask (of *course,* this is on the tip of your tongue), "What happens if you do both — interchange two rows and interchange two columns of the same matrix?" Your instinct may be to say that the two actions cancel one another out. Your instincts are right!

Take a look at matrix E, where I interchange the first and third rows and then, in the new matrix, interchange the first and second columns. I compare the value of the determinant of E with the determinant of the new, revised matrix E'', which has two interchanges performed on it.

$$E = \begin{bmatrix} 1 & -2 & 6 \\ 1 & 0 & 2 \\ -5 & 3 & 1 \end{bmatrix}$$

$$E(R_1 \leftrightarrow R_3) = E' = \begin{bmatrix} -5 & 3 & 1 \\ 1 & 0 & 2 \\ 1 & -2 & 6 \end{bmatrix}$$

$$E'(C_1 \leftrightarrow C_2) = E'' = \begin{bmatrix} 3 & -5 & 1 \\ 0 & 1 & 2 \\ -2 & 1 & 6 \end{bmatrix}$$

$$\det(E) = 0 + 20 + 18 - (0 + 6 - 2) = 38 - 4 = 34$$

$$\det(E'') = 18 + 20 + 0 - (-2 + 6 + 0) = 38 - 4 = 34$$

Zeroing In on Zero Determinants

When computing the value of a determinant, you add and subtract the products of numbers. If you add and subtract the same amount, then the value of the determinant is equal to 0. Being able to recognize, ahead of time, that the value of a determinant is 0 is most helpful — you save the time it would take to do all the computations.

Finding a row or column of zeros

When a matrix has a row or column of zeros, then the value of the determinant is equal to 0. Just consider what you get if you evaluate such a matrix with a row or column of zeros in terms of cofactor expansion (see Chapter 10). If you choose the row or column of zeros to expand upon, then each product would include a zero, so you don't have anything but nothing!

If a matrix has a row or column of zeros, then the determinant associated with that matrix is equal to zero.

Zeroing out equal rows or columns

Another nice property of determinants — one that earns you a big fat 0 for the value — is in terms of elements in two rows or two columns. When the corresponding elements are the same, then the value of the determinant is 0.

If the corresponding elements in two rows (or columns) of a matrix are all equal, then the value of that determinant is zero.

This property involving equal rows or columns may take a bit more convincing, so let me start with an example. Consider the matrix F, in which the elements in the second and fourth columns are equal to one another. I compute the value of the determinant using cofactor expansion along the third row, because that row has a 0 in it.

$$F = \begin{bmatrix} 2 & 5 & 6 & 5 \\ 3 & -2 & 5 & -2 \\ 0 & 6 & 1 & 6 \\ -3 & 1 & 4 & 1 \end{bmatrix}$$

$$\det(F) = 0 * \begin{vmatrix} 5 & 6 & 5 \\ -2 & 5 & -2 \\ 1 & 4 & 1 \end{vmatrix} + (-1)6 * \begin{vmatrix} 2 & 6 & 5 \\ 3 & 5 & -2 \\ -3 & 4 & 1 \end{vmatrix}$$

$$+1 * \begin{vmatrix} 2 & 5 & 5 \\ 3 & -2 & -2 \\ -3 & 1 & 1 \end{vmatrix} + (-1)6 * \begin{vmatrix} 2 & 5 & 6 \\ 3 & -2 & 5 \\ -3 & 1 & 4 \end{vmatrix}$$

$$= 0 - 6(10 + 36 + 60 - (-75 - 16 + 18))$$

$$+1(-4 + 30 + 15 - (30 - 4 + 15)) - 6(-16 - 75 + 18 - (36 + 10 + 60))$$

$$= 0 - 6(106 + 73) + 1(41 - 41) - 6(-73 - 106)$$

$$= 0 - 1{,}074 + 0 + 1{,}074 = 0$$

You might be thinking that I just picked a *convenient* matrix — one that just happens to work for me. So I guess I'd better show you why this particular property holds. I'll use two columns, again, and cofactor expansion over a random row. And this time, I'll use the general terms for the elements in the matrix, letting the elements that are equal be equal to *a, b, c,* and *d* to make them stand out more. So, first look at my new matrix G with columns two and four containing equal elements.

$$G = \begin{bmatrix} g_{11} & a & g_{13} & a \\ g_{21} & b & g_{23} & b \\ g_{31} & c & g_{33} & c \\ g_{41} & d & g_{43} & d \end{bmatrix}$$

I evaluate the determinant of G using cofactor expansion over the first row.

$$\det(G) = \begin{bmatrix} g_{11} & a & g_{13} & a \\ g_{21} & b & g_{23} & b \\ g_{31} & c & g_{33} & c \\ g_{41} & d & g_{43} & d \end{bmatrix}$$

$$\det(G) = g_{11} * \begin{vmatrix} b & g_{23} & b \\ c & g_{33} & c \\ d & g_{43} & d \end{vmatrix} + (-1)a * \begin{vmatrix} g_{21} & g_{23} & b \\ g_{31} & g_{33} & c \\ g_{41} & g_{43} & d \end{vmatrix}$$

$$+g_{13} * \begin{vmatrix} g_{21} & b & b \\ g_{31} & c & c \\ g_{41} & d & d \end{vmatrix} + (-1)a * \begin{vmatrix} g_{21} & b & g_{23} \\ g_{31} & c & g_{33} \\ g_{41} & d & g_{43} \end{vmatrix}$$

$$= g_{11}\left(bg_{33}d + g_{23}cd + bcg_{43} - \left(bg_{33}d + cg_{43}b + dg_{23}c\right)\right)$$

$$-a\left(g_{21}g_{33}d + g_{23}cg_{41} + bg_{31}g_{43} - \left(bg_{33}g_{41} + cg_{43}g_{21} + dg_{23}g_{31}\right)\right)$$

$$+g_{13}\left(g_{21}cd + bcg_{41} + bg_{31}d - \left(bcg_{41} + cdg_{21} + dbg_{31}\right)\right)$$

$$-a\left(g_{21}cg_{43} + bg_{33}g_{41} + g_{23}g_{31}d - \left(g_{23}cg_{41} + g_{33}dg_{21} + g_{44}bg_{31}\right)\right)$$

Look at all those letters with all those terms — and the result is supposed to be zero? Don't despair. You may see some patterns arising. First, look at the two lines starting with g_{11} and g_{13}. If I rearrange the factors in each term so that the letters *b, c,* and *d* come first, in alphabetical order, followed by some g_{ij}, you see that each term is both added and subtracted — the same numbers are in each part of the computation, so the sum is 0.

$$g_{11}\left(bdg_{33} + cdg_{23} + bcg_{43} - \left(bdg_{33} + bcg_{43} + cdg_{23}\right)\right) = g_{11}(0)$$

$$g_{13}\left(cdg_{21} + bcg_{41} + bdg_{31} - \left(bcg_{41} + cdg_{21} + bdg_{31}\right)\right) = g_{13}(0)$$

Now look at the rows in the determinant computation that start with a multiplier of *a*. I've rearranged the factors in each term so that the single letters come first and the elements in the matrix come in order. You see that each element that's added in one row is subtracted in the other row. So, after distributing each *a*, you'll get each term and its opposite, which gives you a 0. In the end, 0 + 0 = 0.

$$-a\left(g_{21}g_{33}d + cg_{23}g_{41} + bg_{31}g_{43} - \left(bg_{33}g_{41} + cg_{21}g_{43} + dg_{23}g_{31}\right)\right)$$

$$-a\left(cg_{21}g_{43} + bg_{33}g_{41} + dg_{23}g_{31} - \left(cg_{23}g_{41} + dg_{21}g_{33} + bg_{31}g_{44}\right)\right)$$

I know that I've only shown you one general example — the determinant of a 4×4 matrix with cofactor expansion over a specific row. I do hope, though, that it's enough to convince you of the property that when you have two rows or columns in a matrix that are equal, then the value of the determinant associated with the matrix is 0.

Manipulating Matrices by Multiplying and Combining

Most matrices and their associated determinants are just fine in their original form, but some matrices need to be tweaked to make them more manageable. The *tweaking* comes, of course, with some particular ground rules. And the reasoning behind the particular manipulations is more clear in the later section "Tracking down determinants of triangular matrices." For now, stick with me and just enjoy the properties that unfold.

Multiplying a row or column by a scalar

You have a perfectly nice matrix with its perfectly swell determinant, and then you decide that the matrix just isn't quite nice enough. The second row has two fractions in it, and you just don't like fractions. So you decide to multiply every term in the second row by 6 to get rid of the fractions. What does that do to the matrix? It changes the matrix, of course. But the change to the determinant associated with the matrix is predictable. The property change applies to multiplying an entire *column* by some number, too.

If every element of a row (or column) of a matrix is multiplied by a particular scalar k, then the determinant of that matrix is k times the determinant of the original matrix. For example, consider the matrix G that has fractions in the second row. You find the determinant of G, slugging through the fractional multiplication:

$$G = \begin{bmatrix} 2 & 3 & 5 \\ \frac{1}{2} & 6 & \frac{5}{6} \\ -2 & 3 & 1 \end{bmatrix}$$

$$\det(G) = 12 - 5 + \frac{15}{2} - \left(-60 + 5 + \frac{3}{2}\right)$$
$$= \frac{29}{2} - \left(-\frac{107}{2}\right) = \frac{136}{2} = 68$$

To be perfectly honest, the multiplication isn't all that bad, but I didn't want to overwhelm you with a truly nasty example — showing you the property with this matrix should be convincing enough. Now multiply each element in the second row of the matrix by 6. The number 6 is the least common multiple of the denominators of the two fractions.

$$6R_2 \to R_2 \begin{bmatrix} 2 & 3 & 5 \\ 6\left(\frac{1}{2}\right) & 6(6) & 6\left(\frac{5}{6}\right) \\ -2 & 3 & 1 \end{bmatrix} = \begin{bmatrix} 2 & 3 & 5 \\ 3 & 36 & 5 \\ -2 & 3 & 1 \end{bmatrix}$$

Now, computing the determinant of the new matrix, you find that the rule holds, and that the value of the determinant is six times that of the determinant of the original matrix: 6(68) = 408.

$$\begin{vmatrix} 2 & 3 & 5 \\ 3 & 36 & 5 \\ -2 & 3 & 1 \end{vmatrix} = 72 - 30 + 45 - (-360 + 30 + 9)$$
$$= 87 - (-321) = 408$$

A common situation in matrices used in practical applications is that you have one or more elements that are decimals. And the decimals often range in the number of places; you even find some matrices using scientific notation to represent the numbers. In matrix H, the entire third column consists of numbers with at least three digits to the right of the decimal point, and the first row has all four numbers written in scientific notation.

$$H = \begin{bmatrix} 2\times10^{-4} & 5\times10^{-5} & 6\times10^{-6} & 5\times10^{-4} \\ 3 & -2 & 0.00005 & -2 \\ 0 & 6 & 0.0001 & 6 \\ -3 & 1 & 0.004 & 1 \end{bmatrix}$$

A number written in scientific notation consists of the product of (a) a number between 0 and 10 and (b) a power of 10. Positive exponents on the 10 are used for numbers greater than 10, and the original numbers are created by moving the decimal point to the right. Negative exponents indicate numbers smaller than 1, and you re-create the original number by moving the decimal point to the left. For example, 5.0×10^5 is another way of saying 500,000, and 5.0×10^{-5} is another way of saying 0.00005.

In matrix H, you multiply each element in the first row by 10^6, or 1,000,000, to eliminate all the negative exponents and decimal numbers.

$$10^6 * R_1 \to R_1 \begin{bmatrix} 200 & 50 & 6 & 500 \\ 3 & -2 & 0.00005 & -2 \\ 0 & 6 & 0.0001 & 6 \\ -3 & 1 & 0.004 & 1 \end{bmatrix}$$

Now, multiply each element in the third column by 100,000 to eliminate all the decimal numbers.

$$100{,}000 * C_3 \to C_3 \begin{bmatrix} 200 & 50 & 600{,}000 & 500 \\ 3 & -2 & 5 & -2 \\ 0 & 6 & 10 & 6 \\ -3 & 1 & 400 & 1 \end{bmatrix}$$

All the decimal numbers and scientific notation are gone, but now the matrix has a lot of large numbers. The determinant of the new matrix is 1,000,000 times 100,000, or 100,000,000,000, times as large as the determinant of the original matrix. Without showing you all the gory details, here are the matrices (beginning and end) and their respective determinants:

$$\begin{vmatrix} 2\times10^{-4} & 5\times10^{-5} & 6\times10^{-6} & 5\times10^{-4} \\ 3 & -2 & 0.00005 & -2 \\ 0 & 6 & 0.0001 & 6 \\ -3 & 1 & 0.004 & 1 \end{vmatrix} = -3.294\times10^{-5}$$

$$\begin{vmatrix} 200 & 50 & 600{,}000 & 500 \\ 3 & -2 & 5 & -2 \\ 0 & 6 & 10 & 6 \\ -3 & 1 & 400 & 1 \end{vmatrix} = -3{,}294{,}000$$

The determinant of the original matrix is -0.00003294, and the determinant of the revised matrix is –3,294,000. When you multiply -0.00003294 by one hundred billion, 100,000,000,000(–0.00003294) = -3,294,000, the decimal point moved 11 places to the right. So, the big question is "Do you prefer decimals and scientific notation or big numbers?"

Adding the multiple of a row or column to another row or column

When solving systems of equations using either the equations themselves or matrices, a common technique used is to add a multiple of one row to another row. The process usually results in reducing the number of variables you have to work with in the problem. (Intrigued? Turn to Chapter 4 to find out more.) The process of adding a multiple of one row to another (or one column to another) in a matrix has a surprising result: The determinant of the matrix doesn't change.

I'm assuming that you're surprised by this turn of events. After all, multiplying a row or column by a scalar *k* makes the determinant of the new matrix *k* times that of the original matrix. And now, by involving an additional row (or column), the effect of the multiplying seems to be negated.

Let me start off with an example. In matrix J, I change the third row by adding two times each element in the second row to the corresponding element in the third row:

$$J = \begin{bmatrix} 2 & 3 & 5 \\ -3 & 6 & 5 \\ -2 & 3 & 1 \end{bmatrix}$$

$$2R_2 + R_3 \to R_3 \begin{bmatrix} 2 & 3 & 5 \\ -3 & 6 & 5 \\ -8 & 15 & 11 \end{bmatrix}$$

The determinants are the same. This property comes in very handy when you're making adjustments to matrices — you don't have to worry about keeping track of the multiples that you've introduced.

$$\begin{vmatrix} 2 & 3 & 5 \\ -3 & 6 & 5 \\ -2 & 3 & 1 \end{vmatrix} = 12 - 30 - 45 - (-60 + 30 - 9)$$

$$= -63 - (-39) = -24$$

$$\begin{vmatrix} 2 & 3 & 5 \\ -3 & 6 & 5 \\ -8 & 15 & 11 \end{vmatrix} = 132 - 120 - 225 - (-240 + 150 - 99)$$

$$= -213 - (-189) = -24$$

Taking on Upper or Lower Triangular Matrices

An upper triangular matrix or lower triangular matrix is a square matrix with all 0s either below or above the main diagonal that runs from the upper left to the lower right. In Chapter 3, I introduce the idea of having the triangular matrices and how the characteristics of triangular matrices make some computations easier. In this section, you see how to create triangular matrices from previously non-triangular matrices, and then you find out just how wonderful it is to have an upper triangular or lower triangular matrix.

Tracking down determinants of triangular matrices

Triangular matrices are full of 0s. All the elements either above or below the main diagonal of a triangular matrix are 0. Other elements in the matrix can be 0, too, but you look at the elements relative to the main diagonal and determine whether you have a triangular matrix.

Why am I getting so excited about these triangular matrices? Why would *anyone* spend so much time talking about them? The answer is simple: The determinant of a triangular matrix is computed by just looking at the elements on the main diagonal.

The main diagonal of a square matrix or determinant runs from the upper-left element down to the lower-right element. It's the main diagonal that's the big player in evaluating the determinants of upper triangular or lower triangular matrices.

If matrix A is an upper (or lower) triangular matrix, then the determinant of A is equal to the product of the elements lying on that main diagonal. Consider, for example, the matrix N.

$$N = \begin{bmatrix} 2 & 1 & 4 & 3 \\ 0 & -2 & 1 & -6 \\ 0 & 0 & 7 & 2 \\ 0 & 0 & 0 & 3 \end{bmatrix}$$

According to the rule, the determinant |N| = 2(–2)(7)(3) = –84. Almost seems too easy. But it's true! Now you see why I go to the trouble to change a matrix to a triangular matrix in the following section, "Cooking up a triangular matrix from scratch."

Just to show you a comparison, I'll compute the determinant of a matrix P. First, I use the tried-and-true method of multiplying on the diagonals. Then I compare the first answer to that obtained by just multiplying along the main diagonal.

$$P = \begin{bmatrix} 2 & 0 & 0 \\ 7 & 1 & 0 \\ 5 & -3 & 1 \end{bmatrix}$$

$$\det(P) = 2 + 0 + 0 - (0 + 0 + 0) = 2$$

$$\det(P) = 2 \text{ using product along main diagonal}$$

This doesn't really prove anything for all sizes of square matrices, but you can see how all the 0s introduce themselves in the products to produce 0 terms.

A very special matrix is considered to be both upper triangular or lower triangular. The matrix I'm referring to is the identity matrix.

The determinant of an identity matrix is equal to 1.

Because you have 0s above and below the main diagonal of any identity matrix, you find the value of the determinant by multiplying along the main diagonal. The only product you're ever going to get — regardless of the size of the identity matrix — is 1.

Cooking up a triangular matrix from scratch

An upper or lower triangular matrix has 0s below or above the main diagonal.

Here are examples of both types:

$$\begin{bmatrix} 2 & 3 & 6 & 5 \\ 0 & -2 & 4 & -2 \\ 0 & 0 & 1 & 6 \\ 0 & 0 & 0 & 1 \end{bmatrix} \quad \begin{bmatrix} 2 & 0 & 0 & 0 \\ 3 & -2 & 0 & 0 \\ 0 & 6 & 1 & 0 \\ -3 & 1 & 4 & 1 \end{bmatrix}$$

If you don't have a matrix that is upper triangular or lower triangular, you can perform row (column) operations to change your original matrix to a triangular matrix. And, it just so happens that the row (column) operations you can perform are exactly some of those described earlier in this chapter: interchanging rows or columns, multiplying a row or column by a scalar, and adding a multiple of a row or column to another row or column. The challenge is in determining just what you need to do to create the desired matrix.

Making one move by adding a multiple of a column to another column

In matrix K, shown next, you see that you almost have a lower triangular matrix. If the 3 in the second row, third column were a 0, then you'd have your triangular matrix.

$$K = \begin{bmatrix} 2 & 0 & 0 \\ -5 & 1 & 3 \\ -2 & 3 & 1 \end{bmatrix}$$

The manipulation that changes the current matrix to a lower triangular matrix is adding the scalar multiple –3 times the second column to the first column.

$$-3C_2 + C_3 \rightarrow C_3 \begin{bmatrix} 2 & 0 & 0 \\ -5 & 1 & 0 \\ -2 & 3 & -8 \end{bmatrix}$$

The process also changed the last element in the third row. The significance (or lack thereof) of the change is made clear in the "Tracking down determinants of triangular matrices" section, found earlier in this chapter.

Planning out more than one manipulation

The matrix K, in the previous section, required just one operation to make it into a triangular matrix. Even though it's more work, you'll find, when computing determinants, that the extra effort to create a triangular matrix is well worth it.

Look at the 4×4 matrix L. You see three 0s on the lower left side. If the 3, –3, and 1 in the lower left portion were 0s, then you'd have an upper triangular matrix.

$$L = \begin{bmatrix} 2 & 1 & 4 & 3 \\ 3 & -2 & 1 & -6 \\ 0 & 0 & 1 & 2 \\ -3 & 1 & 0 & 1 \end{bmatrix}$$

The operations required to change the matrix L to an upper triangular matrix involve multiplying a row by a scalar and adding it to another row.

$$\begin{bmatrix} 2 & 1 & 4 & 3 \\ 3 & -2 & 1 & -6 \\ 0 & 0 & 1 & 2 \\ -3 & 1 & 0 & 1 \end{bmatrix} \quad R_4 + R_2 \to R_2 \begin{bmatrix} 2 & 1 & 4 & 3 \\ 0 & -1 & 1 & -5 \\ 0 & 0 & 1 & 2 \\ -3 & 1 & 0 & 1 \end{bmatrix}$$

$$\frac{3}{2}R_1 + R_4 \to R_4 \begin{bmatrix} 2 & 1 & 4 & 3 \\ 0 & -1 & 1 & -5 \\ 0 & 0 & 1 & 2 \\ 0 & \frac{5}{2} & 6 & \frac{11}{2} \end{bmatrix}$$

In the case of changing the matrix L, I chose to create 0s below the 2 in the first row, first column. That choice accounts for my first two steps. I have 0s below the 2, but, unfortunately, I lost one of the original 0s. But, never fear, row operations are here. Next, concentrate on getting 0s below the –1 on the main diagonal.

$$\begin{bmatrix} 2 & 1 & 4 & 3 \\ 0 & -1 & 1 & -5 \\ 0 & 0 & 1 & 2 \\ 0 & \frac{5}{2} & 6 & \frac{11}{2} \end{bmatrix} \quad \frac{5}{2}R_2 + R_4 \to R_4 \begin{bmatrix} 2 & 1 & 4 & 3 \\ 0 & -1 & 1 & -5 \\ 0 & 0 & 1 & 2 \\ 0 & 0 & \frac{17}{2} & -7 \end{bmatrix}$$

Now the $^{17}/_{2}$ is the only element in the way of having a lower triangular matrix. Just one more operation takes care of it.

$$\begin{bmatrix} 2 & 1 & 4 & 3 \\ 0 & -1 & 1 & -5 \\ 0 & 0 & 1 & 2 \\ 0 & 0 & \frac{17}{2} & -7 \end{bmatrix} \quad -\frac{17}{2}R_3 + R_4 \to R_4 \begin{bmatrix} 2 & 1 & 4 & 3 \\ 0 & -1 & 1 & -5 \\ 0 & 0 & 1 & 2 \\ 0 & 0 & 0 & -24 \end{bmatrix}$$

The last matrix is lower triangular. Ready for action. What action? I tell you in the next section.

Valuing Vandermonde

Alexandre-Théophile Vandermonde was born in Paris in 1735 and died there in 1796. Because Vandermonde's father pretty much insisted that Alexandre pursue a career in music, his contributions to mathematics were not all that great in number. But, what he lacked in number, he made up for in importance. Vandermonde is best known for his work in determinants (although some say that the determinant attributed to him shouldn't have been, because it really wasn't found in his work — someone misread his notations). The Vandermonde matrix, consisting of rows that are powers of fixed numbers, is used in applications such as error-correcting codes and signal processing. And, even though Vandermonde got a late start and was another mathematician (in addition to Carnot, Monge, and others) who was politically involved in the French Revolution, he made significant contributions to the theory of determinants.

Creating an upper triangular or lower triangular matrix

It's fine and dandy to use the rule of multiplying along the main diagonal to evaluate the determinant of a triangular matrix, but what happens if your matrix isn't triangular — or is *almost* triangular? You first determine if the computations and manipulations necessary to change your matrix to one that's triangular are worth it — if it's less hassle and work than just evaluating the determinant of the matrix the way it is. In this section, I show you a few situations where it seems to be worth the necessary manipulations. You can be the judge.

Reflecting on the effects of a reflection

An upper triangular or lower triangular matrix has 0s either in the lower right corner or the upper left corner. But, does the position of the zeros make a difference? What if you have a matrix with 0s in the upper *left* or lower *right* corner? Matrix P, shown next, has zeros in the lower right corner and a diagonal running from upper right to lower left.

$$P = \begin{bmatrix} 2 & 1 & 4 & 3 \\ 6 & -2 & 1 & 0 \\ -1 & 2 & 0 & 0 \\ 4 & 0 & 0 & 0 \end{bmatrix}$$

If you could switch the positions of the columns — exchange the first and the fourth columns and the second and third columns — then you'd have a lower triangular matrix.

$$\begin{matrix} C_1 \leftrightarrow C_4 \\ C_2 \leftrightarrow C_3 \end{matrix} \begin{bmatrix} 3 & 4 & 1 & 2 \\ 0 & 1 & -2 & 6 \\ 0 & 0 & 2 & -1 \\ 0 & 0 & 0 & 4 \end{bmatrix}$$

As I mention earlier, in the "Interchanging Rows or Columns" section, each interchange results in a change of signs for the determinant. Because I performed two interchanges, the sign changed once and then changed back again. So the value of the determinant of P is the same as the value of the determinant of the changed matrix. The determinant is equal to the product along the main diagonal of the new matrix: $3(1)(2)(4) = 24$. In this case, you get the value of the determinant when you multiply along the *reverse* diagonal.

You're probably wondering if this multiplying along *any* diagonal works for all square matrices. The answer: no. For example, a 6×6 square matrix would require three interchanges to make the reverse diagonal a main diagonal — which is an odd number of interchanges, so you'd have three -1 multipliers. The value of the determinant of a 6×6 matrix is the *opposite* of the product along the *reverse* diagonal.

If a square matrix has 0s above or below the *reverse* diagonal (opposite of the main diagonal), then the value of the determinant of the matrix is either the product of the elements along that reverse diagonal or -1 times the product of the elements on that diagonal. If the number of rows and columns, n, is a multiple of 4 or one greater than a multiple of 4 (written $4k$ or $4k + 1$), then the product along the reverse diagonal is used. Multiply the product along the reverse diagonal by -1 in all other cases.

For example, consider matrices Q and R, where Q has dimension 5×5 and R has dimension 6×6. Because 5 is one greater than a multiple of 4, then the determinant of Q is the product along the reverse diagonal. In the case of matrix R, the determinant is equal to -1 times the product along the main diagonal, because the number 6 is not a multiple of 4 or 1 greater than a multiple of 4.

$$Q = \begin{bmatrix} 1 & 2 & 6 & 4 & 3 \\ -3 & 0 & -2 & 4 & 0 \\ 1 & 4 & 5 & 0 & 0 \\ 3 & 2 & 0 & 0 & 0 \\ -1 & 0 & 0 & 0 & 0 \end{bmatrix} \qquad R = \begin{bmatrix} 0 & 0 & 0 & 0 & 0 & 7 \\ 0 & 0 & 0 & 0 & 1 & -1 \\ 0 & 0 & 0 & -1 & 2 & 3 \\ 0 & 0 & 2 & 3 & 4 & 2 \\ 0 & 1 & 4 & -6 & 3 & 5 \\ -2 & 5 & 4 & 3 & -4 & 6 \end{bmatrix}$$

$$|Q| = 3(4)(5)(2)(-1) = -120 \qquad |R| = -1\left[7(1)(-1)(2)(1)(-2)\right] = -28$$

Performing one or more manipulations to create the matrix you want

Several opportunities exist for changing a matrix to your liking. You just need to keep in mind how these *opportunities* affect the value of the determinant — if they do at all. Transposing matrices has no effect on the respective determinants. Multiplying a row or column *may* have an effect, as well as interchanging rows or columns. I give you all the details on the effects to determinants in the earlier sections of this chapter. Here I show you several examples of changing matrices before evaluating determinants.

In the first example, I show you a 3×3 matrix T. To find the determinant of T, I can use the quick rule for a 3×3 matrix, or I can change the corresponding determinant to upper triangular and multiply along the main diagonal. Even though I see a 0 in the second row, third column, I prefer to take advantage of the 1 in the first row, first column. When adding multiples of one row to another, the number 1 comes in very handy.

Adding a multiple of one row to another row doesn't change the value of the determinant (see "Adding the multiple of a row or column to another row or column," earlier in this chapter), so I show the two such manipulations — adding –4 times the first row to the second row and adding –3 times the first row to the third row.

$$\det(T) = \begin{vmatrix} 1 & 2 & 3 \\ 4 & 5 & 0 \\ 3 & 1 & 2 \end{vmatrix}, \quad \begin{matrix} -4R_1 + R_2 \to R_2 \\ -3R_1 + R_3 \to R_3 \end{matrix} \begin{vmatrix} 1 & 2 & 3 \\ 0 & -3 & -12 \\ 0 & -5 & -7 \end{vmatrix} = \det(T)$$

Now all I need for an upper triangular format is to have a 0 for the element in the third row, second column. The process will be easier if I divide each element in the second row by –3; this way, the element in the second row, second column is a 1. Dividing by –3 is the same as multiplying by $-1/3$. And, in the "Multiplying a row or column by a scalar" section, you see that the value of the determinant is changed by the amount of the multiple. Because I want to preserve the value of the original determinant, I need to make an adjustment to compensate for the scalar multiplication.

If the elements of a row or column of a determinant are multiplied by the scalar k (as long as k isn't 0), then the value of the original determinant is equal to $1/k$ times the value of the new determinant.

So, I multiply the elements in the second row by $-1/3$ and adjust for this operation by multiplying the determinant by the reciprocal of $-1/3$, which is -3.

$$-\frac{1}{3}R_2 \to R_2 \begin{vmatrix} 1 & 2 & 3 \\ 0 & 1 & 4 \\ 0 & -5 & -7 \end{vmatrix}$$

$$\det(T) = -3\begin{vmatrix} 1 & 2 & 3 \\ 0 & 1 & 4 \\ 0 & -5 & -7 \end{vmatrix}$$

Now I perform an operation that *doesn't* change the value of the determinant — adding 5 times row two to row three.

$$5R_2 + R_3 \to R_3 \begin{vmatrix} 1 & 2 & 3 \\ 0 & 1 & 4 \\ 0 & 0 & 13 \end{vmatrix}$$

$$\det(T) = -3\begin{vmatrix} 1 & 2 & 3 \\ 0 & 1 & 4 \\ 0 & 0 & 13 \end{vmatrix}$$

And, using the product along the main diagonal, the value of the determinant of matrix T is equal to -3(1)(1)(13) = -39.

Sometimes, you don't need to do much more to a determinant than a couple of interchanges and a minor adding of a scalar multiple to whip the determinant into shape (a triangular shape, of course). The matrix W, which I show you next, has five 0s in it. By interchanging some columns, I almost have a lower triangular format.

$$\det(W) = \begin{vmatrix} 0 & 3 & 3 & 0 \\ 0 & 3 & 1 & 0 \\ 0 & -6 & 5 & -1 \\ 4 & 3 & 7 & 5 \end{vmatrix}$$

$$\begin{matrix} C_1 \leftrightarrow C_3 \\ C_3 \leftrightarrow C_4 \end{matrix} \begin{vmatrix} 3 & 3 & 0 & 0 \\ 1 & 3 & 0 & 0 \\ 5 & -6 & -1 & 0 \\ 7 & 3 & 5 & 4 \end{vmatrix}, \quad \det(W) = \begin{vmatrix} 3 & 3 & 0 & 0 \\ 1 & 3 & 0 & 0 \\ 5 & -6 & -1 & 0 \\ 7 & 3 & 5 & 4 \end{vmatrix}$$

By interchanging columns one and three, and then interchanging the new column three with column four, I almost have a lower triangular format. The two interchanges each introduced a multiplier of -1 to the value of the determinant, but (-1)(-1) = 1, so the determinant has no real change. Now, by multiplying row two by –1 and adding the elements to row one, I finish the manipulations.

$$-1R_2 + R_1 \to R_1 \begin{vmatrix} 2 & 0 & 0 & 0 \\ 1 & 3 & 0 & 0 \\ 5 & -6 & -1 & 0 \\ 7 & 3 & 5 & 4 \end{vmatrix}$$

So now the value of the determinant of W is equal to the product of the elements along the main diagonal: 2(3)(-1)(4) = -24.

Determinants of Matrix Products

You find very specific rules about multiplying matrices — which matrices can multiply which other matrices — that are dependent on the dimensions of the matrices involved. In Chapter 3, you find all you'd ever want to know about multiplying two matrices together. In this chapter, I deal only with matrices that have determinants: square matrices. And, applying the rules involving multiplying matrices, you see that I can only multiply square matrices that have the same dimensions.

When you multiply two $n \times n$ matrices together, you get another $n \times n$ matrix. And the product of the determinants of the two matrices is equal to the determinant of the product matrix.

If A and B are two $n \times n$ matrices, then det(AB) = [det(A)][det(B)].

For example, consider the matrices A and B and the product of A and B.

$$A = \begin{bmatrix} 1 & 2 & 3 \\ 4 & 5 & 0 \\ 3 & 1 & 2 \end{bmatrix}, \; B = \begin{bmatrix} 2 & 0 & -1 \\ 4 & 5 & 3 \\ 1 & 1 & 7 \end{bmatrix}$$

$$AB = \begin{bmatrix} 13 & 13 & 26 \\ 28 & 25 & 11 \\ 12 & 7 & 14 \end{bmatrix}$$

The respective determinants are $|A| = -39$, $|B| = 65$, and $|AB| = -2{,}535 = (-39)(65)$. Go ahead — check my work!

Or, how about a nicer example where I show the advantage of finding the determinant of the product rather than the determinants of the first two matrices and then multiplying them. Look at my matrices C and D:

$$C = \begin{bmatrix} 10 & -2 & 10 \\ -4 & 1 & -4 \\ 12 & -3 & 15 \end{bmatrix}, \quad D = \begin{bmatrix} 1 & 0 & -1 \\ 4 & 5 & 0 \\ 0 & 1 & 1 \end{bmatrix}$$

The product of C and D is a matrix with nonzero elements running down the main diagonal and 0s elsewhere.

$$CD = \begin{bmatrix} 2 & 0 & 0 \\ 0 & 1 & 0 \\ 0 & 0 & 3 \end{bmatrix}$$

The determinant of the product of C and D is 6. This is much easier than finding the determinants of C and D and multiplying them together. You can even use the product and the value of the determinant of either C or D to quickly find the value of the determinant of the other matrix.

Chapter 12

Taking Advantage of Cramer's Rule

In This Chapter

- Joining forces with a matrix adjoint
- Finding matrix inverses using determinants
- Solving systems of equations using Cramer's rule
- Considering the computing possibilities of calculators and computers

Determinants have many applications, and I get to show you one of the applications in this chapter. The application I refer to is that of using determinants to find the inverses of square matrices. And, as a bonus, I get to introduce you to the matrix adjoint — something else that's needed to find matrix inverses using determinants.

Also in this chapter, I compare the good, the bad, and the ugly parts of working with determinants with similar tasks performed using good old algebra or matrix operations. When you're informed about the possibilities, you're better prepared to make a decision. So I give you some possibilities for finding matrix inverses and hope you use them to make the right choices!

Inviting Inverses to the Party with Determined Determinants

Many square matrices have inverses. The way that matrix inverses work is that if matrix M has an inverse, M^{-1}, then the product of M and M^{-1} is an identity matrix, I. (I introduce matrix inverses and how to find them in Chapter 3, if you need a little reminder of how they work.)

Not all square matrices have inverses; when a matrix doesn't have an inverse, it's termed *singular*. In this section, I deal mainly with matrices that *do* have inverses and offer an alternate method for finding that inverse — one that can't be done unless you know how to find a determinant of a matrix.

Setting the scene for finding inverses

In Chapter 3, I show you how to find the inverse of a 2×2 matrix using a quick, down-and-dirty method of switching and dividing, and then I show how to find the inverse of any size square matrix using matrix row operations. With this method, the inverse of matrix A is found by performing row operations to change A to an identity matrix, while carrying along those same row operations into what was already an identity matrix. For example, here's a matrix A, the augmented matrix used to solve for an inverse, and some row operations.

$$A = \begin{bmatrix} 1 & 2 & 3 & 4 \\ 0 & 1 & 2 & 1 \\ 0 & 0 & -1 & -5 \\ 1 & 0 & 0 & 6 \end{bmatrix}$$

$$\left[\begin{array}{cccc|cccc} 1 & 2 & 3 & 4 & 1 & 0 & 0 & 0 \\ 0 & 1 & 2 & 1 & 0 & 1 & 0 & 0 \\ 0 & 0 & -1 & -5 & 0 & 0 & 1 & 0 \\ 1 & 0 & 0 & 6 & 0 & 0 & 0 & 1 \end{array}\right]$$

$$-1R_1 + R_4 \rightarrow R_4 \quad \left[\begin{array}{cccc|cccc} 1 & 2 & 3 & 4 & 1 & 0 & 0 & 0 \\ 0 & 1 & 2 & 1 & 0 & 1 & 0 & 0 \\ 0 & 0 & -1 & -5 & 0 & 0 & 1 & 0 \\ 0 & -2 & -3 & 2 & -1 & 0 & 0 & 1 \end{array}\right]$$

$$\begin{array}{r} -2R_2 + R_1 \rightarrow R_1 \\ 2R_2 + R_4 \rightarrow R_4 \end{array} \quad \left[\begin{array}{cccc|cccc} 1 & 0 & -1 & 2 & 1 & -2 & 0 & 0 \\ 0 & 1 & 2 & 1 & 0 & 1 & 0 & 0 \\ 0 & 0 & -1 & -5 & 0 & 0 & 1 & 0 \\ 0 & 0 & 1 & 4 & -1 & 2 & 0 & 1 \end{array}\right]$$

The row operations are designed to change the left-hand portion of the matrix into an identity matrix. The operations are carried all the way across.

$$\begin{array}{r} -1R_3+R_1 \rightarrow R_1 \\ 2R_3+R_2 \rightarrow R_2 \\ R_3+R_4 \rightarrow R_4 \end{array} \left[\begin{array}{cccc|cccc} 1 & 0 & 0 & 7 & 1 & -2 & -1 & 0 \\ 0 & 1 & 0 & -9 & 0 & 1 & 2 & 0 \\ 0 & 0 & -1 & -5 & 0 & 0 & 1 & 0 \\ 0 & 0 & 0 & -1 & -1 & 2 & 1 & 1 \end{array}\right]$$

$$\begin{array}{r} -1R_3 \rightarrow R_3 \\ -1R_4 \rightarrow R_4 \end{array} \left[\begin{array}{cccc|cccc} 1 & 0 & 0 & 7 & 1 & -2 & -1 & 0 \\ 0 & 1 & 0 & -9 & 0 & 1 & 2 & 0 \\ 0 & 0 & 1 & 5 & 0 & 0 & -1 & 0 \\ 0 & 0 & 0 & 1 & 1 & -2 & -1 & -1 \end{array}\right]$$

$$\begin{array}{r} -7R_4+R_1 \rightarrow R_1 \\ 9R_4+R_2 \rightarrow R_2 \\ -5R_4+R_3 \rightarrow R_3 \end{array} \left[\begin{array}{cccc|cccc} 1 & 0 & 0 & 0 & -6 & 12 & 6 & 7 \\ 0 & 1 & 0 & 0 & 9 & -17 & -7 & -9 \\ 0 & 0 & 1 & 0 & -5 & 10 & 4 & 5 \\ 0 & 0 & 0 & 1 & 1 & -2 & -1 & -1 \end{array}\right]$$

On the left, I now have the identity matrix, and on the right, I have the inverse of the original matrix. The product of A and its inverse, A^{-1}, is an identity matrix.

$$A^*A^{-1} = \begin{bmatrix} 1 & 2 & 3 & 4 \\ 0 & 1 & 2 & 1 \\ 0 & 0 & -1 & -5 \\ 1 & 0 & 0 & 6 \end{bmatrix} * \begin{bmatrix} -6 & 12 & 6 & 7 \\ 9 & -17 & -7 & -9 \\ -5 & 10 & 4 & 5 \\ 1 & -2 & -1 & -1 \end{bmatrix} = \begin{bmatrix} 1 & 0 & 0 & 0 \\ 0 & 1 & 0 & 0 \\ 0 & 0 & 1 & 0 \\ 0 & 0 & 0 & 1 \end{bmatrix}$$

You find more details on inverse matrices in Chapter 3. I use this example to set the tone for what I offer in this chapter as an alternative to finding inverses. And an alternate way of finding the inverse of a matrix is to use determinants.

Introducing the adjoint of a matrix

In Chapters 3 and 4, you find a lot of information about the inverse of a matrix, the transpose of a matrix, the determinant of a matrix, and so on. Now I present you with a whole new character for this play: the *adjoint* of a matrix. The adjoint is also known as an *adjugate*. The adjoint of matrix A, denoted adj(A), is used in computing the inverse of a matrix. In fact, you get the inverse of a matrix by dividing the adjoint by the determinant.

$$A^{-1} = \frac{\text{adj}(A)}{\det(A)}$$

The inverse exists only when det(A) is *not* equal to 0.

Okay, you have the formula now for computing an inverse using the determinant and adjoint of a matrix, but you're probably wondering, "Where in the world do I find this adjoint thing?"

The *adjoint* of matrix A is a matrix of the cofactors of A that has been transposed. So, if C is the matrix of the cofactors of A, then adj(A) = C^T. (You find more on transposing a matrix in Chapter 3.)

In a 4 × 4 matrix, the adjoint is as follows. (Notice that the columns and rows are reversed — like creating the transpose of the original positioning.)

$$\text{adj}(A) = \begin{bmatrix} C_{11} & C_{21} & C_{31} & C_{41} \\ C_{12} & C_{22} & C_{32} & C_{42} \\ C_{13} & C_{23} & C_{33} & C_{43} \\ C_{14} & C_{24} & C_{34} & C_{44} \end{bmatrix}$$

For example, to find the adjoint of matrix A, first recall that the *cofactor* C_{ij} is formed by multiplying the determinant of the minor times -1^{i+j} power — the appropriate power of -1 based on the index of the element. (You can find out all about determinants in Chapters 10 and 11.) Here's matrix A and its cofactors. You identify a cofactor with C_{ij}, where *i* and *j* are the corresponding row and column.

$$A = \begin{bmatrix} 3 & -3 & -1 & 7 \\ 7 & -6 & -3 & 14 \\ 5 & -4 & -2 & 9 \\ -5 & 5 & 2 & -11 \end{bmatrix}$$

$$C_{11} = (-1)^2 \begin{vmatrix} -6 & -3 & 14 \\ -4 & -2 & 9 \\ 5 & 2 & -11 \end{vmatrix} = 1(1) = 1$$

$$C_{12} = (-1)^3 \begin{vmatrix} 7 & -3 & 14 \\ -5 & -2 & 9 \\ 5 & 2 & -11 \end{vmatrix} = -1(-2) = 2$$

$$C_{13} = (-1)^4 \begin{vmatrix} 7 & -6 & 14 \\ 5 & -4 & 9 \\ -5 & 5 & -11 \end{vmatrix} = 1(3) = 3$$

$$C_{14} = (-1)^5 \begin{vmatrix} 7 & -6 & -3 \\ 5 & -4 & -2 \\ -5 & 5 & 2 \end{vmatrix} = -1(-1) = 1$$

$$C_{21} = (-1)^3 \begin{vmatrix} -3 & -1 & 7 \\ -4 & -2 & 9 \\ 5 & 2 & -11 \end{vmatrix} = -1(1) = -1$$

Okay, you get the picture. I don't need to show you all the steps for the rest of the cofactors.

$$C_{22} = 1(2) = 2$$
$$C_{23} = -1(2) = -2$$
$$C_{24} = 1(1) = 1$$
$$C_{31} = 1(2) = 2$$
$$C_{32} = -1(1) = -1$$
$$C_{33} = 1(2) = 2$$
$$C_{34} = -1(1) = -1$$
$$C_{41} = -1(-1) = 1$$
$$C_{42} = 1(3) = 3$$
$$C_{43} = -1(-1) = 1$$
$$C_{44} = 1(1) = 1$$

So now I create the adjoint of matrix A, adj(A), using the cofactors.

$$\text{adj}(A) = \begin{bmatrix} C_{11} & C_{21} & C_{31} & C_{41} \\ C_{12} & C_{22} & C_{32} & C_{42} \\ C_{13} & C_{23} & C_{33} & C_{43} \\ C_{14} & C_{24} & C_{34} & C_{44} \end{bmatrix} = \begin{bmatrix} 1 & -1 & 2 & 1 \\ 2 & 2 & -1 & 3 \\ 3 & -2 & 2 & 1 \\ 1 & 1 & -1 & 1 \end{bmatrix}$$

Instigating steps for the inverse

To find the inverse of matrix A, I need the determinant of A in addition to the adjoint. I find the determinant of a 4×4 matrix using cofactor expansion over a row or column. Because I have all the cofactors computed already, I choose to expand over the first row. (None of the elements is zero, so there's no real advantage to using any particular row; row 1 works as well as any.)

$$\det(A) = 3(1) + (-3)(2) + (-1)(3) + 7(1) = 3 - 6 - 3 + 7 = 1$$

Refer to Chapter 10 if you need a refresher on evaluating determinants using cofactor expansion over a row or column.

Now I'm all set to find the inverse of matrix A using its adjoint and the determinant. Each term in the adjoint is divided by the value of the determinant. Yes, the determinant comes out to have a value of 1, so you don't really change anything by dividing, but I show you the format for all cases and all determinants — those without such a cooperative number.

$$A^{-1} = \frac{\text{adj}(A)}{\det(A)} = \frac{\begin{bmatrix} 1 & -1 & 2 & 1 \\ 2 & 2 & -1 & 3 \\ 3 & -2 & 2 & 1 \\ 1 & 1 & -1 & 1 \end{bmatrix}}{1} = \begin{bmatrix} 1 & -1 & 2 & 1 \\ 2 & 2 & -1 & 3 \\ 3 & -2 & 2 & 1 \\ 1 & 1 & -1 & 1 \end{bmatrix}$$

You may not be completely impressed by this particular method of finding an inverse of a matrix. This method really lends itself more to being programmed in a computer or when using a mathematical computer utility. The biggest benefit comes when your matrix has one or more variables for elements.

Taking calculated steps with variable elements

Finding the inverse of a 3×3 matrix is relatively easy to do using either the row reduction method or the method of dividing the adjoint of the matrix by its determinant. The second method, using the adjoint, is usually the superior method to use when one or more elements are variables.

The matrix B contains the variables a and b. The determinant of B gives you information on the restrictions associated with the values of a and b in this particular matrix:

$$B = \begin{bmatrix} a & b & -b \\ 1 & 1 & 1 \\ 1 & a & a \end{bmatrix}$$

$$\det(B) = \begin{vmatrix} a & b & -b \\ 1 & 1 & 1 \\ 1 & a & a \end{vmatrix} = a^2 + b - ab - (-b + a^2 + ab)$$

$$= a^2 + b - ab + b - a^2 - ab = 2b - 2ab = 2b(1-a)$$

You see from the determinant, $2b(1-a)$, that a cannot be 1 and b cannot be 0, or the determinant would be equal to zero, and these values would make the matrix singular — without an inverse.

Now come the cofactors:

$$\begin{aligned}
C_{11} &= 1(a-a) = 0 \\
C_{12} &= -1(a-1) = 1-a \\
C_{13} &= 1(a-1) = -(1-a) \\
C_{21} &= -1(ab+ab) = -2ab \\
C_{22} &= 1(a^2+b) = a^2+b \\
C_{23} &= -1(a^2-b) = b-a^2 \\
C_{31} &= 1(b+b) = 2b \\
C_{32} &= -1(a+b) = -(a+b) \\
C_{33} &= 1(a-b) = a-b
\end{aligned}$$

The inverse of matrix B is equal to the elements in the adjoint of B divided by the determinant of B.

$$B^{-1} = \frac{\text{adj}(B)}{\det(B)} = \frac{\begin{bmatrix} 0 & -2ab & 2b \\ 1-a & a^2+b & -(a+b) \\ -(1-a) & b-a^2 & a-b \end{bmatrix}}{2b(1-a)}$$

$$= \begin{bmatrix} \frac{0}{2b(1-a)} & \frac{-2ab}{2b(1-a)} & \frac{2b}{2b(1-a)} \\ \frac{1-a}{2b(1-a)} & \frac{a^2+b}{2b(1-a)} & \frac{-(a+b)}{2b(1-a)} \\ \frac{-(1-a)}{2b(1-a)} & \frac{b-a^2}{2b(1-a)} & \frac{a-b}{2b(1-a)} \end{bmatrix}$$

$$= \begin{bmatrix} 0 & -\frac{a}{1-a} & \frac{1}{1-a} \\ \frac{1}{2b} & \frac{a^2+b}{2b(1-a)} & -\frac{a+b}{2b(1-a)} \\ -\frac{1}{2b} & \frac{b-a^2}{2b(1-a)} & \frac{a-b}{2b(1-a)} \end{bmatrix}$$

As values for a and b are chosen in an application, substitutions can then be made into the form for the inverse of the particular matrix. For example, if you have a situation where $a = -1$ and $b = 2$, then you replace the variables in the matrices to determine the matrix and its inverse.

$$B = \begin{bmatrix} a & b & -b \\ 1 & 1 & 1 \\ 1 & a & a \end{bmatrix} = \begin{bmatrix} -1 & 2 & -2 \\ 1 & 1 & 1 \\ 1 & -1 & -1 \end{bmatrix}$$

$$B^{-1} = \begin{bmatrix} 0 & -\frac{a}{1-a} & \frac{1}{1-a} \\ \frac{1}{2b} & \frac{a^2+b}{2b(1-a)} & -\frac{a+b}{2b(1-a)} \\ -\frac{1}{2b} & \frac{b-a^2}{2b(1-a)} & \frac{a-b}{2b(1-a)} \end{bmatrix} = \begin{bmatrix} 0 & \frac{1}{2} & \frac{1}{2} \\ \frac{1}{4} & \frac{3}{8} & -\frac{1}{8} \\ -\frac{1}{4} & \frac{1}{8} & -\frac{3}{8} \end{bmatrix}$$

Solving Systems Using Cramer's Rule

Systems of linear equations are solved in many different ways. You can solve the system of equations shown here in one of many ways:

- You can solve the system algebraically, using elimination or substitution.
- You can use augmented matrices.
- You can graph the lines associated with the equations.
- You can just keep guessing numbers, plugging them into the equations, and stop when you finally get a pair of numbers that works.

$$\begin{cases} x + 2y = 5 \\ 2x + 3y = 6 \end{cases}$$

No matter which way you solve the system, you always get $x = -3$ and $y = 4$. In Chapter 4, I cover the algebraic, matrix, and graphical methods. I leave by-guess-or-by-golly to those who may be so inclined.

In this section, I introduce you to yet another method used to solve systems of equations. The method shown here makes use of the determinants of matrices associated with the system.

Assigning the positions for Cramer's rule

To use Cramer's rule, you write the system of linear equations as the product of matrices and vectors. Then you change one of the matrices by inserting vectors to change whole columns of the original matrices. I show you how this rule works with an example.

Consider the system of linear equations. Using methods from Chapter 4, you get the solution that $x = 3$, $y = -2$, and $z = -1$:

$$\begin{cases} 2x - 3y + z = 11 \\ x + 4y - 2z = -3 \\ 4x - 5y + 3z = 19 \end{cases}$$

The system can be written as a coefficient matrix times a variable matrix/vector and set equal to the constant vector, A*$\mathbf{x} = \mathbf{b}$. (In Chapter 6, I call this equation the *matrix equation.*)

$$A = \begin{bmatrix} 2 & -3 & 1 \\ 1 & 4 & -2 \\ 4 & -5 & 3 \end{bmatrix}, \mathbf{x} = \begin{bmatrix} x \\ y \\ z \end{bmatrix}, \mathbf{b} = \begin{bmatrix} 11 \\ -3 \\ 19 \end{bmatrix}$$

$$\begin{bmatrix} 2 & -3 & 1 \\ 1 & 4 & -2 \\ 4 & -5 & 3 \end{bmatrix} * \begin{bmatrix} x \\ y \\ z \end{bmatrix} = \begin{bmatrix} 11 \\ -3 \\ 19 \end{bmatrix}$$

Now I describe how to create some modified matrices in which a whole column in the coefficient matrix is replaced with the elements in the constant vector. When I write $A_x(\mathbf{b})$, it means to replace the x column (the first column) of the coefficient matrix with the elements in vector **b** to create the new, modified matrix. Likewise, $A_y(\mathbf{b})$ and $A_z(\mathbf{b})$ designate that the y column and the z column of matrix A should be replaced by the elements in the constant vector. So, in the case of the system given here, $A_x(\mathbf{b}) = A_x$, $A_y(\mathbf{b}) = A_y$ and $A_z(\mathbf{b}) = A_z$:

$$A_x = \begin{bmatrix} 11 & -3 & 1 \\ 3 & 4 & -2 \\ 19 & -5 & 3 \end{bmatrix}, A_y = \begin{bmatrix} 2 & 11 & 1 \\ 1 & -3 & -2 \\ 4 & 19 & 3 \end{bmatrix},$$

$$A_z = \begin{bmatrix} 2 & -3 & 11 \\ 1 & 4 & -3 \\ 4 & -5 & 19 \end{bmatrix}$$

The players (the modified matrices) are all in position now. I use the determinants of each of the matrices to solve the system of equations.

Applying Cramer's rule

Cramer's rule for solving systems of linear equations says that, to find the value of each variable, you divide the determinant of each corresponding modified matrix by the determinant of the coefficient matrix.

$$x = \frac{\det(A_x)}{\det(A)}, \; y = \frac{\det(A_y)}{\det(A)}, \; z = \frac{\det(A_z)}{\det(A)}, \; \ldots$$

So I find the determinant of the coefficient of matrix A and the determinants of the three modified matrices.

$$\det(A) = \begin{vmatrix} 2 & -3 & 1 \\ 1 & 4 & -2 \\ 4 & -5 & 3 \end{vmatrix} = 24+24-5-(6+20-9) = 43-27 = 16$$

$$\det(A_x) = \begin{vmatrix} 11 & -3 & 1 \\ -3 & 4 & -2 \\ 19 & -5 & 3 \end{vmatrix} = 132+114+15-(76+110+27) = 261-213 = 48$$

$$\det(A_y) = \begin{vmatrix} 2 & 11 & 1 \\ 1 & -3 & -2 \\ 4 & 19 & 3 \end{vmatrix} = -18-88+19-(-12-76+33) = -87+55 = -32$$

$$\det(A_z) = \begin{vmatrix} 2 & -3 & 11 \\ 1 & 4 & -3 \\ 4 & -5 & 19 \end{vmatrix} = 152+36-55-(176+30-57) = 133-149 = -16$$

Using Cramer's rule and doing the dividing,

$$x = \frac{\det(A_x)}{\det(A)} = \frac{48}{16} = 3$$

$$y = \frac{\det(A_y)}{\det(A)} = \frac{-32}{16} = -2$$

$$z = \frac{\det(A_z)}{\det(A)} = \frac{-16}{16} = -1$$

So, the solution of the system is $x = 3$, $y = -2$, $z = -1$.

Gabriel Cramer

Gabriel Cramer, best known for Cramer's rule, was a Swiss mathematician who lived from 1704 to 1752. His intelligence and mathematical prowess were recognized very early; he received his doctorate at the age of 18, submitting a thesis on the theory of sound, and was co-chairman of the mathematics department at Académie de Clavin in Geneva at the age of 20. While teaching and serving in his administrative position, Cramer introduced a major innovation: teaching his courses in French, rather than Latin.

Cramer published many articles over a wide range of subjects. In addition to his mathematical contributions, he also wrote on philosophy, the date of Easter, the aurora borealis, and law. He also proposed a solution to the St. Petersburg Paradox — a classical situation in game theory and probability where the expected outcome is an infinitely large amount. (Sometimes the actual mathematics flies in the face of common sense.)

Having Cramer's rule as an alternative adds another weapon to your arsenal when solving systems of equations. Creating and evaluating all those determinants necessary with this method takes time and would be extremely cumbersome if the number of variables were large. That's where computers and their programs come in — to perform repeated operations.

Recognizing and Dealing with a Nonanswer

Not all systems of linear equations have solutions, and some systems of equations have an infinite number of solutions. When solving systems of equations algebraically or with row operations in matrices, you have indicators to tell you when the situation is infinitely many or none.

Taking clues from algebraic and augmented matrix solutions

When no solution exists for a system of equations, you get *impossible statements* — statements that are *never* true — when working algebraically, and you get 0s equaling a nonzero number in a matrix. For example, consider solving the following system of linear equations:

$$\begin{cases} x + y - z = 2 \\ 2x - y + z = 1 \\ 3x - y + z = 1 \end{cases}$$

Working algebraically, when you add the first and second equations together, you get $3x = 3$, which has a solution of $x = 1$. When you add the first and fourth equations together, you get $4x = 3$, which tells you that $x = {}^{3}/_{4}$. You have a contradiction. The two equations can't possibly provide solutions to the same system. For example, because the equations say that both variable products are equal to 3, then you can set $3x = 4x$. The only solution to this equation is $x = 0$, and that doesn't satisfy $3x = 3$. This is an example of the *impossible statement.* (For more on solving systems of equations, refer to Chapter 4.)

Perhaps you prefer to solve systems of linear equations using matrices and row operations. Using the same system of linear equations, you start with the augmented matrix I show you here, and, after several operations, you end up with the last matrix.

$$\left[\begin{array}{ccc|c} 1 & 1 & -1 & 2 \\ 2 & -1 & 1 & 1 \\ 3 & -1 & 1 & 1 \end{array}\right] \rightarrow \cdots \rightarrow \left[\begin{array}{ccc|c} 1 & 1 & -1 & 2 \\ 0 & 1 & -1 & 1 \\ 0 & 0 & 0 & 1 \end{array}\right]$$

The last line in the matrix translates into $0 = 1$. The statement is impossible as a solution of the system.

In contrast, when a system of equations has an infinite number of solutions, you end up with equations and statements that are *always* true. For example, the system shown here has an infinite number of solutions:

$$\begin{cases} x + y - 2z = 1 \\ 2x - y + z = 3 \\ 4x + y - 3z = 5 \end{cases}$$

Solving algebraically, when you add the first and second equations together, you get $3x - z = 4$. You get the same equation when you add -1 times the first equation to the third equation: $3x - z = 4$. Further algebra gives you $0 = 0$. The statement $0 = 0$ is always true. So the system of equations has an infinite number of solutions, all in the form of the ordered triple: $(k, 5k - 7, 3k - 4)$. Using an augmented matrix, you'd end up with the bottom row all being 0s — very much like $0 = 0$. Both are indicators of the fact that you have many, many solutions of the system of equations.

Now that I've outlined what to look for when solving systems of equations using algebraic techniques or augmented matrices, let me show you what the indicators are when using Cramer's rule.

Cramming with Cramer for non-solutions

Cramer's rule for solving systems of equations deals with dividing by a determinant. The determinant is the greatest determiner of a situation of *infinitely many or none.* If you evaluate the determinant first, before doing any other work on the problem, you'll save yourself a lot of time by recognizing the situation upfront.

In both cases — having no solution or an infinite number of solutions — the determinant of the coefficient matrix is equal to zero. Using Cramer's rule, you're to divide by the coefficient matrix, so the zero prevents that step. After you've determined that the determinant of the coefficient matrix is equal to zero, then you investigate further to find out which situation you're dealing with: none or infinitely many.

A good plan of attack to find the answer to your question is to go to the augmented matrix of the system of equations and perform row operations to determine if you have a system with no solutions (getting a row of 0s ending in a nonzero number) or a system with infinitely many solutions (getting all 0s in a row).

Making a Case for Calculators and Computer Programs

When matrices and their determinants and adjoints are of a reasonable size (though some would say that anything larger than 2×2 is *not* reasonable), the computations and manipulations necessary to perform the necessary tasks are relatively easy to do. By just jumping to a 4×4 matrix, you suddenly have to resort to cofactor expansion. Just imagine how much fun you'd have with a 6×6 matrix or larger! But, in real life and with real-life applications, the matrices needed get larger and larger to handle all the different aspects of a problem.

What you find in this section is not only how to handle the smaller matrices, but also the basis for handling any size matrix — if you have the strength and stamina to do it.

Calculating with a calculator

Hand-held graphing calculators have more power to compute than the first huge computers did — the ones that took up several rooms and required all those cards with holes in them. So, no wonder you can quickly and efficiently work with matrices using calculators.

The main problem with matrices and calculators is that you have to enter all the elements of the matrix by hand, keep track of which matrix you put where, and (often) interpret the decimal values that result — even when you carefully put in fractions (not decimals). The larger matrices can't all be seen on the screen at once — you have to scroll right and left and up and down to see portions of the matrix. But, if you're willing to put up with these little inconveniences, a graphing calculator is the tool for you.

Most graphing calculators have buttons specifically designated for entering matrices and elements in the matrices. A matrix is usually denoted with brackets around the name, such as [A]. The bracket notation is consistent with your writing the elements of a matrix by hand within that same grouping symbol.

Graphing calculators vary by manufacturer, but they all have similar types of functions when it comes to matrices. Here are some of the more common functions and a representative type of notation indicating the function:

[A] 3×4	Naming or editing a matrix and entering the dimension
det ([A])	Calculating the determinant of a particular matrix
$[A]^T$	Performing a matrix transpose
dim ([A])	Giving the dimension of the matrix
identity (*k*)	Creating an identity matrix of dimension $k \times k$
rowSwap	Interchanging two rows
row+	Adding one row to another
*row	Multiplying a row by a scalar
*row+	Multiplying a row by a scalar and adding it to another row

The row-operation capabilities of calculators are wonderful, but they're very specifically scripted. You need to enter matrix names and rows and scalar multiples in an exact fashion, usually with commas separating portions of the operation and no helpful prompts. When I use my calculator to do these operations, I usually tape a listing of the commands to the inside cover of my calculator, because they're a challenge to remember (and, yes, I let my students do this, too).

Sometimes you enter perfectly nice integers or fractions and end up with long, indistinguishable decimals. One way to combat this problem is to change the *mode* on your calculator to a specific number of decimal places — you can have the calculator round to two or three decimal places.

Another option is to try to change the decimals back to fractions. You won't see the fractions on the calculator screen — they'll keep reverting back to the decimals — but you can record the fractions on paper. The calculators usually have a function that changes repeating fractions back to decimals, but the function fails if the repeating part is too long or the number of digits entered is too short. You just have to experiment to see what it takes to get a result.

Computing with a computer

One very nice thing about matrices, determinants, and Cramer's rule is that the processes used are repeated and programmable. Just consider the way the elements of a matrix are identified:

$$A = \begin{bmatrix} a_{11} & a_{12} & \cdots & a_{1n} \\ a_{21} & a_{22} & \cdots & a_{2n} \\ \vdots & \vdots & \ddots & \vdots \\ a_{m1} & a_{m2} & \cdots & a_{mn} \end{bmatrix}$$

Each element in matrix A is identified with a_{ij}, where the i represents the row and j represents the column. By creating commands using the i and j, starting with a particular number and increasing by one more, over and over, you can perform operations on any size matrix — letting the computer do all the computations. And computers are notoriously more accurate and reliable than humans. Of course, garbage in, garbage out: If you don't program the computer correctly, you won't get your answers.

You'll find many computer utilities that already have the matrix operations built in. For instance, Derive and Mathematica are popular and widespread in educational institutions. But they take all the fun out of matrices! You just plop in the matrix values and get the answer! (You don't see my problem with that? Oh, well.) Even Excel spreadsheets have operations that deal with matrices. One of the biggest advantages to computer programs and spreadsheets is that you get a printout of your work, too.

Part IV
Involving Vector Spaces

In this part . . .

This isn't Yankees vs. Red Sox or Democrats vs. Republicans. Rather than being for or against something, you get to embrace and be *for* the vector space and all the fine points that come with it. This part ends with eigenvalues vs. eigenvectors — and you're the victor!

Chapter 13

Promoting the Properties of Vector Spaces

In This Chapter

- Defining a vector space in terms of elements and operations
- Looking for closure with operations performed on elements
- Investigating the associative and commutative properties in terms of vector operations
- Zeroing in on zeros and inverses in vector spaces

In mathematics, you find many instances where a collection of objects is identified or described and the objects in the collection are tied to some rules so that they can actually be a part of that collection. The simplest example is a set of real numbers and the operation of addition. The set A = {1, 2, 3, . . . } contains all the counting numbers. The rule for belonging to that set is that the smallest possible element is a 1, and all the other numbers are obtained by adding one more and then one more and then. . . . If you add any two counting numbers together, you get another counting number. The addition part is an operation performed on elements in the set.

In this chapter, I introduce you to the collection called a *vector space.* The qualifications for becoming a member of a vector space and the rules to be abided by are stated and explained in this chapter. And, to further pique your interest, I show you how a vector space doesn't even have to contain vectors — as you know them.

Delving into the Vector Space

A *vector space* is a collection of objects called vectors; the name *vector space* came from the properties of vectors and how they behave under two specific operations. For now, just think of the vectors in a vector space in terms of the objects described in Chapter 2 and found throughout the other earlier chapters. (I introduce other types of objects later in the chapter.)

You can think of a vector space as a space that contains vectors — it's where they all gather together (to do heaven only knows what). Those who occupy this vector space have to abide by some specific rules and cooperate when involved in the two operations.

Just to get the vision of *vector spaces* dancing in your head, I'll show you the two simplest versions of vectors and vector operations. I use 2×1 vectors, with vector addition, $\oplus$, being the traditional addition process as described in Chapter 2. I also define the vector multiplication, $\otimes$, as the scalar multiplication also found in Chapter 2. The vectors A and B are added together to give me vector C. I then multiply vector C by the scalar 6.

$$A = \begin{bmatrix} 3 \\ -2 \end{bmatrix}, B = \begin{bmatrix} 4 \\ 8 \end{bmatrix}$$

$$A \oplus B = \begin{bmatrix} 3 \\ -2 \end{bmatrix} \oplus \begin{bmatrix} 4 \\ 8 \end{bmatrix} = \begin{bmatrix} 3+4 \\ -2+8 \end{bmatrix} = \begin{bmatrix} 7 \\ 6 \end{bmatrix} = C$$

$$6 \otimes C = 6 \otimes \begin{bmatrix} 7 \\ 6 \end{bmatrix} = \begin{bmatrix} 6 \cdot 7 \\ 6 \cdot 6 \end{bmatrix} = \begin{bmatrix} 42 \\ 36 \end{bmatrix}$$

Let me give you a formal definition of a vector space: Consider a set of elements, *V*, with vectors **u, v,** and **w** belonging to *V*, and the real numbers *k* and *l*, in which the operations of $\oplus$ and $\otimes$ have the following properties:

- $\mathbf{u} \oplus \mathbf{v}$ is also in *V*. (The set *V* is *closed* under the operation $\oplus$.)
- $\mathbf{u} \oplus \mathbf{v} = \mathbf{v} \oplus \mathbf{u}$. (There exists commutativity under the operation $\oplus$.)
- $\mathbf{u} \oplus (\mathbf{v} \oplus \mathbf{w}) = (\mathbf{u} \oplus \mathbf{v}) \oplus \mathbf{w}$. (There exists associativity under the operation $\oplus$.)
- There exists an element **0** in *V* such that $\mathbf{u} \oplus \mathbf{0} = \mathbf{0} \oplus \mathbf{u} = \mathbf{u}$ for any element **u.** (There exists an identity under the operation $\oplus$.)
- For every element **u** in *V*, there exists an element **–u** such that $\mathbf{u} \oplus -\mathbf{u} = -\mathbf{u} \oplus \mathbf{u} = \mathbf{0}$. (There exist inverses under the operation $\oplus$.)
- $k \otimes \mathbf{u}$ is also in *V*. (The set *V* is *closed* under the operation $\otimes$.) For simplicity, the expression $k \otimes \mathbf{u}$ is often written $k\mathbf{u}$ with the operation implied.
- $k \otimes (\mathbf{u} \oplus \mathbf{v}) = k \otimes \mathbf{u} \oplus k \otimes \otimes \mathbf{v}$, also written $k(\mathbf{u} \oplus \mathbf{v}) = k\mathbf{u} \oplus k\mathbf{v}$. (The scalar *k* distributes over the operation $\oplus$.)
- $(k + l) \otimes \mathbf{u} = k \otimes \mathbf{u} \oplus l \otimes \mathbf{u}$ (also written $[k + l]\mathbf{u} = k\mathbf{u} \oplus l\mathbf{u}$.)
- $k \otimes (l \otimes \mathbf{u}) = (kl) \otimes \mathbf{u}$.
- $1 \otimes \mathbf{u} = \mathbf{u}$.

Yes, you're probably wondering about these strange $\oplus$ and $\otimes$ symbols and asking yourself if I couldn't just have used + and ×. The answer is a simple "No." The operations of $\oplus$ and $\otimes$ sometimes act in a manner similar to our usual addition and multiplication, but they're often quite different. The rules are very strict and a bit different in vector spaces. But one bright note is that sets containing all the matrices of a particular dimension, $m \times n$, will form a vector space. In the next sections of this chapter, I describe the vector space operations and rules in more detail.

Describing the Two Operations

The two operations associated with a vector space are *vector addition* (denoted $\oplus$) and *scalar multiplication* (with the symbol $\otimes$). The two operations may behave exactly as you'd expect, considering your vast experience with adding and multiplying. But the two operations might take a wild swing one way or another to a completely different type of compounding of activities and operations.

Letting vector spaces grow with vector addition

A vector space is tied to the operation of vector addition. The operation itself may vary, depending on the set of objects involved, but, in all cases, vector addition is performed on two vectors (unlike vector multiplication). For example, consider the vector space, *V*, consisting of all ordered pairs (x,y) which represent a 2×1 vector in standard position. (Refer to Chapter 2 for more on how vectors come in all sizes.) In the case of these ordered pairs, (x,y), vector addition is defined as $(x_1,y_1) \oplus (x_2,y_2) = (x_1 + x_2, y_1 + y_2)$. You add the respective coordinates together to get a new set of coordinates. So the vector addition has some adding in it, but the format is just a tad different from adding 2 + 3.

Another example of a vector space that uses a familiar method of vector addition is the set of all 2×2 matrices that have a *trace* equal to zero.

The *trace* of a square matrix is the sum of the elements that lie along the main diagonal. The matrices shown here have a trace of 0:

$$C = \begin{bmatrix} 7 & 3 \\ -5 & -7 \end{bmatrix}, B = \begin{bmatrix} 0 & 7 \\ -1 & 0 \end{bmatrix}, A = \begin{bmatrix} a_{11} & a_{12} \\ a_{21} & a_{22} \end{bmatrix}$$

where trace $a_{11} + a_{22} = 0$.

So vector addition using elements from the set of all 2×2 matrices whose trace is equal to zero is defined:

$$\begin{bmatrix} a_{11} & a_{12} \\ a_{21} & a_{22} \end{bmatrix} \oplus \begin{bmatrix} b_{11} & b_{12} \\ b_{21} & b_{22} \end{bmatrix} = \begin{bmatrix} a_{11}+b_{11} & a_{12}+b_{12} \\ a_{21}+b_{21} & a_{22}+b_{22} \end{bmatrix}$$

and $(a_{11} + b_{11}) + (a_{22} + b_{22}) = 0$.

Any square matrix can have a trace of 0. For example, matrices D and E, shown here, have sums of 0 along the main diagonal.

$$D = \begin{bmatrix} 2 & 8 & 0 \\ -6 & -1 & 5 \\ 4 & 2 & -1 \end{bmatrix}, E = \begin{bmatrix} -5 & 6 & 0 & 4 \\ 1 & -3 & 1 & -5 \\ 0 & 7 & 0 & -7 \\ 2 & 2 & 3 & 8 \end{bmatrix}$$

Now let me depart from having vectors or matrices in my vector space and describe *Z*, a vector space whose elements are all positive real numbers. I define the vector addition in this vector space as follows:

Let x and y be positive real numbers. Then $x \oplus y = xy$.

For example, in this vector space, if I use vector addition on the numbers 2 and 5, I get $2 \oplus 5 = 10$. Yes, I can do that. I can define my vector addition in any way that it works with my particular application. Of course, other qualifications have to be met for my vector addition to qualify in the vector space. I cover the other qualifications in the next few sections of this chapter.

Making vector multiplication meaningful

Vector multiplication is often referred to as *scalar multiplication,* because the operation involves just one vector and a real number, not two vectors as in vector addition. For example, I describe vector multiplication on that set of 2×2 matrices whose trace is equal to 0. Letting A be a matrix in the set and k be a real number, then:

$$k \otimes A = k \otimes \begin{bmatrix} a_{11} & a_{12} \\ a_{21} & a_{22} \end{bmatrix} = \begin{bmatrix} ka_{11} & ka_{12} \\ ka_{21} & ka_{22} \end{bmatrix}$$

and $ka_{11} + ka_{22} = k(a_{11} + a_{22}) = 0$.

Also, going for another nonstandard vector space (not using vectors or matrices for the elements), I could define the operation of vector multiplication on the set of positive real numbers as follows:

If x is a positive real number and k is any real number, then $k \otimes x = x^k$.

So, if you perform vector multiplication on the number 2 where k is 5, you get $5 \otimes 2 = 2^5 = 32$.

Looking for closure with vector operations

One requirement of a vector space and its vector operations is that the space must be *closed* under both of the operations. The designation *closed* is used whenever any results of performing a particular operation on any elements in a set are always more elements in the original set. For example, the counting numbers 1, 2, 3, . . . are closed under addition because, when you add two counting numbers together, you always get another counting number. The counting numbers are *not* closed under subtraction, because sometimes you get negative numbers as results when you subtract (not counting numbers).

Consider the vector space containing the 2×2 matrices with a trace of 0. For the set of matrices to be *closed* under vector addition, every time you use the addition operation on two of the matrices, you must always get another matrix in the set. For example, consider the matrices A and B that I show you here, and the resulting matrix under vector addition. The resulting matrices all have a trace of 0.

$$A = \begin{bmatrix} 3 & -2 \\ 9 & -3 \end{bmatrix}, B = \begin{bmatrix} -6 & 11 \\ 8 & 6 \end{bmatrix}$$

$$A \oplus B = \begin{bmatrix} 3 & -2 \\ 9 & -3 \end{bmatrix} \oplus \begin{bmatrix} -6 & 11 \\ 8 & 6 \end{bmatrix} = \begin{bmatrix} 3-6 & -2+11 \\ 9+8 & -3+6 \end{bmatrix} = \begin{bmatrix} -3 & 9 \\ 17 & 3 \end{bmatrix}$$

Did I just pull out a set of matrices that appear to work under these circumstances? Can you be sure that the resulting trace will always be equal to 0, no matter what matrices I start with? Here I show you that the set of matrices is closed under vector addition by using generalized vectors:

$$A = \begin{bmatrix} a_{11} & a_{12} \\ a_{21} & a_{22} \end{bmatrix}, B = \begin{bmatrix} b_{11} & b_{12} \\ b_{21} & b_{22} \end{bmatrix}$$

$$\text{Traces: } a_{11} + a_{22} = 0 \text{ and } b_{11} + b_{22} = 0$$

$$A \oplus B = \begin{bmatrix} a_{11} + b_{11} & a_{12} + b_{12} \\ a_{21} + b_{21} & a_{22} + b_{22} \end{bmatrix}$$

$$\text{Trace: } a_{11} + b_{11} + a_{22} + b_{22} = (a_{11} + a_{22}) + (b_{11} + b_{22}) = 0 + 0 = 0$$

Any matrix resulting from this vector addition still has a trace equal to 0.

Because the vector addition in the previous example is so closely tied to matrix addition, you may want to see something a little different to be completely convinced about this closure stuff. Consider the vector space consisting of real numbers and with vector addition defined: $x \oplus y = xy$ (where x and y are two of those real numbers). For example, replace the x and the y with the numbers 2 and 5. Performing vector addition, you get a result of 10, which is another positive real number. Multiplying two positive real numbers together will always result in a positive real number (although you're probably still wondering why I call this operation *addition*).

I now continue with the positive real numbers and my definition of vector multiplication given in "Making vector multiplication meaningful." If you define the vector multiplication $k \otimes x = x^k$, the result of raising a positive real number to a power is always another positive real number. For example, $3 \otimes 4 = 4^3 = 64$, and $-3 \otimes 4 = 4^{-3} = 1/64$. The result will always be a positive real number (even if it's a fraction).

Ferreting out the failures to close

I keep showing you examples of vector spaces and their operations and how there always seems to be closure. Instead of having you believe that, no matter what the elements or what the operations, you always have closure, I'd better set the record straight right now.

For example, consider the set of 2 × 2 matrices of the form:

$$\begin{bmatrix} a & -b \\ b & 1 \end{bmatrix}$$

where a and b are real numbers. The elements b and $-b$ are opposites of one another, and the number 1 is always in the a_{22} position. I'll let the vector addition be just normal addition of matrices — where you add the elements in the corresponding positions. I show you a lack of closure with a simple example of using the vector addition on two matrices:

$$A = \begin{bmatrix} 3 & -2 \\ 2 & 1 \end{bmatrix}, B = \begin{bmatrix} -6 & 11 \\ -11 & 1 \end{bmatrix}$$

$$A \oplus B = \begin{bmatrix} 3 & -2 \\ 2 & 1 \end{bmatrix} \oplus \begin{bmatrix} -6 & 11 \\ -11 & 1 \end{bmatrix} = \begin{bmatrix} 3-6 & -2+11 \\ 2-11 & 1+1 \end{bmatrix} = \begin{bmatrix} -3 & 9 \\ -9 & 2 \end{bmatrix}$$

William Hamilton

Sir William Rowan Hamilton was an Irish mathematician, physicist, and astronomer. He was born in Dublin, Ireland, in 1805, and died there in 1865. William was pretty much a child prodigy, able to read Greek, Hebrew, and Latin by the age of 5, and familiar with half a dozen Asian languages by the age of 10. He was introduced to formal mathematics at the age of 13 — the mathematics written in French, with which he was then fluent.

Hamilton was interested in poetry as a young man, but his poetical endeavors weren't particularly good. Perhaps his tendency to write letters that were 50 to 100 pages long (covering every minute detail of a subject) were his outlet for lack of success in poetry.

Hamilton was appointed Royal Astronomer of Ireland at the age of 22 and was knighted for his mathematical and other scientific contributions at the age of 30. Hamilton is well known for his discovery of *quaternions,* an extension of complex numbers. This discovery led the way to a freedom that mathematicians could then enjoy — being able to describe new algebras that weren't held to the formal rules. The new algebras could be collections of specific objects held to operations and rules dictated by the application. Quaternions are used today in computer graphics, signal processing, and control theory.

In the final matrix, the elements in all but the second row, second column position are what they should be. All the elements are real numbers, with the opposites still in the a_{12} and a_{21} positions, but I have a 2 where there should be a 1. The resulting matrix is not in the original set. The set is not closed under $\oplus$.

An example of a lack of closure involving vector multiplication occurs when the set of elements consists of the vectors $(x,y,2)$. The vectors with this format have real numbers, x and y for the first two coordinates, and the number 2 for the third coordinate. Letting the vector multiplication be defined $k \otimes (x,y,2) = (kx,ky,2k)$, the resulting vectors would not be in the original set, because the last coordinate is not a 2 unless the multiplier, k, is equal to 1.

Singling Out the Specifics of Vector Space Properties

The various properties and requirements of the elements and operations involved in a vector space include closure, commutativity, associativity, opposites, and identities. I cover closure in the earlier section, "Looking for closure with vector operations." In this section, I address the other properties by explaining the properties more fully and showing you examples of success and failure (to meet the properties).

Changing the order with commutativity of vector addition

The vector addition operation, $\oplus$, is performed on two vectors — resulting in another vector in the vector space. One property of the operation of vector addition that is very nice and very helpful is *commutativity.* When a vector operation provides commutativity, you don't have to worry about what order the vectors are in when you perform the operation on the two of them. You get the same result when you add vectors $\mathbf{u} \oplus \mathbf{v}$ as when you add $\mathbf{v} \oplus \mathbf{u}$. For example, consider the vector space containing 2×2 matrices of the form:

$$\begin{bmatrix} a & b \\ -b & 0 \end{bmatrix}$$

where a and b are real numbers.

When performing vector addition, the order doesn't matter.

$$\begin{bmatrix} 4 & 9 \\ -9 & 0 \end{bmatrix} \oplus \begin{bmatrix} -5 & -6 \\ 6 & 0 \end{bmatrix} = \begin{bmatrix} -5 & -6 \\ 6 & 0 \end{bmatrix} \oplus \begin{bmatrix} 4 & 9 \\ -9 & 0 \end{bmatrix}$$
$$\begin{bmatrix} 4+(-5) & 9+(-6) \\ -9+6 & 0+0 \end{bmatrix} = \begin{bmatrix} -5+4 & -6+9 \\ 6+(-9) & 0+0 \end{bmatrix}$$
$$\begin{bmatrix} -1 & 3 \\ -3 & 0 \end{bmatrix} = \begin{bmatrix} -1 & 3 \\ -3 & 0 \end{bmatrix}$$

The commutativity holds for this example, because the definition of vector addition is based on our addition of real numbers, which is also commutative.

Now let me show you an example of what you might think should illustrate commutativity of addition, but what doesn't pan out in the end. Let the set X consist of the six 2×2 matrices that represent reflections and rotations of points in the coordinate plane. This set is closed under $\oplus$, where adding means to perform the one transformation and then the second transformation on the results of the first. The matrices and their transformations are as follows:

$$X = \left\{ \begin{bmatrix} -1 & 0 \\ 0 & 1 \end{bmatrix}, \begin{bmatrix} 1 & 0 \\ 0 & -1 \end{bmatrix}, \begin{bmatrix} 1 & 0 \\ 0 & 1 \end{bmatrix}, \begin{bmatrix} 0 & -1 \\ 1 & 0 \end{bmatrix}, \begin{bmatrix} -1 & 0 \\ 0 & -1 \end{bmatrix}, \begin{bmatrix} 0 & 1 \\ -1 & 0 \end{bmatrix} \right\}$$

$$\begin{bmatrix} -1 & 0 \\ 0 & 1 \end{bmatrix}: \text{reflection about } y\text{-axis}$$

$$\begin{bmatrix} 1 & 0 \\ 0 & -1 \end{bmatrix}: \text{reflection about } x\text{-axis}$$

$$\begin{bmatrix} 1 & 0 \\ 0 & 1 \end{bmatrix}: \text{rotation of } 360°$$

$$\begin{bmatrix} 0 & -1 \\ 1 & 0 \end{bmatrix}: \text{rotation of } 90°$$

$$\begin{bmatrix} -1 & 0 \\ 0 & -1 \end{bmatrix}: \text{rotation of } 180°$$

$$\begin{bmatrix} 0 & 1 \\ -1 & 0 \end{bmatrix}: \text{rotation of } 270°$$

Refer to Chapter 8 for more on matrices representing reflections and rotations.

When you perform vector addition using these matrices, you find that the vector addition over the set X is *not* commutative. For example, if you take the point (2,3) and first reflect it over the y-axis and then rotate it 90 degrees, you don't get the same result as rotating it 90 degrees followed by reflecting it over the y-axis. First, let me write the vector addition as $T_1(\mathbf{v}) \oplus T_2(\mathbf{v})$, which, in this case, is $T_y(\mathbf{v}) \oplus T_{90}(\mathbf{v})$ for the first way and $T_{90}(\mathbf{v}) \oplus T_y(\mathbf{v})$ for the second way. The vector that's first input has coordinates x and y. The result of that first transformation gives you vector $\mathbf{v}'$, with coordinates x' and y' — which are input into the second transformation.

Using the matrix transformations as described previously,

$$T_y(\mathbf{v}) = \begin{bmatrix} -1 & 0 \\ 0 & 1 \end{bmatrix} \begin{bmatrix} x \\ y \end{bmatrix}, \; T_{90}(\mathbf{v}) = \begin{bmatrix} 0 & -1 \\ 1 & 0 \end{bmatrix} \begin{bmatrix} x \\ y \end{bmatrix}$$

and showing the two operations performed in opposite orders,

$$
\begin{aligned}
T_y(\mathbf{v}) \oplus T_{90}(\mathbf{v}') = T_{90}\left(T_y(\mathbf{v})\right) \\
= T_{90}\left(\begin{bmatrix} -1 & 0 \\ 0 & 1 \end{bmatrix} * \begin{bmatrix} 2 \\ 3 \end{bmatrix}\right) \\
= T_{90}\left(\begin{bmatrix} -2 \\ 3 \end{bmatrix}\right) \\
= \begin{bmatrix} 0 & -1 \\ 1 & 0 \end{bmatrix} * \begin{bmatrix} -2 \\ 3 \end{bmatrix} = \begin{bmatrix} -3 \\ -2 \end{bmatrix}
\end{aligned}
$$

$$
\begin{aligned}
T_{90}(\mathbf{v}) \oplus T_y(\mathbf{v}) = T_y\left(T_{90}(\mathbf{v})\right) \\
= T_y\left(\begin{bmatrix} 0 & -1 \\ 1 & 0 \end{bmatrix} * \begin{bmatrix} 2 \\ 3 \end{bmatrix}\right) \\
= T_y\left(\begin{bmatrix} -3 \\ 2 \end{bmatrix}\right) \\
= \begin{bmatrix} -1 & 0 \\ 0 & 1 \end{bmatrix} * \begin{bmatrix} -3 \\ 2 \end{bmatrix} = \begin{bmatrix} 3 \\ 2 \end{bmatrix}
\end{aligned}
$$

The results of the different ordering of the operations are completely different. The first ordering has you end up at the point (-3,-2) and the second ordering ends up at (3,2). Commutativity does *not* exist for $\oplus$ in the situation.

Regrouping with addition and scalar multiplication

The associative property refers to groupings rather than order. When you add three numbers together, such as 47 + 8 + 2, you get the same answer if you add (47 + 8) + 2 as you do when you add 47 + (8 + 2). The second version is easier, because you add 47 to 10 and don't have to stumble with any *carrying* in addition. A vector space has associativity under addition, and you also find some associativity at work when multiplying by more than one scalar.

First, let me show you how the vector space consisting of 2 × 2 vectors of the following form illustrates the property of associativity over vector addition:

$$
\mathbf{v} = \begin{bmatrix} a & b \\ -b & c \end{bmatrix}
$$

I add three general vectors of the same form together and observe that the results are the same whether you perform vector addition on the first two followed by the third or add the first to the result of vector addition on the second two. The vector addition, $\oplus$, is just normal matrix addition.

$$\left(\begin{bmatrix} a & b \\ -b & c \end{bmatrix} \oplus \begin{bmatrix} d & e \\ -e & f \end{bmatrix}\right) \oplus \begin{bmatrix} g & h \\ -h & i \end{bmatrix} = \begin{bmatrix} a & b \\ -b & c \end{bmatrix} \oplus \left(\begin{bmatrix} d & e \\ -e & f \end{bmatrix} \oplus \begin{bmatrix} g & h \\ -h & i \end{bmatrix}\right)$$

$$\left(\begin{bmatrix} a+d & b+e \\ -(b+e) & c+f \end{bmatrix}\right) \oplus \begin{bmatrix} g & h \\ -h & i \end{bmatrix} = \begin{bmatrix} a & b \\ -b & c \end{bmatrix} \oplus \left(\begin{bmatrix} d+g & e+h \\ -(e+h) & f+i \end{bmatrix}\right)$$

$$\begin{bmatrix} a+d+g & b+e+h \\ -(b+e+h) & c+f+i \end{bmatrix} = \begin{bmatrix} a+d+g & b+e+h \\ -(b+e+h) & c+f+i \end{bmatrix}$$

Now look at what happens when you use the associative property with two scalar multiplications over a vector. I have to be a little free with the process of multiplying two scalars together — I assume regular multiplication — but you should see the structure here:

$$k \otimes (l \otimes \mathbf{u}) = (kl) \otimes \mathbf{u}$$

You see that using vector multiplication of the scalar k over the result of multiplying l times the vector **u** is the same as first multiplying the two scalars together and then using vector multiplication of the result over the vector.

For example, using a 2×2 vector and normal scalar multiplication of a matrix, I show the results of the two different methods. I choose to use some specific numbers, although any will work.

$$2 \otimes \left(-5 \otimes \begin{bmatrix} 3 & 7 \\ -7 & 4 \end{bmatrix}\right) = 2 \otimes \begin{bmatrix} -15 & -35 \\ 35 & -20 \end{bmatrix} = \begin{bmatrix} -30 & -70 \\ 70 & -40 \end{bmatrix}$$

$$\left(2(-5)\right) \otimes \begin{bmatrix} 3 & 7 \\ -7 & 4 \end{bmatrix} = -10 \otimes \begin{bmatrix} 3 & 7 \\ -7 & 4 \end{bmatrix} = \begin{bmatrix} -30 & -70 \\ 70 & -40 \end{bmatrix}$$

The results are the same. You choose one method over another when one method is more convenient — or when the computation is easier to do.

Distributing the wealth of scalars over vectors

The distributive property of algebra reads $a(b + c) = ab + ac$. The property says that you get the same result when multiplying a number times the sum of two values as you do when multiplying the number times each of the two values and then finding the sum of the products. You see two properties of vector spaces that appear to act like the distributive property.

First, $k \otimes (\mathbf{u} \oplus \mathbf{v}) = k \otimes \mathbf{u} \oplus k \otimes \otimes \mathbf{v}$. This statement says that the scalar k distributes or multiplies over the vector operation $\oplus$. You can either add the two vectors together and then perform scalar multiplication on the result, or perform the scalar multiplication first. Let me show you an example of where adding the two vectors together first makes more sense. I show you two 1×3 vectors from a vector space and the execution of the property.

$$\begin{aligned}&13\otimes\left(\begin{bmatrix}14 & 87 & -23\end{bmatrix}\oplus\begin{bmatrix}-14 & -86 & 22\end{bmatrix}\right)\\&\quad=13\otimes\left(\begin{bmatrix}0 & 1 & -1\end{bmatrix}\right)=\begin{bmatrix}0 & 13 & -13\end{bmatrix}\end{aligned}$$

Compare the preceding computations with those that are necessary if you don't perform the addition first.

$$\begin{aligned}&13\otimes\left(\begin{bmatrix}14 & 87 & -23\end{bmatrix}\oplus\begin{bmatrix}-14 & -86 & 22\end{bmatrix}\right)\\&\quad=\left(13\otimes\begin{bmatrix}14 & 87 & -23\end{bmatrix}\right)\oplus\left(13\otimes\begin{bmatrix}-14 & -86 & 22\end{bmatrix}\right)\\&\quad=\begin{bmatrix}182 & 1{,}131 & -299\end{bmatrix}\oplus\begin{bmatrix}-182 & -1{,}118 & 286\end{bmatrix}\\&\quad=\begin{bmatrix}0 & 13 & -13\end{bmatrix}\end{aligned}$$

You want to take advantage of the nicer computations. Because the vectors and operations are a part of a vector space, you can take the more sensible route.

Now look at the other form of distributive property. In this property, you see that multiplying a vector by the sum of two scalars is equal to distributing each of the scalars over the vector by multiplying and then adding the products together: $(k + l) \otimes \mathbf{u} = k \otimes \mathbf{u} \oplus l \otimes \mathbf{u}$. Compare the two equal results with the computations that are required.

$$\begin{aligned}(-8+9)\otimes\begin{bmatrix}14 & 87 & -23\end{bmatrix}&=-8\begin{bmatrix}14 & 87 & -23\end{bmatrix}\oplus9\begin{bmatrix}14 & 87 & -23\end{bmatrix}\\1\otimes\begin{bmatrix}14 & 87 & -23\end{bmatrix}&=\begin{bmatrix}-112 & 696 & 184\end{bmatrix}\oplus\begin{bmatrix}126 & 783 & -207\end{bmatrix}\\\begin{bmatrix}14 & 87 & -23\end{bmatrix}&=\begin{bmatrix}14 & 87 & -23\end{bmatrix}\end{aligned}$$

Zeroing in on the idea of a zero vector

A vector space must have an identity element or a zero. The additive identity of the real numbers is 0. And the way a zero behaves is that it doesn't *change the identity* of any value it's added to or that's added to it. The same property of zero exists with vector spaces and vector addition. In a vector space, you have a special vector, designated **0**, where $\mathbf{0} \oplus \mathbf{v} = \mathbf{v}$, and $\mathbf{v} \oplus \mathbf{0} = \mathbf{v}$.

For example, the vector space containing 2 × 2 matrices of the following form has a zero vector whose elements are all zeros.

$$\mathbf{v} = \begin{bmatrix} a & b \\ -b & 0 \end{bmatrix}$$

$$\begin{bmatrix} a & b \\ -b & 0 \end{bmatrix} \oplus \begin{bmatrix} 0 & 0 \\ 0 & 0 \end{bmatrix} = \begin{bmatrix} 0 & 0 \\ 0 & 0 \end{bmatrix} \oplus \begin{bmatrix} a & b \\ -b & 0 \end{bmatrix} = \begin{bmatrix} a & b \\ -b & 0 \end{bmatrix}$$

The zero vector also plays a role with inverses of vectors, as you see in the next section.

Adding in the inverse of addition

In a vector space, every element has an additive inverse. What this means is that for every vector, **u,** you also have the inverse of **u,** which I designate **–u.** Combining **u** and its inverse using vector addition, you get the zero vector. So $\mathbf{u} \oplus -\mathbf{u} = -\mathbf{u} \oplus \mathbf{u} = \mathbf{0}$.

For example, the set of 2 × 1 vectors, where the second element is -3 times the first element, forms a vector space under the usual definitions of matrix addition and scalar multiplication. So the general vector and its inverse are

$$\mathbf{u} = \begin{bmatrix} a \\ -3a \end{bmatrix}, \quad -\mathbf{u} = \begin{bmatrix} -a \\ 3a \end{bmatrix}$$

Adding the two vectors together gives you the zero vector. What's the inverse of the zero vector? Why, the zero vector is its own inverse! How cool!

Delighting in some final details

Vector spaces have specific requirements and properties. I offer three more statements or properties that arise from the original properties — or are just clarifications of what goes on:

- $0 \otimes \mathbf{u} = \mathbf{0}$: Whenever you multiply a vector in a vector space by the scalar 0, you end up with the zero vector of that vector space.
- $k \otimes \mathbf{0} = \mathbf{0}$: No matter what scalar you multiply the zero vector by, you still have the zero vector — the scalar can't change that.
- $-\mathbf{u} = (-1) \otimes \mathbf{u}$: The opposite of some vector **u**, is written as $-\mathbf{u}$, which is equivalent to multiplying the original vector, **u**, by the scalar -1.

Chapter 14

Seeking Out Subspaces of a Vector Space

In This Chapter

- Determining whether a subset is a subspace of a vector space
- Spanning the horizon with spanning sets
- Getting in line with column spaces and null spaces

You go into a local sandwich shop and order a sub. What goes into your sub? Why, anything you want — or, at least, anything that's available in the shop. One person's sub is usually different from another person's sub, but they all contain ingredients that are available on the menu. Now that I have you drooling, let me connect this culinary meandering to vector spaces and subspaces.

The relationship between a *subspace* and its vector space is somewhat like that sandwich made up of some of the available ingredients. Or, if you prefer my leaving the world of food and moving to a mathematical analogy, think of the subset of a set as it's related to its *superset.* Mathematically speaking, for a set to be a subset of another set (the superset), the subset has to be made up of only elements found in that superset. For example, the set $M = \{1, 2, 3, a, b, c\}$ is a subset of $N = \{1, 2, 3, 4, a, b, c, d, e, \$, \#\}$, because all of the elements in M are also in N. The set N has many other possible subsets that can be made of anywhere from 0 to 11 elements. And now, to get to the subject at hand, I tackle vector spaces and subspaces.

In order for a particular set of objects to be termed a *subspace* of a vector space, that set of objects has to meet some rigid qualifications or standards. The qualifications have to do with being a particular type of subset and with using some particular operations. I give you all the details on the terms of qualifying for *subspacehood* in this chapter.

Investigating Properties Associated with Subspaces

For a set W to be a subspace of vector space V, then W must be a non-empty subset of V. Having only non-empty sets is the first difference between the set-subset relationship and the vector-subspace relationship. The other qualification is that subset W must itself be a vector space with respect to the same operations as those in V.

The set W is a subspace of vector space V if all the following apply:

- W is a non-empty subset of V.
- W is a vector space.
- The operations of vector addition and scalar multiplication as defined in W are also those same operations found in V.

It's actually sufficient, when determining if W is a subspace of V, to just show that W is a subset of V and that W is *closed* with respect to the two operations of $\oplus$ and $\otimes$ that apply to V. In other words, the sum $\mathbf{v}_1 \oplus \mathbf{v}_2$ belongs to the vector space V, if both $\mathbf{v}_1$ and $\mathbf{v}_2$ are in V, and $c \otimes \mathbf{v}_1$ is also in V. If you need a refresher on the operations associated with a vector space, then refer to Chapter 13, where you find all the information you need on the operations.

In general, vector spaces have an infinite number of vectors. The only real vector space with a finite (countable) number of vectors is the vector space containing only the zero vector.

Determining whether you have a subset

If a set W is even to be considered as a subspace of a vector space V, then you first have to determine if W even qualifies as a subset of V. For example, here's a set of 2 × 1 vectors, A, where the pairs, x and y, represent all the points in the coordinate plane and x and y are real numbers.

$$A = \left\{ \begin{bmatrix} x \\ y \end{bmatrix} \right\}, B = \left\{ \begin{bmatrix} x \\ 0 \end{bmatrix} \right\}$$

Set B is a subset of A, because B contains all the points in the coordinate plane that have a y-coordinate of 0. Another way of describing the points represented in B is to say that they're all the points that lie on the x-axis.

Building a subset with a rule

It's easy to see, with sets A and B, that B is a subset of A. When you have just one vector and one format to look at, you easily spot the relationship between all the vectors in the two sets.

Now think about a new set; I'll name the set C. I don't show you all the elements, in C, but I tell you that C is *spanned* by the vectors in the set D, shown here:

$$D = \left\{ \begin{bmatrix} 1 \\ 0 \\ 0 \end{bmatrix}, \begin{bmatrix} 1 \\ 0 \\ 1 \end{bmatrix} \right\}$$

Given a set of vectors $\{\mathbf{v}_1, \mathbf{v}_2, \ldots, \mathbf{v}_k\}$, the set of all *linear combinations* of this set is called its *span*. (You find lots of information on the span of a set in Chapter 5.)

What do the vectors in set C look like? They're also 3×1 vectors, and they're all linear combinations of the vectors in set D. For example, two vectors from set C — $\mathbf{c}_1$ and $\mathbf{c}_2$ — are shown here (with linear combinations that produce those vectors):

$$\mathbf{c}_1 = \begin{bmatrix} 1 \\ 0 \\ -2 \end{bmatrix} = 3\begin{bmatrix} 1 \\ 0 \\ 0 \end{bmatrix} - 2\begin{bmatrix} 1 \\ 0 \\ 1 \end{bmatrix} \qquad \mathbf{c}_2 = \begin{bmatrix} 10 \\ 0 \\ 4 \end{bmatrix} = 6\begin{bmatrix} 1 \\ 0 \\ 0 \end{bmatrix} + 4\begin{bmatrix} 1 \\ 0 \\ 1 \end{bmatrix}$$

Putting the two vectors $\mathbf{c}_1$ and $\mathbf{c}_2$ in a set E, I produce a subset of set C; E contains two of the vectors that are in set C.

$$E = \left\{ \begin{bmatrix} 1 \\ 0 \\ -2 \end{bmatrix}, \begin{bmatrix} 10 \\ 0 \\ 4 \end{bmatrix} \right\}$$

In general, if C is the set of vectors created by the span of set D, then you create any vector in C by writing a linear combination using a_1 and a_2 as multipliers. The elements of the resulting vector in C are c_1, c_2, and c_3.

$$a_1\begin{bmatrix} 1 \\ 0 \\ 0 \end{bmatrix} + a_2\begin{bmatrix} 1 \\ 0 \\ 1 \end{bmatrix} = \begin{bmatrix} c_1 \\ c_2 \\ c_3 \end{bmatrix} = \begin{bmatrix} a_1 + a_2 \\ 0 \\ a_2 \end{bmatrix}$$

The vectors in C do have a particular format. The second element, c_2, is always equal to 0. Also, since $a_1 + a_2 = c_1$ and $a_2 = c_3$, you always have a relationship between the first and third elements and the first multiplier: $a_1 = c_1 - c_3$.

As it turns out, even though set E is a subset of set C, set E isn't a subspace of C. I show you why in the "Getting spaced out with a subset being a vector space" section, later in this chapter.

Determining if a set is a subset

In the previous example, I demonstrate one way to create a subset — constructing vectors that belong in the subset. But what if you're given a set of vectors and you need to see if they all belong to a particular subset?

Here I show you two sets. The vectors in set A span a vector space, and set B is up for consideration: Is set B a subset of the span of set A?

$$A = \left\{ \begin{bmatrix} 1 \\ -1 \\ -2 \end{bmatrix}, \begin{bmatrix} 2 \\ -1 \\ 3 \end{bmatrix} \right\}, B = \left\{ \begin{bmatrix} 2 \\ -3 \\ -9 \end{bmatrix}, \begin{bmatrix} 1 \\ 0 \\ -2 \end{bmatrix} \right\}$$

First, I write an expression for the linear combinations of the vectors in set A (which constitutes the span of A).

$$a_1 \begin{bmatrix} 1 \\ -1 \\ -2 \end{bmatrix} + a_2 \begin{bmatrix} 2 \\ -1 \\ 3 \end{bmatrix} = \begin{bmatrix} a_1 + 2a_2 \\ -a_1 - a_2 \\ -2a_1 + 3a_2 \end{bmatrix}$$

Now I determine if the two vectors in B are the result of some linear combination of the vectors in A. Taking each vector separately:

$$\begin{bmatrix} a_1 + 2a_2 \\ -a_1 - a_2 \\ -2a_1 + 3a_2 \end{bmatrix} = \begin{bmatrix} 2 \\ -3 \\ -11 \end{bmatrix} \quad \begin{bmatrix} a_1 + 2a_2 \\ -a_1 - a_2 \\ -2a_1 + 3a_2 \end{bmatrix} = \begin{bmatrix} 1 \\ 0 \\ -2 \end{bmatrix}$$

$$\begin{cases} a_1 + 2a_2 = 2 \\ -a_1 - a_2 = -3 \\ -2a_1 + 3a_2 = -11 \end{cases} \quad \begin{cases} a_1 + 2a_2 = 1 \\ -a_1 - a_2 = 0 \\ -2a_1 + 3a_2 = -2 \end{cases}$$

The system of equations on the left, corresponding to the first vector in B, has a solution when $a_1 = 4$ and $a_2 = -1$. (Don't remember how to solve systems of equations? Refer to Chapter 4.) So far, so good. But the system of equations on the right, corresponding to the second vector in B, doesn't have a solution. There are no values of a_1 and a_2 that give you all three elements of the vector. So set B can't be a subset of the span of set A.

Getting spaced out with a subset being a vector space

Earlier, in the "Building a subset with a rule" section, I show a subset of a vector space and then declare that even though you have a subset, you don't necessarily have a subset that's a vector space. A subspace of a vector space must not only be a non-empty subset, but it also must be a vector space itself.

Set E consists of two vectors and is a subset of the set C that I describe in the earlier section.

$$E = \left\{ \begin{bmatrix} 1 \\ 0 \\ -2 \end{bmatrix}, \begin{bmatrix} 10 \\ 0 \\ 4 \end{bmatrix} \right\}$$

But set E is not a vector space. For one thing, E is not closed under vector addition. (In Chapter 13, I explain what it means to have closure under the vector operations.) If you use vector addition on the two vectors in set E, you get the following result:

$$\begin{bmatrix} 1 \\ 0 \\ -2 \end{bmatrix} + \begin{bmatrix} 10 \\ 0 \\ 4 \end{bmatrix} = \begin{bmatrix} 11 \\ 0 \\ 2 \end{bmatrix}$$

The vector resulting from the vector addition isn't in the set E, so the set isn't closed under that operation. Also, E cannot be a vector space because it has no zero vector. And, as if this weren't enough to convince you, look at the number of vectors in E: You see only two vectors. The only vector space with fewer than an infinite number of vectors is the vector space containing only the zero vector.

Now I show you an example of a subset of a vector space that *is* a vector space itself — and, thus, a subspace of the vector space. First, consider the vector space, G, consisting of all possible 2×4 matrices. G is a vector space under the usual matrix addition and scalar multiplication.

$$G = \begin{bmatrix} g_{11} & g_{12} & g_{13} & g_{14} \\ g_{21} & g_{22} & g_{23} & g_{24} \end{bmatrix}$$

Now I introduce you to set H, a subset of G:

$$H = \begin{bmatrix} h_{11} & h_{12} & h_{13} & h_{14} \\ 0 & h_{22} & 0 & 0 \end{bmatrix} \text{ where } h_{14} = h_{11} + h_{12} + h_{13}$$

I need to show you that H is a subspace of G. To show that H is a subspace, all I have to do is show you that, as well as being a subset of G, H is also closed under matrix addition and scalar multiplication. I first demonstrate the closure using an example, and then I show you the generalized form.

Choosing two random vectors in H, I add them together. Then I multiply the first vector by the scalar 3.

$$\begin{bmatrix} 2 & 3 & 4 & 9 \\ 0 & -5 & 0 & 0 \end{bmatrix} \oplus \begin{bmatrix} -2 & 5 & -8 & -5 \\ 0 & 1 & 0 & 0 \end{bmatrix} = \begin{bmatrix} 0 & 8 & -4 & 4 \\ 0 & -4 & 0 & 0 \end{bmatrix}$$

$$3 \otimes \begin{bmatrix} 2 & 3 & 4 & 9 \\ 0 & -5 & 0 & 0 \end{bmatrix} = \begin{bmatrix} 6 & 9 & 12 & 27 \\ 0 & -15 & 0 & 0 \end{bmatrix}$$

With this simple example, it appears that the vectors in the set are closed under the vector operations. But, to be sure, I show you the vector addition in general.

$$\begin{bmatrix} h_{11} & h_{12} & h_{13} & h_{14} \\ 0 & h_{22} & 0 & 0 \end{bmatrix} \oplus \begin{bmatrix} k_{11} & k_{12} & k_{13} & k_{14} \\ 0 & k_{22} & 0 & 0 \end{bmatrix}$$
$$= \begin{bmatrix} h_{11} + k_{11} & h_{12} + k_{12} & h_{13} + k_{13} & h_{14} + k_{14} \\ 0 & h_{22} + k_{22} & 0 & 0 \end{bmatrix}$$

Because the two matrices are in set H, then $h_{11} + h_{12} + h_{13} = h_{14}$ and $k_{11} + k_{12} + k_{13} = k_{14}$. The vector sum of the two matrices has, as its element in the first row, fourth column, $h_{14} + k_{14}$. Substituting in the equivalent values, you get

$$\begin{aligned} h_{14} + k_{14} &= h_{11} + h_{12} + h_{13} + k_{11} + k_{12} + k_{13} \\ &= (h_{11} + k_{11}) + (h_{12} + k_{12}) + (h_{13} + k_{13}) \end{aligned}$$

The subset H is closed under vector addition. Likewise, with scalar multiplication,

$$c \otimes \begin{bmatrix} h_{11} & h_{12} & h_{13} & h_{14} \\ 0 & h_{22} & 0 & 0 \end{bmatrix} = \begin{bmatrix} ch_{11} & ch_{12} & ch_{13} & ch_{14} \\ 0 & ch_{22} & 0 & 0 \end{bmatrix}$$

Checking the relationship between the elements in the first row, you find that $ch_{11} + ch_{12} + ch_{13} = c(h_{11} + h_{12} + h_{13}) = ch_{14}$. So the subset is also closed under the multiplication operation. Therefore, H is a subspace of G.

Finding a Spanning Set for a Vector Space

Vector spaces have an infinite number of vectors (except for the vector space that contains only the zero vector). You determine if a particular set spans a vector space or a subspace of a vector space by determining if every vector in the vector space can be obtained from a linear combination of the vectors in the *spanning set.* A vector space usually has more than one spanning set. What I show you in this section is how to find one or more of those spanning sets.

Checking out a candidate for spanning

The simplest spanning set of a vector space with elements that are matrices is the *natural basis.* In Chapter 7, I talk about natural bases for sets of vectors; in this chapter, I expand that discussion to vector spaces in general.

The vector space consisting of all 2×2 matrices is spanned by the set of vectors T. T is also a *natural basis* for the vector space.

$$T = \left\{ \begin{bmatrix} 1 & 0 \\ 0 & 0 \end{bmatrix}, \begin{bmatrix} 0 & 1 \\ 0 & 0 \end{bmatrix}, \begin{bmatrix} 0 & 0 \\ 1 & 0 \end{bmatrix}, \begin{bmatrix} 0 & 0 \\ 0 & 1 \end{bmatrix} \right\}$$

But the set T isn't the only spanning set for all 2×2 matrices. Look at set U.

$$U = \left\{ \begin{bmatrix} 1 & 0 \\ 0 & 0 \end{bmatrix}, \begin{bmatrix} 1 & 1 \\ 0 & 0 \end{bmatrix}, \begin{bmatrix} 1 & 1 \\ 1 & 0 \end{bmatrix}, \begin{bmatrix} 1 & 0 \\ 1 & 1 \end{bmatrix} \right\}$$

Writing a general linear combination of the matrices in set U,

$$a_1\begin{bmatrix}1 & 0\\0 & 0\end{bmatrix}+a_2\begin{bmatrix}1 & 1\\0 & 0\end{bmatrix}+a_3\begin{bmatrix}1 & 1\\1 & 0\end{bmatrix}+a_4\begin{bmatrix}1 & 0\\1 & 1\end{bmatrix}=\begin{bmatrix}x_1 & x_2\\x_3 & x_4\end{bmatrix}$$

I now write the system of equations associated with the elements in the matrices after multiplying by the scalars.

$$\begin{cases}a_1+a_2+a_3+a_4=x_1\\ \quad a_2+a_3 \qquad =x_2\\ \qquad\quad a_3+a_4=x_3\\ \qquad\qquad\quad a_4=x_4\end{cases}$$

Solving for the values of the scalars (each a_i) in terms of the elements of the 2×2 matrix, I find that any 2×2 matrix can be formed. Once I've chosen the elements of the matrix I want, I can write a linear combination of the matrices in U to create that matrix. So set U is also a spanning set for the 2×2 matrices.

$$\begin{bmatrix}a_1\\a_2\\a_3\\a_4\end{bmatrix}=\begin{bmatrix}x_1-a_2-a_3-a_4\\x_2-a_3\\x_3-a_4\\x_4\end{bmatrix}=\begin{bmatrix}x_1-x_2-x_4\\x_2-x_3+x_4\\x_3-x_4\\x_4\end{bmatrix}$$

Putting polynomials into the spanning mix

Another type of vector space involves polynomials of varying degrees. The symbol P^2 represents all polynomials of degree 2 or less. And the standard spanning set of such a vector space is $\{x^2, x, 1\}$.

A polynomial is a function written in the form:

$$f(x) = c_n x^n + c_{n-1} x^{n-1} + \ldots + c_1 x^1 + c_0$$

where each c_i is a real number and each n is a whole number.

Another spanning set for P^2 is the set $R = \{x^2 + 1, x^2 + x, x + 1\}$. R isn't the only spanning set for P^2, but I show you that R works by writing the linear combinations of the elements of R and determining how they create a second-degree polynomial.

$$\begin{aligned}&a_1(x^2+1)+a_2(x^2+x)+a_3(x+1)\\&=(a_1+a_2)x^2+(a_2+a_3)x+(a_1+a_3)1\end{aligned}$$

If the coefficients and constant of the polynomial are c_1, c_2, and c_3, as shown, then those coefficients and their relationship with the scalar multiples are

$$(a_1+a_2)x^2+(a_2+a_3)x+(a_1+a_3)1=c_2x^2+c_1x^1+c_0$$

$$\begin{cases}a_1+a_2 \qquad = c_2\\ \qquad a_2+a_3 = c_1\\ a_1 \qquad +a_3 = c_0\end{cases}$$

Solving for the values of the scalars,

$$a_1=\frac{c_2-c_1+c_0}{2},\ a_2=\frac{c_2+c_1-c_0}{2},\ a_3=\frac{-c_2+c_1+c_0}{2}$$

For example, the linear combination necessary to write the second degree polynomial $5x^2+2x+1$ is found as follows:

$$\begin{aligned}a_1&=\frac{c_2-c_1+c_0}{2}=\frac{5-2+1}{2}=2\\a_2&=\frac{c_2+c_1-c_0}{2}=\frac{5+2-1}{2}=3\\a_3&=\frac{-c_2+c_1+c_0}{2}=\frac{-5+2+1}{2}=-1\end{aligned}$$

Substituting into the equation involving the scalar multiples,

$$\begin{aligned}&(a_1+a_2)x^2+(a_2+a_3)x+(a_1+a_3)\\&=(2+3)x^2+(3-1)x+(2-1)1\\&=5x^2+2x+1\end{aligned}$$

Skewing the results with a skew-symmetric matrix

A square matrix is called *skew-symmetric* if its transpose exchanges each element with that element's opposite (negative). You create the transpose of a square matrix by exchanging elements on the opposite side of the main diagonal. And the opposite of a matrix has each element changed to the opposite sign. (Turn to Chapter 3 if these procedures seem strange to you; I cover matrices in great detail there.) This *skew-symmetric* matrix takes a bit of each process, transposing and negating, to produce a new matrix.

The properties of a *skew-symmetric* matrix A are

- A is a square matrix.
- The main diagonal of A is all 0s.
- The transpose of A is equal to the opposite of A, $A^T = -A$.
- For every element in A, $a_{ji} = -a_{ij}$.

Here's an example of a 4 × 4 skew-symmetric matrix:

$$\begin{bmatrix} 0 & 1 & 2 & -3 \\ -1 & 0 & 4 & 6 \\ -2 & -4 & 0 & -8 \\ 3 & -6 & 8 & 0 \end{bmatrix}$$

The set of all 4 × 4 skew-symmetric matrices is a subspace of the vector space of all 4 × 4 matrices. If you're so inclined, you can easily show that the set exhibits closure under addition and scalar multiplication. To create a spanning set for the skew-symmetric matrices, first consider a generalized format for the skew-symmetric matrix:

$$\begin{bmatrix} 0 & a_{12} & a_{13} & a_{14} \\ -a_{12} & 0 & a_{23} & a_{24} \\ -a_{13} & -a_{23} & 0 & a_{34} \\ -a_{14} & -a_{24} & -a_{34} & 0 \end{bmatrix}$$

Here are the six scalar multiples and their respective matrices as linear combinations. The six matrices make up a spanning set for the 4 × 4 skew-symmetric matrices.

$$a_{12}\begin{bmatrix} 0 & 1 & 0 & 0 \\ -1 & 0 & 0 & 0 \\ 0 & 0 & 0 & 0 \\ 0 & 0 & 0 & 0 \end{bmatrix} + a_{13}\begin{bmatrix} 0 & 0 & 1 & 0 \\ 0 & 0 & 0 & 0 \\ -1 & 0 & 0 & 0 \\ 0 & 0 & 0 & 0 \end{bmatrix} + a_{14}\begin{bmatrix} 0 & 0 & 0 & 1 \\ 0 & 0 & 0 & 0 \\ 0 & 0 & 0 & 0 \\ -1 & 0 & 0 & 0 \end{bmatrix}$$
$$+ a_{23}\begin{bmatrix} 0 & 0 & 0 & 0 \\ 0 & 0 & 1 & 0 \\ 0 & -1 & 0 & 0 \\ 0 & 0 & 0 & 0 \end{bmatrix} + a_{24}\begin{bmatrix} 0 & 0 & 0 & 0 \\ 0 & 0 & 0 & 1 \\ 0 & 0 & 0 & 0 \\ 0 & -1 & 0 & 0 \end{bmatrix} + a_{34}\begin{bmatrix} 0 & 0 & 0 & 0 \\ 0 & 0 & 0 & 0 \\ 0 & 0 & 0 & 1 \\ 0 & 0 & -1 & 0 \end{bmatrix}$$

Defining and Using the Column Space

The *column space* of a matrix is the set of all linear combinations of the columns of the matrix. The column space of a matrix is often used to describe a subspace of R^m, when a matrix has dimension $m \times n$.

For example, consider the matrix A, which has dimension 3×4. The column space for matrix A consists of four 3×1 vectors. And the four vectors constitute a spanning set.

$$A = \begin{bmatrix} 1 & -1 & 2 & 0 \\ 3 & 2 & 0 & 1 \\ -1 & 0 & 1 & -2 \end{bmatrix}$$

$$Col\ A = Span\left\{\begin{bmatrix} 1 \\ 3 \\ -1 \end{bmatrix}, \begin{bmatrix} -1 \\ 2 \\ 0 \end{bmatrix}, \begin{bmatrix} 2 \\ 0 \\ 1 \end{bmatrix}, \begin{bmatrix} 0 \\ 1 \\ -2 \end{bmatrix}\right\}$$

The notation *Col* A means *column space of matrix* A, and *Span* { } means that the vectors are a spanning set. Another way to describe the column space of matrix A is to name the columns vector $\mathbf{a}_1, \mathbf{a}_2, \mathbf{a}_3, \ldots$ and write $Col\ A = Span\{\mathbf{a}_1, \mathbf{a}_2, \mathbf{a}_3, \ldots\}$.

Because *Col* A is the set of all the linear combinations of the column vectors $\mathbf{a}^i$, you can assign **x** to be some vector in R_n and describe the resulting vector, **b**, as a product of the column space and **x**.

$Col\ A = Span\{\mathbf{a}_1, \mathbf{a}_2, \mathbf{a}_3, \ldots\} = \{\mathbf{b}: \mathbf{b} = A\mathbf{x}\}$.

If a vector **b** is in *Col* A, then **b** is the result of a linear combination of the column vectors in A. For example, I look at matrix A, and want to determine if vector **b,** shown here, is in *Col* A.

$$A = \begin{bmatrix} 1 & -1 & 2 & 0 \\ 3 & 2 & 0 & 1 \\ -1 & 0 & 1 & -2 \end{bmatrix}, \mathbf{b} = \begin{bmatrix} 13 \\ 1 \\ 0 \end{bmatrix}$$

$$A\mathbf{x} = \begin{bmatrix} 1 & -1 & 2 & 0 \\ 3 & 2 & 0 & 1 \\ -1 & 0 & 1 & -2 \end{bmatrix}\begin{bmatrix} x_1 \\ x_2 \\ x_3 \\ x_4 \end{bmatrix} = \begin{bmatrix} 13 \\ 1 \\ 0 \end{bmatrix} = \mathbf{b}$$

In other words, I need to find a linear combination of the column vectors that results in **b.**

$$x_1\begin{bmatrix}1\\3\\-1\end{bmatrix}+x_2\begin{bmatrix}-1\\2\\0\end{bmatrix}+x_3\begin{bmatrix}2\\0\\1\end{bmatrix}+x_4\begin{bmatrix}0\\1\\-2\end{bmatrix}=\begin{bmatrix}13\\1\\0\end{bmatrix}$$

To solve for the elements (each x_i), I use an augmented matrix and do row reductions. (A little rusty on solving systems of equations using augmented matrices? Just refer to Chapter 4, and you'll get a full explanation.)

$$\left[\begin{array}{cccc|c}1&-1&2&0&13\\3&2&0&1&1\\-1&0&1&-2&0\end{array}\right]$$

$$\begin{array}{r}-3R_1+R_2\rightarrow R_2\\R_1+R_3\rightarrow R_3\end{array}\left[\begin{array}{cccc|c}1&-1&2&0&13\\0&5&-6&1&-38\\0&-1&3&-2&13\end{array}\right]$$

$$R_2 \quad R_3\left[\begin{array}{cccc|c}1&-1&2&0&13\\0&-1&3&-2&13\\0&5&-6&1&-38\end{array}\right]$$

$$-R_2\rightarrow R_2\left[\begin{array}{cccc|c}1&-1&2&0&13\\0&1&-3&2&-13\\0&5&-6&1&-38\end{array}\right]$$

$$\begin{array}{r}R_2+R_1\rightarrow R_1\\-5R_2+R_3\rightarrow R_3\end{array}\left[\begin{array}{cccc|c}1&0&-1&2&0\\0&1&-3&2&-13\\0&0&9&-9&27\end{array}\right]$$

$$\frac{1}{9}R_3\rightarrow R_3\left[\begin{array}{cccc|c}1&0&-1&2&0\\0&1&-3&2&-13\\0&0&1&-1&3\end{array}\right]$$

$$\begin{array}{r}R_3+R_1\rightarrow R_1\\3R_3+R_2\rightarrow R_2\end{array}\left[\begin{array}{cccc|c}1&0&0&1&3\\0&1&0&-1&-4\\0&0&1&-1&3\end{array}\right]$$

The row reductions leave the following relationships: $x_1 + x_4 = 3$, $x_2 - x_4 = -4$, and $x_3 - x_4 = 3$. Many different combinations of values can be used to create the vector **b** from the column vectors. For example, if I choose to let $x_4 = 1$, I get $x_1 = 2$, $x_2 = -3$, and $x_3 - 4$. Or I could let $x_4 = -2$ and get $x_1 = 5$, $x_2 = -6$, and $x_3 = 1$. Here are the two linear combinations — and you can find many, many more.

$$2\begin{bmatrix}1\\3\\-1\end{bmatrix}-3\begin{bmatrix}-1\\2\\0\end{bmatrix}+4\begin{bmatrix}2\\0\\1\end{bmatrix}+1\begin{bmatrix}0\\1\\-2\end{bmatrix}=\begin{bmatrix}13\\1\\0\end{bmatrix}$$

$$5\begin{bmatrix}1\\3\\-1\end{bmatrix}-6\begin{bmatrix}-1\\2\\0\end{bmatrix}+1\begin{bmatrix}2\\0\\1\end{bmatrix}-2\begin{bmatrix}0\\1\\-2\end{bmatrix}=\begin{bmatrix}13\\1\\0\end{bmatrix}$$

Earlier, I show you how to determine if a particular vector is in *Col* A by writing the general linear combination and solving for the necessary multipliers. In terms of the matrix A and vector **x**, I essentially solved $A\mathbf{x} = \mathbf{b}$. With so many possibilities for **x**, let me create a format or pattern for all the possible solutions when you're given a particular vector **b** to create.

The elements of a vector **b** are b_1, b_2, and b_3. I still need the same linear combinations of the vectors to create a desired vector.

$$x_1\begin{bmatrix}1\\3\\-1\end{bmatrix}+x_2\begin{bmatrix}-1\\2\\0\end{bmatrix}+x_3\begin{bmatrix}2\\0\\1\end{bmatrix}+x_4\begin{bmatrix}0\\1\\-2\end{bmatrix}=\begin{bmatrix}b_1\\b_2\\b_3\end{bmatrix}$$

Now, going through all the same row operations as I did for the previous example and using the general elements for vector **b**,

$$\left[\begin{array}{cccc:c} 1 & -1 & 2 & 0 & b_1 \\ 3 & 2 & 0 & 1 & b_2 \\ -1 & 0 & 1 & -2 & b_3 \end{array}\right]$$

$$\begin{array}{r} -3R_1 + R_2 \to R_2 \\ R_1 + R_3 \to R_3 \end{array} \left[\begin{array}{cccc:c} 1 & -1 & 2 & 0 & b_1 \\ 0 & 5 & -6 & 1 & b_2 - 3b_1 \\ 0 & -1 & 3 & -2 & b_1 + b_3 \end{array}\right]$$

$$R_2 \quad R_3 \left[\begin{array}{cccc:c} 1 & -1 & 2 & 0 & b_1 \\ 0 & -1 & 3 & -2 & b_1 + b_3 \\ 0 & 5 & -6 & 1 & b_2 - 3b_1 \end{array}\right]$$

$$-R_2 \to R_2 \left[\begin{array}{cccc:c} 1 & -1 & 2 & 0 & b_1 \\ 0 & 1 & -3 & 2 & -b_1 - b_3 \\ 0 & 5 & -6 & 1 & b_2 - 3b_1 \end{array}\right]$$

$$\begin{array}{r} R_2 + R_1 \to R_1 \\ -5R_2 + R_3 \to R_3 \end{array} \left[\begin{array}{cccc:c} 1 & 0 & -1 & 2 & -b_3 \\ 0 & 1 & -3 & 2 & -b_1 - b_3 \\ 0 & 0 & 9 & -9 & 2b_1 + b_2 + 5b_3 \end{array}\right]$$

$$\frac{1}{9}R_3 \to R_3 \left[\begin{array}{cccc:c} 1 & 0 & -1 & 2 & -b_3 \\ 0 & 1 & -3 & 2 & -b_1 - b_3 \\ 0 & 0 & 1 & -1 & \frac{2}{9}b_1 + \frac{1}{9}b_2 + \frac{5}{9}b_3 \end{array}\right]$$

$$\begin{array}{r} R_3 + R_1 \to R_1 \\ 3R_3 + R_2 \to R_2 \end{array} \left[\begin{array}{cccc:c} 1 & 0 & 0 & 1 & \frac{2}{9}b_1 + \frac{1}{9}b_2 - \frac{4}{9}b_3 \\ 0 & 1 & 0 & -1 & -\frac{1}{3}b_1 + \frac{1}{3}b_2 + \frac{2}{3}b_3 \\ 0 & 0 & 1 & -1 & \frac{2}{9}b_1 + \frac{1}{9}b_2 + \frac{5}{9}b_3 \end{array}\right]$$

Don't be put off by all the fractions. You'll still have combinations of numbers that result in integers for the end results. I'll choose wisely when showing you the rest of the story.

First, though, look at the results of the row reductions in terms of the relationships between the multipliers in the linear combinations.

$$x_1 + x_4 = \frac{2}{9}b_1 + \frac{1}{9}b_2 - \frac{4}{9}b_3$$

$$x_2 - x_4 = -\frac{1}{3}b_1 + \frac{1}{3}b_2 + \frac{2}{3}b_3$$

$$x_3 - x_4 = \frac{2}{9}b_1 + \frac{1}{9}b_2 + \frac{5}{9}b_3$$

The variable x_4 appears in each equation. Replace the x_4 with the constant k. After a vector has been chosen to be created, I can let k be any real number, substitute it into the equations, and produce the desired linear combinations.

$$\begin{aligned} x_1 &= -k + \frac{2}{9}b_1 + \frac{1}{9}b_2 - \frac{4}{9}b_3 \\ x_2 &= k - \frac{1}{3}b_1 + \frac{1}{3}b_2 + \frac{2}{3}b_3 \\ x_3 &= k + \frac{2}{9}b_1 + \frac{1}{9}b_2 + \frac{5}{9}b_3 \end{aligned}$$

For example, you need the linear combinations necessary to create the vector:

$$\mathbf{b} = \begin{bmatrix} 3 \\ -2 \\ 1 \end{bmatrix}$$

Substituting the elements of the vector into the equations:

$$\mathbf{b} = \begin{bmatrix} 3 \\ -2 \\ 1 \end{bmatrix} = \begin{bmatrix} b_1 \\ b_2 \\ b_3 \end{bmatrix}$$

$$\begin{aligned} x_1 &= -k + \frac{6}{9} - \frac{2}{9} - \frac{4}{9} = -k \\ x_2 &= k - 1 - \frac{2}{3} + \frac{2}{3} = k - 1 \\ x_3 &= k + \frac{2}{3} - \frac{2}{9} + \frac{5}{9} = k + 1 \\ x_4 &= k \end{aligned}$$

Now, letting k be the constant value 4 (I've just chosen a number at random), you get values for the multipliers in the linear combination.

$$\begin{aligned} x_1 &= -4 \\ x_2 &= 4 - 1 = 3 \\ x_3 &= 4 + 1 = 5 \\ x_4 &= 4 \end{aligned}$$

$$-4\begin{bmatrix} 1 \\ 3 \\ -1 \end{bmatrix} + 3\begin{bmatrix} -1 \\ 2 \\ 0 \end{bmatrix} + 5\begin{bmatrix} 2 \\ 0 \\ 1 \end{bmatrix} + 4\begin{bmatrix} 0 \\ 1 \\ -2 \end{bmatrix} = \begin{bmatrix} 3 \\ -2 \\ 1 \end{bmatrix}$$

You'll get the same result, no matter what you choose for k.

Connecting Null Space and Column Space

I talk about a column space in the preceding section. And now I get to introduce you to yet another designation for interaction with a matrix: *null space.*

The *null space* of a matrix A is the set of all the vectors, $\mathbf{x}_i$, for which A**x** = **0.** The notation for the null space of A is *Nul* A.

You may recognize the equation A**x** = **0** as being a *homogeneous* system. (If this doesn't sound familiar, refer back to Chapter 7.)

In the "Defining and Using the Column Space" section, earlier, I show you how to find linear combinations that produce a particular vector. What's special about this section is that vector **x** produces the zero vector when multiplied by A. To find a null space, you isolate all the vectors $\mathbf{x}_i$ so that A**x** produces the zero vector. I will be using linear combinations with the elements of **x.**

For example, consider the matrix A and vector **x.** I picked **x** because it's in the null space of matrix A.

$$A = \begin{bmatrix} 3 & 4 & 11 & -4 \\ 3 & 2 & 1 & 10 \\ -2 & -3 & -9 & 5 \end{bmatrix}, \mathbf{x} = \begin{bmatrix} -8 \\ 7 \\ 0 \\ 1 \end{bmatrix}$$

The vector **x** is in *Nul* A, because A**x** is the zero vector.

$$\begin{aligned} A\mathbf{x} &= -8\begin{bmatrix} 3 \\ 3 \\ -2 \end{bmatrix} + 7\begin{bmatrix} 4 \\ 2 \\ -3 \end{bmatrix} + 0\begin{bmatrix} 11 \\ 1 \\ -9 \end{bmatrix} + 1\begin{bmatrix} -4 \\ 10 \\ 5 \end{bmatrix} \\ &= \begin{bmatrix} -24+28+0-4 \\ -24+14+0+10 \\ 16-21+0+5 \end{bmatrix} = \begin{bmatrix} 0 \\ 0 \\ 0 \end{bmatrix} \end{aligned}$$

The null space is a set of vectors. The vectors usually have a particular format and are described in terms of the basis for the set. For example, here's matrix B and a general format for a vector **x,** which is in *Nul* B.

$$\mathbf{Bx} = \begin{bmatrix} 1 & 0 & -1 \\ -1 & 1 & 2 \\ 0 & 2 & 2 \end{bmatrix} \begin{bmatrix} x_1 \\ x_2 \\ x_3 \end{bmatrix} = \begin{bmatrix} 0 \\ 0 \\ 0 \end{bmatrix}$$

For this particular matrix B, the vectors in *Nul* B are all of the form:

$$\begin{bmatrix} k \\ -k \\ k \end{bmatrix} = k \begin{bmatrix} 1 \\ -1 \\ 1 \end{bmatrix}$$

So the basis for *Nul* B is the single 3×1 vector of 1s and a -1. For example, if you let k be the number 3, then you have:

$$\mathbf{Bx} = \begin{bmatrix} 1 & 0 & -1 \\ -1 & 1 & 2 \\ 0 & 2 & 2 \end{bmatrix} \begin{bmatrix} 3 \\ -3 \\ 3 \end{bmatrix} = \begin{bmatrix} 0 \\ 0 \\ 0 \end{bmatrix}$$

Chapter 15

Scoring Big with Vector Space Bases

In This Chapter

- Determining alternate bases for vector spaces
- Reducing spanning sets to bases
- Writing orthogonal and orthonormal bases for vector spaces
- Finding the scalars when changing bases

In Chapter 14, I discuss subspaces of vector spaces and spanning sets for vector spaces. In this chapter, you find topics from many earlier chapters all tied together with a common goal of investigating the bases of vector spaces. Of course, I refer you back to appropriate chapters, in case you need a little nudge, encouragment, or review.

So, what more can be said about the basis of a vector space? My answer: *plenty.* Vector spaces usually have more than one basis, but all the bases of a vector space have something in common: The bases of a particular vector space all have the same dimension. Also, a vector space can have orthogonal bases and orthonormal bases. And then you can switch bases (like a baseball player?) by adjusting the scalar multipliers.

It's important to be able to work with a particular vector space in terms of your choice of the kind of basis. The applications work better when you can use a simpler, but equivalent, set of vectors — the circumstances dictate the format that you want. In this chapter, I acquaint you with options for the basis of a vector space.

Going Geometric with Vector Spaces

Some lines and planes are subspaces of R^n. One reason that I discuss lines and planes at this time is that I can show you, with a sketch, what a basis looks like.

Not all lines and planes form vector spaces, though. If I limit the discussion to just those lines and planes that go through the origin, then I'm safe — I have vector spaces.

Lining up with lines

The *slope-intercept form* of the equation of a line is $y = mx + b$. The x and y represent the coordinates x and y as they're graphed on the coordinate plane. The m represents the slope of the line, and the b represents the y-intercept. (If you're not familiar with this equation, brush up on the basics of algebra with *Algebra For Dummies,* published by Wiley.)

The general vector equation of a line is written as follows:

$$v(x) = \begin{bmatrix} x \\ xm+b \end{bmatrix} = x\begin{bmatrix} 1 \\ m \end{bmatrix} + \begin{bmatrix} 0 \\ b \end{bmatrix}$$

where the 2×1 vectors represent the points (x,y) on the coordinate plane. In Figure 15-1, I show you the vector representations for two different lines: $y = 2x + 3$ and $y = -2x$.

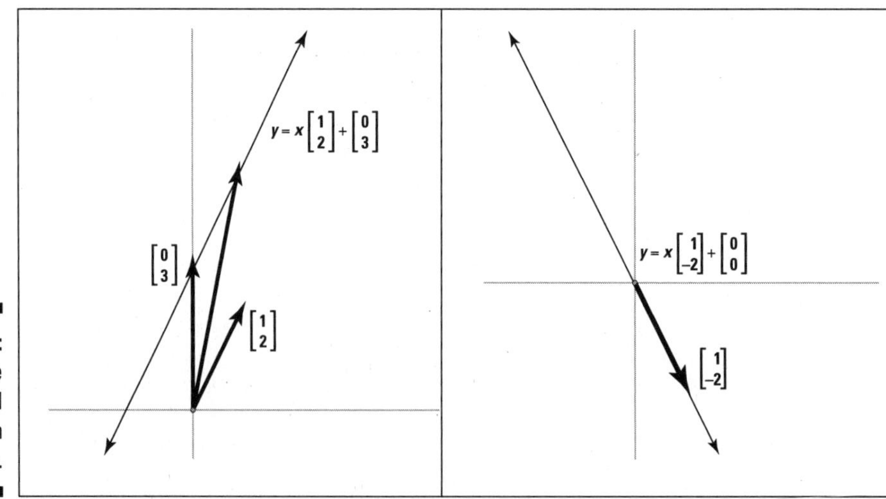

Figure 15-1: Lines are represented by the sum of vectors.

The two lines I show are of two different types. The first type, represented by the $y = 2x + 3$, are the lines that are *not* vector spaces. The second type, represented by $y = -2x$, are those lines that *are* vector spaces. The main difference between the two lines is the constant value of 3 that is added to the first equation. By adding the constant, you can no longer write a spanning set — and then a basis — for the points on the line.

So, lines on the coordinate plane that are vector spaces are spanned by all the linear combinations (multiples, in this case) of the real numbers, x, multiplying the following 2×1 vector:

$$\begin{bmatrix} 1 \\ m \end{bmatrix}$$

And this span is a basis. Actually, any scalar multiple of the vector can also form a basis.

Providing plain talk for planes

A plane is represented by a flat surface. In geometry, planes extend infinitely in all directions, determined by three *distinct* (the points are all different), *non-collinear points* (only two points at a time share the same line — they don't all three lie on the same line).

As with lines, only planes passing through the origin constitute vector spaces. The planes of a vector space are spanned by two nonparallel vectors. Here's the general vector equation of a plane:

$$v(a_1, a_2) = a_1\mathbf{v}_1 + a_2\mathbf{v}_2 = a_1\begin{bmatrix} x_1 \\ y_1 \\ z_1 \end{bmatrix} + a_2\begin{bmatrix} x_2 \\ y_2 \\ z_2 \end{bmatrix}$$

To show you an example of a plane with a vector equation and graph, I'll choose two vectors to create a plane — one vector ends behind the vertical *yz* plane, and the other vector ends in the first, front octant:

$$\mathbf{v}_1 = \begin{bmatrix} -1 \\ 2 \\ 1 \end{bmatrix}, \mathbf{v}_2 = \begin{bmatrix} 1 \\ 6 \\ 2 \end{bmatrix}$$

Figure 15-2 shows a plane described by those vectors $\mathbf{v}_1$ and $\mathbf{v}_2$. Try to imagine the plane extending in all directions from the flat surface defined by the vectors. When the vectors used are linearly independent, you have a plane. If the vectors are not linearly independent, you have a single line through the origin.

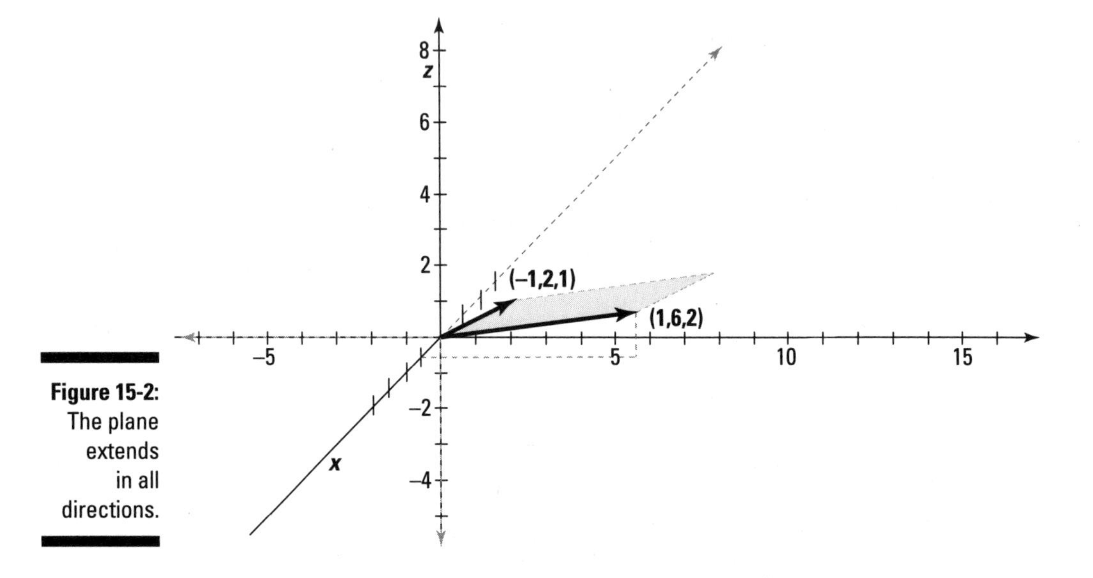

Figure 15-2: The plane extends in all directions.

Creating Bases from Spanning Sets

All bases of vector spaces are spanning sets, but not all spanning sets are bases. The one thing that differentiates spanning sets from bases is linear independence. (I talk about linear independence in greater detail in Chapters 7 and 14.) The vectors in a basis are linearly independent. You can have more than one basis for a particular vector space, but each basis will have the same number of linearly independent vectors in it. The number of vectors in the basis of a vector space is the *dimension* of the vector space. The basis of a vector space has no *redundancy* — you don't have more vectors than you need.

If you have a vector space with a basis containing n vectors, then any subset of the vector space with more than n vectors has some linearly dependent vectors in it.

In this section, I show you a spanning set with too many vectors — some are linear combinations of the others. I then show you how to delete the redundant vectors and create a span for the vector space.

Consider the spanning set, S, with five 4×1 vectors. $S = \{\mathbf{s}_1, \mathbf{s}_2, \mathbf{s}_3, \mathbf{s}_4, \mathbf{s}_5\}$. The span of S is a subspace of R^4.

$$S = \left\{ \begin{bmatrix} 1 \\ 1 \\ -2 \\ 1 \end{bmatrix}, \begin{bmatrix} 1 \\ 2 \\ 1 \\ 1 \end{bmatrix}, \begin{bmatrix} 2 \\ 1 \\ 4 \\ -1 \end{bmatrix}, \begin{bmatrix} -1 \\ 6 \\ 1 \\ 5 \end{bmatrix}, \begin{bmatrix} 0 \\ 3 \\ -2 \\ 3 \end{bmatrix} \right\}$$

The vectors of S are linearly dependent. Writing the augmented matrix for the relationship $a_1\mathbf{s}_1 + a_2\mathbf{s}_2 + a_3\mathbf{s}_3 + a_4\mathbf{s}_4 + a_5\mathbf{s}_5 = 0$ and reducing to echelon form (I don't show all the steps — just the beginning and end), you get

$$\left[\begin{array}{ccccc|c} 1 & 1 & 2 & -1 & 0 & 0 \\ 1 & 2 & 1 & 6 & 3 & 0 \\ -2 & 1 & 4 & 1 & -2 & 0 \\ 1 & 1 & -1 & 5 & 3 & 0 \end{array}\right] \rightarrow \left[\begin{array}{ccccc|c} 1 & 0 & 0 & -2 & 0 & 0 \\ 0 & 1 & 0 & 5 & -2 & 0 \\ 0 & 0 & 1 & -2 & -1 & 0 \\ 0 & 0 & 0 & 0 & 0 & 0 \end{array}\right]$$

From the echelon form, you get the following relationships between the variables: $a_1 = 2a_4$, $a_2 = -5a_4 + 2a_5$, and $a_3 = 2a_4 + a_5$. If the vectors were linearly independent, then the only solutions making the equations in the system equal to 0 would be for each a_i to be 0. As you see, with the relationships I found in the echelon form, you could make $a_3 = 0$ if $a_4 = 1$ and $a_5 = -2$. Other such combinations arise, too. In any case, the vectors are linearly dependent and don't form a basis.

Okay, you say, I've made my point. So how do we eliminate enough vectors to do away with linear dependency but keep enough to form a basis? Here's the plan:

1. **Write the vectors in a 4×5 matrix.**
2. **Change the matrix to its transpose.**
3. **Row-reduce the transposed matrix until it's triangular.**
4. **Transpose back to a 4×5 matrix.**
5. **Collect the nonzero vectors as the basis.**

If you need a refresher on transposing matrices and triangular matrices, refer to Chapter 3.

So, writing the original vectors from S in a single matrix and then writing the transpose, you get:

$$\begin{bmatrix} 1 & 1 & 2 & -1 & 0 \\ 1 & 2 & 1 & 6 & 3 \\ -2 & 1 & 4 & 1 & -2 \\ 1 & 1 & -1 & 5 & 3 \end{bmatrix}^T = \begin{bmatrix} 1 & 1 & -2 & 1 \\ 1 & 2 & 1 & 1 \\ 2 & 1 & 4 & -1 \\ -1 & 6 & 1 & 5 \\ 0 & 3 & -2 & 3 \end{bmatrix}$$

Now, you perform row reductions:

$$\begin{matrix} \\ -1R_1 + R_2 \rightarrow R_2 \\ -2R_1 + R_3 \rightarrow R_3 \\ R_1 + R_4 \rightarrow R_4 \\ \\ \end{matrix} \begin{bmatrix} 1 & 1 & -2 & 1 \\ 0 & 1 & 3 & 0 \\ 0 & -1 & 8 & -3 \\ 0 & 7 & -1 & 6 \\ 0 & 3 & -2 & 3 \end{bmatrix}$$

$$\begin{matrix} \\ R_2 + R_3 \rightarrow R_3 \\ -7R_2 + R_4 \rightarrow R_4 \\ -3R_2 + R_5 \rightarrow R_5 \\ \\ \end{matrix} \begin{bmatrix} 1 & 1 & -2 & 1 \\ 0 & 1 & 3 & 0 \\ 0 & 0 & 11 & -3 \\ 0 & 0 & -22 & 6 \\ 0 & 0 & -11 & 3 \end{bmatrix}$$

$$\begin{matrix} 2R_3 + R_4 \rightarrow R_4 \\ R_3 + R_5 \rightarrow R_5 \end{matrix} \begin{bmatrix} 1 & 1 & -2 & 1 \\ 0 & 1 & 3 & 0 \\ 0 & 0 & 11 & -3 \\ 0 & 0 & 0 & 0 \\ 0 & 0 & 0 & 0 \end{bmatrix}$$

Next, you transpose the last matrix:

$$\begin{bmatrix} 1 & 1 & -2 & 1 \\ 0 & 1 & 3 & 0 \\ 0 & 0 & 11 & -3 \\ 0 & 0 & 0 & 0 \\ 0 & 0 & 0 & 0 \end{bmatrix}^T \rightarrow \begin{bmatrix} 1 & 0 & 0 & 0 & 0 \\ 1 & 1 & 0 & 0 & 0 \\ -2 & 3 & 11 & 0 & 0 \\ 1 & 0 & -3 & 0 & 0 \end{bmatrix}$$

You collect the nonzero columns as vectors and put them in a set Z. The vectors shown in set Z are a basis for the subset S.

$$Z = \left\{ \begin{bmatrix} 1 \\ 1 \\ -2 \\ 1 \end{bmatrix}, \begin{bmatrix} 0 \\ 1 \\ 3 \\ 0 \end{bmatrix}, \begin{bmatrix} 0 \\ 0 \\ 11 \\ -3 \end{bmatrix} \right\}$$

The set Z isn't the only basis; there are many different bases possible.

Making the Right Moves with Orthogonal Bases

When you have 2 × 1 vectors that are *orthogonal,* another way of describing the relationship between the two vectors is to say that they're perpendicular. In Chapter 2, I show you how to check for *orthogonality* of vectors by finding their *inner product.* And, just to remind you about vectors being perpendicular or orthogonal, I offer the following:

Two vectors **u** and **v** are *orthogonal* if their inner product, $\mathbf{u}^T\mathbf{v}$, is equal to 0.

The two vectors **u** and **v** that I show here are orthogonal. Graphing vector **u** on the coordinate plane, it has its point in the second quadrant. At right angles to vector **u** is vector **v,** with its point in the third quadrant. (In Chapter 2, you see a sketch of two perpendicular vectors.) The inner product is, of course, 0.

$$\mathbf{u} = \begin{bmatrix} -3 \\ 7 \end{bmatrix}, \mathbf{v} = \begin{bmatrix} -14 \\ -6 \end{bmatrix}$$

$$\mathbf{u}^T\mathbf{v} = \begin{bmatrix} -3 & 7 \end{bmatrix} \begin{bmatrix} -14 \\ -6 \end{bmatrix} = 42 - 42 = 0$$

You aren't limited to 2 × 1 vectors or even 3 × 1 vectors when it comes to orthogonality. Even though the vectors can't be represented graphically, the vectors are still orthogonal if the inner product is 0. For example, consider the two vectors **w** and **z:**

$$\mathbf{w} = \begin{bmatrix} 1 \\ 0 \\ -3 \\ 2 \end{bmatrix}, \mathbf{z} = \begin{bmatrix} -7 \\ 8 \\ -5 \\ -4 \end{bmatrix}$$

$$\mathbf{w}^T\mathbf{z} = \begin{bmatrix} 1 & 0 & -3 & 2 \end{bmatrix} \begin{bmatrix} -7 \\ 8 \\ -5 \\ -4 \end{bmatrix} = -7 + 0 + 15 - 8 = 0$$

In fact, **w** and **z** have many other vectors that are orthogonal to either one or both of them.

An *orthogonal set* is a set of vectors in which each pair of vectors is orthogonal (every vector is orthogonal to every other vector in the set). For example, the set P is orthogonal. Each pair of vectors has an inner product of 0.

$$P = \left\{ \begin{bmatrix} 1 \\ 0 \\ 2 \end{bmatrix}, \begin{bmatrix} -2 \\ 3 \\ 1 \end{bmatrix}, \begin{bmatrix} -6 \\ -5 \\ 3 \end{bmatrix} \right\}$$

$$\begin{bmatrix} 1 & 0 & 2 \end{bmatrix} \begin{bmatrix} -2 \\ 3 \\ 1 \end{bmatrix} = -2 + 0 + 2 = 0$$

$$\begin{bmatrix} 1 & 0 & 2 \end{bmatrix} \begin{bmatrix} -6 \\ -5 \\ 3 \end{bmatrix} = -6 + 0 + 6 = 0$$

$$\begin{bmatrix} -2 & 3 & 1 \end{bmatrix} \begin{bmatrix} -6 \\ -5 \\ 3 \end{bmatrix} = 12 - 15 + 3 = 0$$

So what good are orthogonal sets of vectors? They're very helpful when working with various bases of vector spaces.

An *orthogonal set* is linearly independent, as long as the set doesn't contain the **0** vector. Furthermore, an orthogonal set containing n vectors is a basis for R^n.

Creating an orthogonal basis

A vector space can have more than one basis. And, an especially useful basis of a vector space is its *orthogonal basis.* To create an orthogonal basis from an existing basis, you use the *Gram-Schmidt orthogonalization process.*

If a subspace, W, has a basis consisting of the vectors $\{\mathbf{w}_1, \mathbf{w}_2, \mathbf{w}_3, \ldots\}$, then an orthogonal basis for W, $U = \{\mathbf{u}_1, \mathbf{u}_2, \mathbf{u}_3, \ldots\}$, is found with the following:

$$\mathbf{u}_1 = \mathbf{w}_1$$

$$\mathbf{u}_2 = \mathbf{w}_2 - \frac{\mathbf{u}_1^T\mathbf{w}_2}{\mathbf{u}_1^T\mathbf{u}_1}\mathbf{u}_1$$

$$\mathbf{u}_3 = \mathbf{w}_3 - \frac{\mathbf{u}_1^T\mathbf{w}_3}{\mathbf{u}_1^T\mathbf{u}_1}\mathbf{u}_1 - \frac{\mathbf{u}_2^T\mathbf{w}_3}{\mathbf{u}_2^T\mathbf{u}_2}\mathbf{u}_2$$

$$\vdots$$

$$\mathbf{u}_i = \mathbf{w}_i - \sum_{k-1}^{i-1}\frac{\mathbf{u}_k^T\mathbf{w}_i}{\mathbf{u}_k^T\mathbf{u}_k}\mathbf{u}_k,\ 2 \le i \le p$$

Using this formula in the process, the number of terms in the computation of a particular vector is one more than for the previous vector. The computations can get pretty hairy, but computers can come to the rescue if needed. You can use a spreadsheet and enter the formulas; it's easier to keep track of everything with a computer program than with a calculator. The variable p in the formula represents the *dimension* (number of vectors) in the subspace.

Here's how to create an orthogonal basis from the given basis of a set, starting with set W, a basis for R^3.

$$W = \left\{\begin{bmatrix}1\\-1\\0\end{bmatrix}, \begin{bmatrix}-2\\3\\1\end{bmatrix}, \begin{bmatrix}1\\2\\4\end{bmatrix}\right\}$$

According to the Gram-Schmidt process, the first vector in the orthogonal basis is the same as the first vector in the given basis. I now apply the process to find the other two vectors.

$$W = \begin{matrix} \mathbf{w}_1 & \mathbf{w}_2 & \mathbf{w}_3 \\ \begin{bmatrix} 1 \\ -1 \\ 0 \end{bmatrix}, & \begin{bmatrix} -2 \\ 3 \\ 1 \end{bmatrix}, & \begin{bmatrix} 1 \\ 2 \\ 4 \end{bmatrix} \end{matrix}$$

$$\mathbf{u}_1 = \mathbf{w}_1 = \begin{bmatrix} 1 \\ -1 \\ 0 \end{bmatrix}$$

$$\mathbf{u}_2 = \mathbf{w}_2 - \frac{\mathbf{u}_1^T\mathbf{w}_2}{\mathbf{u}_1^T\mathbf{u}_1}\mathbf{u}_1 = \begin{bmatrix} -2 \\ 3 \\ 1 \end{bmatrix} - \frac{-5}{2}\begin{bmatrix} 1 \\ -1 \\ 0 \end{bmatrix}$$

$$= \begin{bmatrix} -2+\frac{5}{2} \\ 3-\frac{5}{2} \\ 1+0 \end{bmatrix} = \begin{bmatrix} \frac{1}{2} \\ \frac{1}{2} \\ 1 \end{bmatrix}$$

$$\mathbf{u}_3 = \mathbf{w}_3 - \frac{\mathbf{u}_1^T\mathbf{w}_3}{\mathbf{u}_1^T\mathbf{u}_1}\mathbf{u}_1 - \frac{\mathbf{u}_2^T\mathbf{w}_3}{\mathbf{u}_2^T\mathbf{u}_2}\mathbf{u}_2$$

$$= \begin{bmatrix} 1 \\ 2 \\ 4 \end{bmatrix} - \frac{-1}{2}\begin{bmatrix} 1 \\ -1 \\ 0 \end{bmatrix} - \frac{11}{3}\begin{bmatrix} \frac{1}{2} \\ \frac{1}{2} \\ 1 \end{bmatrix} = \begin{bmatrix} 1+\frac{1}{2}-\frac{11}{6} \\ 2-\frac{1}{2}-\frac{11}{6} \\ 4+0-\frac{11}{3} \end{bmatrix} = \begin{bmatrix} -\frac{1}{3} \\ -\frac{1}{3} \\ \frac{1}{3} \end{bmatrix}$$

Whew! Now, to make future computations easier, multiply the second vector in U by 2 and the third vector by 3 to change all the elements to integers.

$$U = \left\{ \begin{bmatrix} 1 \\ -1 \\ 0 \end{bmatrix}, \begin{bmatrix} 1 \\ 1 \\ 2 \end{bmatrix}, \begin{bmatrix} -1 \\ -1 \\ 1 \end{bmatrix} \right\}$$

Using the orthogonal basis to write the linear combination

One advantage of the orthogonal basis is that you have an easier task when writing a linear combination of a particular vector from the vectors in a set. You choose your vector, you line up the vectors in your orthogonal basis, and then you plug values into a formula for finding the scalar multiples for each vector in the basis.

If the vectors $\{\mathbf{u}_1, \mathbf{u}_2, \mathbf{u}_3, \ldots\}$ are the orthogonal basis for some subspace U, then a vector $\mathbf{v}$ belonging to U is written as the linear combination:

$\mathbf{v} = a_1\mathbf{u}_1 + a_2\mathbf{u}_2 + a_3\mathbf{u}_3 + \ldots$ where

$$a_i = \frac{\mathbf{u}_i{}^T\mathbf{v}}{\mathbf{u}_i{}^T\mathbf{u}_i}$$

For example, using the orthogonal basis U, I determine the scalars needed to write the vector $\mathbf{v}$ as a linear combination of the vectors in U.

$$\text{U} = \left\{ \begin{bmatrix} 1 \\ -1 \\ 0 \end{bmatrix}, \begin{bmatrix} 1 \\ 1 \\ 2 \end{bmatrix}, \begin{bmatrix} -1 \\ -1 \\ 1 \end{bmatrix} \right\},\ \mathbf{v} = \begin{bmatrix} -6 \\ 12 \\ 5 \end{bmatrix}$$

$$a_1 = \frac{\mathbf{u}_1{}^T\mathbf{v}}{\mathbf{u}_1{}^T\mathbf{u}_1} = \frac{-18}{2} = -9,\ a_2 = \frac{\mathbf{u}_2{}^T\mathbf{v}}{\mathbf{u}_2{}^T\mathbf{u}_2} = \frac{16}{6} = \frac{8}{3}$$

$$a_3 = \frac{\mathbf{u}_3{}^T\mathbf{v}}{\mathbf{u}_3{}^T\mathbf{u}_3} = \frac{-1}{3} = -\frac{1}{3}$$

$$a_1\mathbf{u}_1 + a_2\mathbf{u}_2 + a_3\mathbf{u}_3 = -9\begin{bmatrix} 1 \\ -1 \\ 0 \end{bmatrix} + \frac{8}{3}\begin{bmatrix} 1 \\ 1 \\ 2 \end{bmatrix} - \frac{1}{3}\begin{bmatrix} -1 \\ -1 \\ 1 \end{bmatrix}$$

$$= \begin{bmatrix} -9 + \frac{8}{3} + \frac{1}{3} \\ 9 + \frac{8}{3} + \frac{1}{3} \\ 0 + \frac{16}{3} - \frac{1}{3} \end{bmatrix} = \begin{bmatrix} -6 \\ 12 \\ 5 \end{bmatrix}$$

Neato! Makes a great case for orthogonal bases!

Making orthogonal orthonormal

Just when you thought it couldn't get any better, I step in to add to your repertoire by showing you one more process or procedure to change the vectors of a basis. An *orthonormal* basis consists of vectors that all have a magnitude of 1. (I discuss magnitude in greater detail in Chapter 2.)

The magnitude of a vector, designated $\|\mathbf{v}\|$ is the square root of the sum of the squares of the elements of the vector.

For example, the magnitude of vector **w,** shown here, is 7.

$$\mathbf{w}=\begin{bmatrix}-2\\0\\6\\-3\end{bmatrix},\ \|\mathbf{w}\|=\sqrt{(-2)^2+0^2+6^2+(-3)^2}=\sqrt{49}=7$$

In "Creating an orthogonal basis," earlier in this chapter, I show you how to take any old basis and change it to an orthogonal basis. Now I go one step further and change an orthogonal basis to an orthonormal basis — all the vectors in the basis have magnitude 1.

The process needed to change to an orthonormal basis is really relatively simple. You just multiply each vector in the orthogonal basis by the reciprocal of the square root of its inner product. (Actually, the words describing the multiplier are worse than the actual number.)

To change each vector $\mathbf{v}_i$ to a vector with magnitude 1, multiply each $\mathbf{v}_i$ by

$$a_i=\frac{1}{\sqrt{\mathbf{v}_i^T\mathbf{v}_i}}$$

So, to change the orthogonal basis U to an orthonormal basis:

$$U=\left\{\overset{\mathbf{u}_1}{\begin{bmatrix}1\\-1\\0\end{bmatrix}},\overset{\mathbf{u}_2}{\begin{bmatrix}1\\1\\2\end{bmatrix}},\overset{\mathbf{u}_3}{\begin{bmatrix}-1\\-1\\1\end{bmatrix}}\right\}$$

$$a_1\mathbf{u}_1=\frac{1}{\sqrt{2}}\begin{bmatrix}1\\-1\\0\end{bmatrix}=\begin{bmatrix}\frac{1}{\sqrt{2}}\\-\frac{1}{\sqrt{2}}\\0\end{bmatrix},\ a_2\mathbf{u}_2=\frac{1}{\sqrt{3}}\begin{bmatrix}-1\\-1\\1\end{bmatrix}=\begin{bmatrix}-\frac{1}{\sqrt{3}}\\-\frac{1}{\sqrt{3}}\\\frac{1}{\sqrt{3}}\end{bmatrix}$$

$$a_3\mathbf{u}_3=\frac{1}{\sqrt{6}}\begin{bmatrix}1\\1\\2\end{bmatrix}=\begin{bmatrix}\frac{1}{\sqrt{6}}\\\frac{1}{\sqrt{6}}\\\frac{2}{\sqrt{6}}\end{bmatrix}$$

The vectors are orthogonal *and* the magnitude of each vector is 1.

Writing the Same Vector after Changing Bases

Two bases for R^3 are W and U:

$$W = \left\{ \begin{bmatrix} 1 \\ -1 \\ 0 \end{bmatrix}, \begin{bmatrix} -2 \\ 3 \\ 1 \end{bmatrix}, \begin{bmatrix} 1 \\ 2 \\ 4 \end{bmatrix} \right\}, \; U = \left\{ \begin{bmatrix} 1 \\ -1 \\ 0 \end{bmatrix}, \begin{bmatrix} 1 \\ 1 \\ 2 \end{bmatrix}, \begin{bmatrix} -1 \\ -1 \\ 1 \end{bmatrix} \right\}$$

You may recognize the two bases if you read the earlier section on orthogonal bases. The set U contains the vectors of an orthogonal basis.

Because both sets W and U are bases for R^3, you can write any 3×1 vector as a linear combination of the vectors in each set. For example, the vector **v** shown here is written as linear combinations of the vectors:

$$\mathbf{v} = \begin{bmatrix} 0 \\ 19 \\ 24 \end{bmatrix}$$

$$W = 3\begin{bmatrix} 1 \\ -1 \\ 0 \end{bmatrix} + 4\begin{bmatrix} -2 \\ 3 \\ 1 \end{bmatrix} + 5\begin{bmatrix} 1 \\ 2 \\ 4 \end{bmatrix} = \begin{bmatrix} 0 \\ 19 \\ 24 \end{bmatrix}$$

$$U = -\frac{19}{2}\begin{bmatrix} 1 \\ -1 \\ 0 \end{bmatrix} + \frac{67}{6}\begin{bmatrix} 1 \\ 1 \\ 2 \end{bmatrix} + \frac{5}{3}\begin{bmatrix} -1 \\ -1 \\ 1 \end{bmatrix} = \begin{bmatrix} 0 \\ 19 \\ 24 \end{bmatrix}$$

You may be a bit skeptical about the scalar multiples for the vectors in U, but go ahead, check it out! Skeptical or not, though, you're probably wondering where in the world those multipliers came from. The scalars here aren't from the formula in "Using the orthogonal basis to write the linear combination."

One method you use to solve for scalar multiples is to set up augmented matrices for each set of vectors, use row reduction, and solve for the multipliers. Another method is to solve for the scalar multiples used in one of the sets — those used in a linear combination that you already know — and then use a *change of basis* matrix to compute other scalars for the new set of vectors.

I want to construct a *transition matrix* that will take the scalars from a linear combination of the vectors in W and give me the scalars needed on the vectors of U. I may prefer to use the basis described by the vectors in U for some reason or another — maybe because U is the orthogonal basis. You can create a transition matrix to go in either direction — U to W or W to U. The process for finding the transition matrix from W to U is as follows:

1. **Write an augmented matrix that has the vectors from set U on the left and the vectors from set W on the right.**

$$\left[\begin{array}{rrr|rrr} 1 & 1 & -1 & 1 & -2 & 1 \\ -1 & 1 & -1 & -1 & 3 & 2 \\ 0 & 2 & 1 & 0 & 1 & 4 \end{array}\right]$$

   The augmented matrix represents three different equations involving vectors and scalar multiples:

$$\begin{aligned} a_1\mathbf{u}_1 + a_2\mathbf{u}_2 + a_3\mathbf{u}_3 &= \mathbf{w}_1 \\ b_1\mathbf{u}_1 + b_2\mathbf{u}_2 + b_3\mathbf{u}_3 &= \mathbf{w}_2 \\ c_1\mathbf{u}_1 + c_2\mathbf{u}_2 + c_3\mathbf{u}_3 &= \mathbf{w}_3 \end{aligned}$$

   Each $\mathbf{u}_i$ represents a vector in set U, and each $\mathbf{w}_i$ represents a vector in set W.

2. **Perform row reductions on the augmented matrix to put the matrix in reduced row echelon form.**

$$\left[\begin{array}{ccc|ccc} 1 & 1 & -1 & 1 & -2 & 1 \\ -1 & 1 & -1 & -1 & 3 & 2 \\ 0 & 2 & 1 & 0 & 1 & 4 \end{array}\right]$$

$$R_1 + R_2 \rightarrow R_2 \left[\begin{array}{ccc|ccc} 1 & 1 & -1 & 1 & -2 & 1 \\ 0 & 2 & -2 & 0 & 1 & 3 \\ 0 & 2 & 1 & 0 & 1 & 4 \end{array}\right]$$

$$\frac{1}{2}R_2 \rightarrow R_2 \left[\begin{array}{ccc|ccc} 1 & 1 & -1 & 1 & -2 & 1 \\ 0 & 1 & -1 & 0 & \frac{1}{2} & \frac{3}{2} \\ 0 & 2 & 1 & 0 & 1 & 4 \end{array}\right]$$

$$\begin{array}{l} -1R_2 + R_1 \rightarrow R_1 \\ -2R_2 + R_3 \rightarrow R_3 \end{array} \left[\begin{array}{ccc|ccc} 1 & 0 & 0 & 1 & -\frac{5}{2} & -\frac{1}{2} \\ 0 & 1 & -1 & 0 & \frac{1}{2} & \frac{3}{2} \\ 0 & 0 & 3 & 0 & 0 & 1 \end{array}\right]$$

$$\frac{1}{3}R_3 \rightarrow R_3 \left[\begin{array}{ccc|ccc} 1 & 0 & 0 & 1 & -\frac{5}{2} & -\frac{1}{2} \\ 0 & 1 & -1 & 0 & \frac{1}{2} & \frac{3}{2} \\ 0 & 0 & 1 & 0 & 0 & \frac{1}{3} \end{array}\right]$$

$$R_3 + R_2 \rightarrow R_2 \left[\begin{array}{ccc|ccc} 1 & 0 & 0 & 1 & -\frac{5}{2} & -\frac{1}{2} \\ 0 & 1 & 0 & 0 & \frac{1}{2} & \frac{11}{6} \\ 0 & 0 & 1 & 0 & 0 & \frac{1}{3} \end{array}\right]$$

The matrix on the left is the transition matrix that you use to find the scalars needed to write the linear combination of the vectors in set U.

Here's the transition matrix from W to U:

$$\begin{bmatrix} 1 & -\frac{5}{2} & -\frac{1}{2} \\ 0 & \frac{1}{2} & \frac{11}{6} \\ 0 & 0 & \frac{1}{3} \end{bmatrix}$$

For example, consider the following vector, **v.** I show the vector and the corresponding linear combinations of the vectors in the basis W.

$$\mathbf{v} = \begin{bmatrix} -13 \\ -1 \\ -20 \end{bmatrix}$$

$$W: 1\begin{bmatrix} 1 \\ -1 \\ 0 \end{bmatrix} + 4\begin{bmatrix} -2 \\ 3 \\ 1 \end{bmatrix} - 6\begin{bmatrix} 1 \\ 2 \\ 4 \end{bmatrix} = \begin{bmatrix} -13 \\ -1 \\ -20 \end{bmatrix}$$

3. **Now I construct a 3 × 1 vector of the scalars used in the linear combination.**

$$\begin{bmatrix} 1 \\ 4 \\ -6 \end{bmatrix}$$

4. **To find the scalars needed for a linear combination of the vectors in the basis U, multiply the transition matrix times the vector of scalars from the linear combination of W.**

$$\begin{bmatrix} 1 & -\frac{5}{2} & -\frac{1}{2} \\ 0 & \frac{1}{2} & \frac{11}{6} \\ 0 & 0 & \frac{1}{3} \end{bmatrix} * \begin{bmatrix} 1 \\ 4 \\ -6 \end{bmatrix} = \begin{bmatrix} 1-\frac{5}{2}(4)-\frac{1}{2}(-6) \\ 0(1)+\frac{1}{2}(4)+\frac{11}{6}(-6) \\ 0(1)+0(4)+\frac{1}{3}(-6) \end{bmatrix} = \begin{bmatrix} -6 \\ -9 \\ -2 \end{bmatrix}$$

Using the elements of the resulting matrix as scalar multipliers in the linear combination with the vectors in U, the resulting matrix is the same.

$$U: -6\begin{bmatrix} 1 \\ -1 \\ 0 \end{bmatrix} - 9\begin{bmatrix} 1 \\ 1 \\ 2 \end{bmatrix} - 2\begin{bmatrix} -1 \\ -1 \\ 1 \end{bmatrix} = \begin{bmatrix} -6-9+2 \\ 6-9+2 \\ 0-18-2 \end{bmatrix} = \begin{bmatrix} -13 \\ -1 \\ -20 \end{bmatrix}$$

Chapter 16

Eyeing Eigenvalues and Eigenvectors

In This Chapter

- Showing how eigenvalues and eigenvectors are related
- Solving for eigenvalues and their corresponding eigenvectors
- Describing eigenvectors under special circumstances
- Digging into diagonalization

The earlier chapters of this book are full of fun things to do with vectors and matrices. Of course, they're *fun* things! All math is a blast — you must think so if you're reading this book, right?

What I show you in this chapter is how to take some basic mathematical ingredients, mix them up judiciously, heat up that mixture, and come up with something even more intriguing than before. You take a cup of vectors and a pinch of scalars, add a matrix, combine them all up with a determinant and a subtraction, and then heat up the mixture with the solution of an equation using algebra. I know that you just can't wait to get a taste of what's coming, so I won't hold you back any longer.

Putting the mix into a nutshell (something I'm sure many cooks do), what I show you in this chapter is how to find special vectors for matrices. When performing matrix operations, the matrices transform the vectors into multiples of themselves.

Defining Eigenvalues and Eigenvectors

Eigenvalues and eigenvectors are related to one another through a matrix. You actually start with a matrix and say to yourself, "I wonder what I get when I multiply this matrix times a vector?" Of course, if you're familiar with the material in Chapter 3, you immediately say that the result of multiplying a matrix times a vector is another vector. What's involved in this chapter that's so special is the *kind* of vector that you get as a result.

Demonstrating eigenvectors of a matrix

Take a look at matrix A and vector **v**. Matrix A looks like any other 2×2 matrix. It doesn't appear to be special at all. And vector **v** is nothing spectacular — unless, of course, you're fond of that particular vector.

$$A = \begin{bmatrix} -1 & 2 \\ 4 & 6 \end{bmatrix}, \mathbf{v} = \begin{bmatrix} 1 \\ 4 \end{bmatrix}$$

Now I perform the multiplication of matrix A times vector **v**. (You can find out how the multiplication works in Chapter 3.)

$$A\mathbf{v} = \begin{bmatrix} -1 & 2 \\ 4 & 6 \end{bmatrix} \begin{bmatrix} 1 \\ 4 \end{bmatrix} = \begin{bmatrix} -1+8 \\ 4+24 \end{bmatrix} = \begin{bmatrix} 7 \\ 28 \end{bmatrix}$$

The multiplication was relatively uneventful, but did you notice something about the product? The two elements in the resulting vector are multiples of the original vector **v**.

$$\begin{bmatrix} 7 \\ 28 \end{bmatrix} = 7 \begin{bmatrix} 1 \\ 4 \end{bmatrix}$$

The curious thing happening here is that multiplying matrix A by the vector **v** gives you the same result that you'd get by just multiplying vector **v** by the scalar 7.

Want to see me do that again? Okay, I can do it once more with a new vector, **w**.

$$A = \begin{bmatrix} -1 & 2 \\ 4 & 6 \end{bmatrix}, \mathbf{w} = \begin{bmatrix} -2 \\ 1 \end{bmatrix}$$

$$A\mathbf{w} = \begin{bmatrix} -1 & 2 \\ 4 & 6 \end{bmatrix} \begin{bmatrix} -2 \\ 1 \end{bmatrix} = \begin{bmatrix} 2+2 \\ -8+6 \end{bmatrix} = \begin{bmatrix} 4 \\ -2 \end{bmatrix} = -2 \begin{bmatrix} -2 \\ 1 \end{bmatrix}$$

This time, multiplying matrix A times the vector **w** has the same effect as multiplying vector **w** by the scalar -2.

Okay, I'm all out of magic tricks — or, actually, I'm all out of eigenvalues.

Coming to grips with the eigenvector definition

All square matrices have eigenvectors and their related eigenvalues. But not all square matrices will have eigenvectors that are real (some involve imaginary numbers). When real eigenvalues and eigenvectors exist, the effect of multiplying the eigenvector by the matrix is the same as multiplying that same vector by a scalar.

An *eigenvector* of an $n \times n$ matrix is a nonzero vector **x** such that $A\mathbf{x} = \lambda\mathbf{x}$, where λ is some scalar. The scalar λ is called the *eigenvalue* of A, but only if there is a *nontrivial* solution (one that doesn't involve all zeros) for **x** of $A\mathbf{x} = \lambda\mathbf{x}$. You say that **x** is an *eigenvector* corresponding to λ.

The matrix A in "Demonstrating eigenvectors of a matrix" has two eigenvectors, **v** and **w.** The eigenvector **v** corresponds to the eigenvalue 7, and the eigenvector **w** corresponds to the eigenvalue -2.

Illustrating eigenvectors with reflections and rotations

In Chapter 8, I show you matrices that represent rotations and dilations of points in the coordinate plane. When you rotate a point about the origin, the resulting point is always the same distance from the origin as the original point and sometimes a multiple of the original point. Dilating or contracting points or segments changes distances — making distances larger or smaller. I choose to use these two geometric-type transformations to illustrate eigenvectors.

Rotating with eigenvectors

In general, rotations of points about the origin are performed using the following 2×2 matrix as the linear operator:

$$\begin{bmatrix} \cos\theta & -\sin\theta \\ \sin\theta & \cos\theta \end{bmatrix}$$

If you let θ be π (180 degrees), then you get the matrix:

$$\begin{bmatrix} \cos\pi & -\sin\pi \\ \sin\pi & \cos\pi \end{bmatrix} = \begin{bmatrix} -1 & 0 \\ 0 & -1 \end{bmatrix}$$

A rotation of 180 degrees takes points in a quadrant to the opposite quadrant and reverses the signs of the coordinates. Rotating a point that lies on one of the axes by 180 degrees results in another point on the same axis — but on the other side of the origin. For example, I multiply the matrix for a rotation of 180 degrees times vectors representing the points (3,4), (-2,7), (-5,-2), and (0,3).

$$\begin{bmatrix} -1 & 0 \\ 0 & -1 \end{bmatrix}\begin{bmatrix} 3 \\ 4 \end{bmatrix} = \begin{bmatrix} -3 \\ -4 \end{bmatrix} \quad \begin{bmatrix} -1 & 0 \\ 0 & -1 \end{bmatrix}\begin{bmatrix} -2 \\ 7 \end{bmatrix} = \begin{bmatrix} 2 \\ -7 \end{bmatrix}$$
$$\begin{bmatrix} -1 & 0 \\ 0 & -1 \end{bmatrix}\begin{bmatrix} -5 \\ -2 \end{bmatrix} = \begin{bmatrix} 5 \\ 2 \end{bmatrix} \quad \begin{bmatrix} -1 & 0 \\ 0 & -1 \end{bmatrix}\begin{bmatrix} 0 \\ 3 \end{bmatrix} = \begin{bmatrix} 0 \\ -3 \end{bmatrix}$$

Each resulting vector is a multiple of the original vector, and the multiplier is the number -1. This particular transformation works on any 2×2 vector, so, technically, for the matrix that performs 180-degree rotations, any vector representing a point on the coordinate plane is an eigenvector with an eigenvalue of -1. In Figure 16-1, I show you the four original points and their corresponding points under a rotation of 180 degrees.

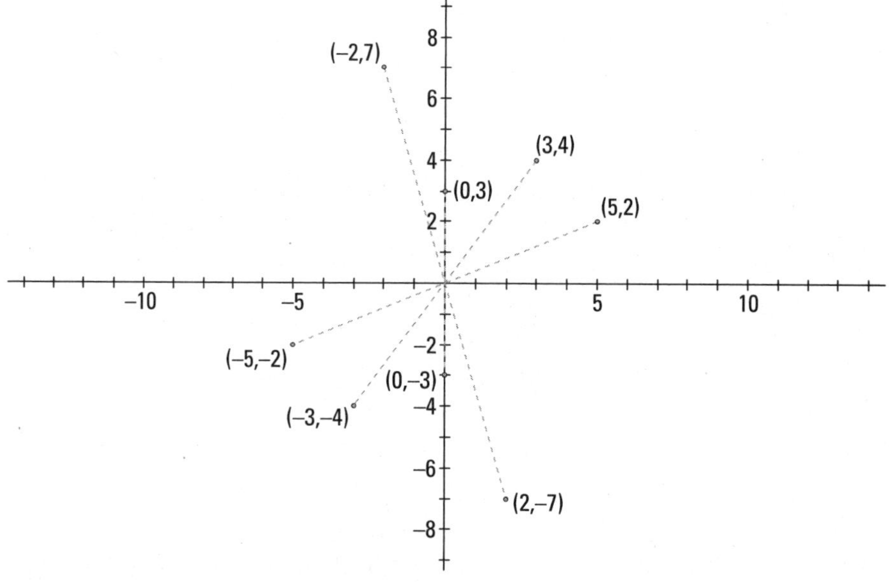

Figure 16-1: Points and images under a rotation of 180 degrees.

Dilating and contracting with eigenvectors

Two other related types of geometric transformations are dilations and contractions. In Chapter 8, I show you how these transformations work. In this section, I show you how these transformations are related to eigenvectors and eigenvalues.

Consider the following matrix and what happens when it multiplies three vectors that represent coordinates in the plane.

$$\begin{bmatrix} 2 & 0 \\ 0 & 2 \end{bmatrix}\begin{bmatrix} -1 \\ 2 \end{bmatrix} = \begin{bmatrix} -2 \\ 4 \end{bmatrix} \quad \begin{bmatrix} 2 & 0 \\ 0 & 2 \end{bmatrix}\begin{bmatrix} -1 \\ 1 \end{bmatrix} = \begin{bmatrix} -2 \\ 2 \end{bmatrix}$$

$$\begin{bmatrix} 2 & 0 \\ 0 & 2 \end{bmatrix}\begin{bmatrix} 3 \\ -1 \end{bmatrix} = \begin{bmatrix} 6 \\ -2 \end{bmatrix}$$

In Figure 16-2, I show you the original three points — all connected with segments. Then I connect the resulting points with segments.

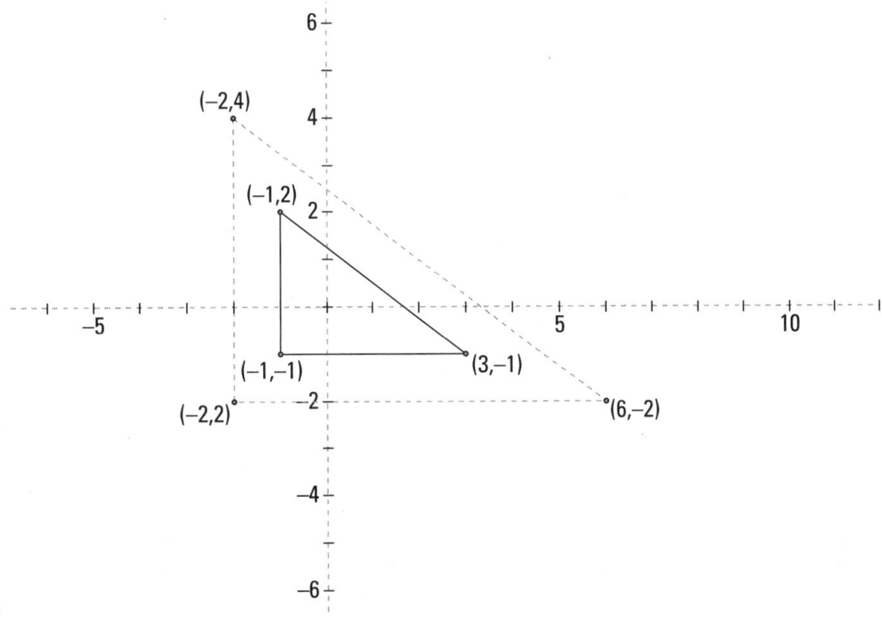

Figure 16-2: The triangle expands with the dilation.

Each side of the new triangle is twice that of the original triangle, which is no surprise, because each coordinate doubled in size. Also, each new point is twice as far from the origin as the original point. The matrix represents a dilation of 2 (which is also the eigenvalue). Using a number between 0 and 1 instead of the 2 in the matrix would result in a contraction. In any case, the matrix performs the operation on any vector representing a point in the coordinate plane.

Solving for Eigenvalues and Eigenvectors

The eigenvectors of a matrix are the special vectors that, when multiplied by the matrix, result in multiples of themselves. An eigenvalue is a constant multiplier that is associated with the vector that gets multiplied.

A process used to determine the eigenvalues and corresponding eigenvectors of a matrix involves solving the equation: $\det(A - \lambda\mathbf{I})\mathbf{x} = 0$. You solve for the values that λ assumes to make the determinant of a matrix be equal to 0. The matrix whose determinant is used is formed by subtracting the multiple of the identity matrix from the target matrix, A. Let me put all this in steps to explain it better.

To solve for the eigenvalues, λ_i, and corresponding eigenvectors, $\mathbf{x}_i$, of $n \times n$ matrix A, you:

1. **Multiply an $n \times n$ identity matrix by the scalar λ.**
2. **Subtract the identity matrix multiple from the matrix A.**
3. **Find the determinant of the matrix of the difference.**
4. **Solve for the values of λ that satisfy the equation $\det(A - \lambda I) = 0$.**
5. **Solve for the vector corresponding to each λ.**

Determining the eigenvalues of a 2 × 2 matrix

In this section, I show you how to find eigenvalues and corresponding eigenvectors using, first, a 2 × 2 matrix and, then, a 3 × 3 matrix.

For this example, start with matrix A.

$$A = \begin{bmatrix} 7 & 3 \\ 3 & -1 \end{bmatrix}$$

1. **Multiply the 2 × 2 identity matrix by the scalar λ.**

$$\lambda\begin{bmatrix} 1 & 0 \\ 0 & 1 \end{bmatrix} = \begin{bmatrix} \lambda & 0 \\ 0 & \lambda \end{bmatrix}$$

2. Subtract the multiple of the identity matrix from matrix A.

$$A - \lambda \mathbf{I} = \begin{bmatrix} 7 & 3 \\ 3 & -1 \end{bmatrix} - \begin{bmatrix} \lambda & 0 \\ 0 & \lambda \end{bmatrix} = \begin{bmatrix} 7-\lambda & 3 \\ 3 & -1-\lambda \end{bmatrix}$$

3. Find the determinant of the matrix found in Step 2 by computing the difference of the cross-products.

$$\begin{aligned} \det\begin{bmatrix} 7-\lambda & 3 \\ 3 & -1-\lambda \end{bmatrix} &= (7-\lambda)(-1-\lambda) - 3(3) \\ &= -7 - 6\lambda + \lambda^2 - 9 \\ &= \lambda^2 - 6\lambda - 16 \end{aligned}$$

If you don't remember how to compute determinants, turn to Chapter 10.

4. Solve for the values of λ satisfying the equation found by setting the expression equal to 0.

$$\begin{aligned} &\lambda^2 - 6\lambda - 16 = 0 \\ &(\lambda - 8)(\lambda + 2) = 0 \\ &\lambda = 8 \text{ or } \lambda = -2 \end{aligned}$$

The eigenvalues for matrix A are 8 and -2.

5. Solve for the corresponding eigenvectors.

Begin with $\lambda = 8$, and replace each λ in the difference matrix with an 8.

$$\begin{aligned} \begin{bmatrix} 7-\lambda & 3 \\ 3 & -1-\lambda \end{bmatrix} &= \begin{bmatrix} 7-8 & 3 \\ 3 & -1-8 \end{bmatrix} \\ &= \begin{bmatrix} -1 & 3 \\ 3 & -9 \end{bmatrix} \end{aligned}$$

Now you essentially solve the matrix equation $A\mathbf{x} = \mathbf{0}$ for the x values. Performing a row reduction, the row of nonzero elements is used to form the equation $-x_1 + 3x_2 = 0$ or that $x_1 = 3x_2$. This relationship describes the eigenvector corresponding to the eigenvalue of 8.

$$\begin{bmatrix} -1 & 3 \\ 3 & -9 \end{bmatrix}\begin{bmatrix} x_1 \\ x_2 \end{bmatrix}=\begin{bmatrix} 0 \\ 0 \end{bmatrix}$$

$$\left[\begin{array}{cc|c} -1 & 3 & 0 \\ 3 & -9 & 0 \end{array}\right]$$

simplifies to:

$$\begin{bmatrix} -1 & 3 \\ 3 & -9 \end{bmatrix} \rightarrow \begin{matrix} x_1 & x_2 \\ \begin{bmatrix} -1 & 3 \\ 0 & 0 \end{bmatrix} \end{matrix}$$

$$-x_1+3x_2=0$$

$$x_1=3x_2$$

$$\begin{bmatrix} x_1 \\ x_2 \end{bmatrix}=\begin{bmatrix} 3 \\ 1 \end{bmatrix}$$

So, when the eigenvalue is 8, the product of matrix A and the corresponding eigenvector is equal to 8 times the eigenvector.

$$\begin{bmatrix} 7 & 3 \\ 3 & -1 \end{bmatrix}\begin{bmatrix} 3 \\ 1 \end{bmatrix}=\begin{bmatrix} 24 \\ 8 \end{bmatrix}=8\begin{bmatrix} 3 \\ 1 \end{bmatrix}$$

Similarly, when determining the eigenvector corresponding to the eigenvalue -2,

$$\begin{bmatrix} 7-(-2) & 3 \\ 3 & -1-(-2) \end{bmatrix}=\begin{bmatrix} 9 & 3 \\ 3 & 1 \end{bmatrix}$$

$$\begin{bmatrix} 9 & 3 \\ 3 & 1 \end{bmatrix} \rightarrow \begin{matrix} x_1 & x_2 \\ \begin{bmatrix} 3 & 1 \\ 0 & 0 \end{bmatrix} \end{matrix}$$

$$3x_1+x_2=0$$

$$x_2=-3x_1$$

$$\begin{bmatrix} x_1 \\ x_2 \end{bmatrix}=\begin{bmatrix} 1 \\ -3 \end{bmatrix}$$

and multiplying the eigenvector by matrix A, you get a -2 multiple of the vector.

$$\begin{bmatrix} 7 & 3 \\ 3 & -1 \end{bmatrix}\begin{bmatrix} 1 \\ -3 \end{bmatrix}=\begin{bmatrix} -2 \\ 6 \end{bmatrix}=-2\begin{bmatrix} 1 \\ -3 \end{bmatrix}$$

Getting in deep with a 3 × 3 matrix

At first glance, you might not think there'd be too much more to do when solving for eigenvalues of a 3×3 matrix. After all, that's just one up from a 2×2 matrix. Well, I hate to burst your bubble, but unless the matrix you're working with contains a fair amount of 0s, the algebra involved in solving the equation found from the determinant can be a bit of a challenge.

For example, here are a relatively well-behaved 3×3 matrix A and the first steps toward solving for the eigenvalues:

$$A = \begin{bmatrix} 4 & -3 & 0 \\ 4 & -1 & -2 \\ 1 & -3 & 3 \end{bmatrix}$$

$$A - \lambda \mathbf{I} = \begin{bmatrix} 4 & -3 & 0 \\ 4 & -1 & -2 \\ 1 & -3 & 3 \end{bmatrix} - \begin{bmatrix} \lambda & 0 & 0 \\ 0 & \lambda & 0 \\ 0 & 0 & \lambda \end{bmatrix}$$

$$= \begin{bmatrix} 4-\lambda & -3 & 0 \\ 4 & -1-\lambda & -2 \\ 1 & -3 & 3-\lambda \end{bmatrix}$$

Now, evaluating the determinant of the difference matrix, I use the technique shown in Chapter 10:

$$\det \begin{bmatrix} 4-\lambda & -3 & 0 \\ 4 & -1-\lambda & -2 \\ 1 & -3 & 3-\lambda \end{bmatrix}$$
$$= (4-\lambda)(-1-\lambda)(3-\lambda) + 6 + 0$$
$$-(0 + 6(4-\lambda) - 12(3-\lambda))$$
$$= -\lambda^3 + 6\lambda^2 - 11\lambda + 6$$

Yes, I know that I skipped all the lovely algebra steps and didn't show you the product of the three binomials and the distribution of factors over the other terms. I thought you'd prefer to ignore all the gory details and get to the resulting polynomial. So, assuming that it's okay with you, I now factor the polynomial after setting it equal to 0.

How did I factor the polynomial? I used synthetic division. If you're not familiar with synthetic division, check out *Algebra II For Dummies* (Wiley).

$$-\lambda^3+6\lambda^2-11\lambda+6=0$$
$$-(\lambda-1)(\lambda-2)(\lambda-3)=0$$

The eigenvalues are 1, 2, and 3. Solving for the corresponding eigenvectors, I start with λ = 1.

$$\lambda=1,\begin{bmatrix}4-1 & -3 & 0\\ 4 & -1-1 & -2\\ 1 & -3 & 3-1\end{bmatrix}$$
$$=\begin{bmatrix}3 & -3 & 0\\ 4 & -2 & -2\\ 1 & -3 & 2\end{bmatrix}$$

Performing row reductions on the resulting matrix, I get

$$\begin{bmatrix}3 & -3 & 0\\ 4 & -2 & -2\\ 1 & -3 & 2\end{bmatrix}\to\begin{bmatrix}1 & -1 & 0\\ 0 & 2 & -2\\ 0 & -2 & 2\end{bmatrix}\to\begin{bmatrix}1 & -1 & 0\\ 0 & 1 & -1\\ 0 & 0 & 0\end{bmatrix}\to\begin{matrix}x_1 & x_2 & x_3\\ \begin{bmatrix}1 & 0 & -1\\ 0 & 1 & -1\\ 0 & 0 & 0\end{bmatrix}\end{matrix}$$

From the rows of the matrix, I write the equations $x_1 - x_3 = 0$ and $x_2 - x_3 = 0$. The vector corresponding to the relationships between the elements is

$$\begin{bmatrix}1\\ 1\\ 1\end{bmatrix}$$

The corresponding eigenvectors for λ = 2 are found as follows. I've left out steps in the row reduction, but feel free to perform them yourself for the pure pleasure of it.

$$\lambda=2,\begin{bmatrix}2 & -3 & 0\\ 4 & -3 & -2\\ 1 & -3 & 1\end{bmatrix}\to\begin{bmatrix}1 & -3 & 1\\ 0 & 3 & -2\\ 0 & 0 & 0\end{bmatrix}\to\begin{bmatrix}1 & 0 & -1\\ 0 & 3 & -2\\ 0 & 0 & 0\end{bmatrix}$$

The equations I get from the final matrix are $x_1 - x_3 = 0$ and $3x_2 - 2x_3 = 0$. From the equations, I get the eigenvector:

$$\begin{bmatrix} 1 \\ \frac{2}{3} \\ 1 \end{bmatrix} \text{ or } \begin{bmatrix} 3 \\ 2 \\ 3 \end{bmatrix}$$

I multiplied each element by 3 to get rid of the fraction.

Finally, when $\lambda = 3$, I get

$$\lambda = 3, \begin{bmatrix} 1 & -3 & 0 \\ 4 & -4 & -2 \\ 1 & -3 & 0 \end{bmatrix} \rightarrow \begin{bmatrix} 1 & -3 & 0 \\ 0 & 8 & -2 \\ 0 & 0 & 0 \end{bmatrix} \rightarrow \begin{bmatrix} 1 & 0 & -\frac{3}{4} \\ 0 & 1 & -\frac{1}{4} \\ 0 & 0 & 0 \end{bmatrix}$$

The equations from the matrix are $x_1 - {}^3/_4\, x_3 = 0$ and $x_2 - {}^1/_4\, x_3 = 0$. The resulting eigenvector is

$$\begin{bmatrix} \frac{3}{4} \\ \frac{1}{4} \\ 1 \end{bmatrix} \text{ or } \begin{bmatrix} 3 \\ 1 \\ 4 \end{bmatrix}$$

As you can see, the computations and algebra get more difficult as the size of the matrix increases. Thank goodness for modern technology. However, some $n \times n$ matrices have eigenvalues that are easy to determine; the matrices in point are the triangular matrices.

Circling Around Special Circumstances

Eigenvalues of matrices are determined either by guessing or by using a tried-and-true method. Some nice properties of the eigenvalues of matrices and various related matrices make the computations easier and, sometimes, unnecessary. (One of the properties, involving powers of matrices, is discussed even further in "Getting It Straight with Diagonalization," later in this chapter.)

Transforming eigenvalues of a matrix transpose

One property of matrices and their respective eigenvalues is that you don't change those eigenvalues when you transpose the matrix. In Chapter 3, I define a matrix transpose for you and show you examples. But just to define transposition quickly, *transposing a matrix* means to switch all the rows and columns (row one becomes column one, and so on).

The eigenvalues of matrix A and matrix A^T are the same.

So, if you know the eigenvalues of matrix A, shown here, then you know the eigenvalues of matrix A^T.

$$A = \begin{bmatrix} a & b & c \\ d & e & f \\ g & h & i \end{bmatrix}, \ A^T = \begin{bmatrix} a & d & g \\ b & e & h \\ c & f & i \end{bmatrix}$$

Just to demonstrate this property with the general 3×3 matrices, here are the terms in the respective difference determinants:

$$\begin{aligned} \det(A - \lambda \mathbf{I}) &= \det \begin{bmatrix} a-\lambda & b & c \\ d & e-\lambda & f \\ g & h & i-\lambda \end{bmatrix} \\ &= (a-\lambda)(e-\lambda)(i-\lambda) + bfg + cdh \\ &\quad - \left(cg(e-\lambda) + fh(a-\lambda) + bd(i-\lambda)\right) \\ \det(A^T - \lambda \mathbf{I}) &= \det \begin{bmatrix} a-\lambda & d & g \\ b & e-\lambda & h \\ c & f & i-\lambda \end{bmatrix} \\ &= (a-\lambda)(e-\lambda)(i-\lambda) + bfg + cdh \\ &\quad - \left(cg(e-\lambda) + fh(a-\lambda) + bd(i-\lambda)\right) \end{aligned}$$

The values of the determinants are the same. The same holds true, of course, for any $n \times n$ matrices.

Reciprocating with the eigenvalue reciprocal

A matrix doesn't really have a reciprocal. (The *reciprocal* of a number is equal to 1 divided by the number.) But a matrix may have an *inverse,* which is another matrix that, when it multiplies the first matrix, gives you an identity matrix. I go into a lot more detail about matrices and their inverses in Chapter 3. A number or scalar multiple *does* have a reciprocal (assuming that you aren't silly enough to have the scalar be 0). So, how do inverses of matrices and reciprocals of scalars tie together? I'm here to show you how.

If matrix A is nonsingular (has an inverse) and λ is an eigenvalue of matrix A, then $^1/_\lambda$ is an eigenvalue of the inverse of matrix A, matrix A^{-1}. For example, here's a matrix A:

$$A = \begin{bmatrix} 1 & 2 & 0 \\ 2 & 2 & 2 \\ 0 & 2 & 3 \end{bmatrix}$$

And here are matrix A's eigenvalues: $\lambda = 2, \lambda = 5, \lambda = -1$.

The inverse of matrix A is

$$A^{-1} = \begin{bmatrix} -0.2 & 0.6 & -0.4 \\ 0.6 & -0.3 & 0.2 \\ -0.4 & 0.2 & 0.2 \end{bmatrix}$$

And the eigenvalues of A^{-1} are $\lambda = {}^1/_2, \lambda = {}^1/_5$, and $\lambda = -1$.

Just to demonstrate, I show you the eigenvectors corresponding to the eigenvalue 2 for matrix A and the eigenvalue corresponding to $^1/_2$ for matrix A^{-1}.

$$A = \begin{bmatrix} 1 & 2 & 0 \\ 2 & 2 & 2 \\ 0 & 2 & 3 \end{bmatrix}, \lambda = 2,$$

$$\begin{bmatrix} 1 & 2 & 0 \\ 2 & 2 & 2 \\ 0 & 2 & 3 \end{bmatrix} \begin{bmatrix} 2 \\ 1 \\ -2 \end{bmatrix} = \begin{bmatrix} 4 \\ 2 \\ -4 \end{bmatrix} = 2 \begin{bmatrix} 2 \\ 1 \\ -2 \end{bmatrix}$$

$$A^{-1} = \begin{bmatrix} -0.2 & 0.6 & -0.4 \\ 0.6 & -0.3 & 0.2 \\ -0.4 & 0.2 & 0.2 \end{bmatrix}, \lambda = \frac{1}{2}$$

$$\begin{bmatrix} -0.2 & 0.6 & -0.4 \\ 0.6 & -0.3 & 0.2 \\ -0.4 & 0.2 & 0.2 \end{bmatrix} \begin{bmatrix} 2 \\ 1 \\ -2 \end{bmatrix} = \begin{bmatrix} 1 \\ \frac{1}{2} \\ -1 \end{bmatrix} = \frac{1}{2} \begin{bmatrix} 2 \\ 1 \\ -2 \end{bmatrix}$$

Notice that the eigenvectors corresponding to $\lambda = 2$ and $\lambda = 1/2$ are the same.

Triangulating with triangular matrices

In Chapter 3, you see many different ways of describing matrices. Matrices can be square, triangular, zero, identity, singular, nonsingular, and so on. Many matrices fit into more than one classification.

The matrices I discuss in this section are all square and triangular or square and diagonal — essentially, they're triangular or diagonal. Now, if this was a geometry discussion, you could tell me that being both square and triangular just doesn't happen. Aren't you glad this isn't geometry?

A *square matrix* has an equal number of rows and columns.

A *triangular matrix* has all 0s either above or below the main diagonal. If all the elements are 0s above the main diagonal, then the matrix is *lower triangular;* if the 0s are all below the main diagonal, the matrix is *upper triangular* (you're identifying where the nonzero elements *are*).

Now let me show you something nice about triangular matrices. I start with the upper triangular matrix A and the lower triangular matrix B:

$$A = \begin{bmatrix} 1 & 4 & 2 \\ 0 & 2 & -1 \\ 0 & 0 & -3 \end{bmatrix} \quad B = \begin{bmatrix} 3 & 0 & 0 \\ 1 & -4 & 0 \\ 0 & 1 & 2 \end{bmatrix}$$

What's so special about triangular matrices is their eigenvalues.

The eigenvalues of a triangular matrix all lie along the main diagonal.

So, if you look at matrices A and B, you see that the eigenvalues for matrix A are λ = 1, 2, and -3, and the eigenvalues for matrix B are λ = 3, -4, and 2. Are you skeptical? I'll give you a quick demonstration using matrix A to show why the rule works for triangular matrices. I set up the matrix A - λ**I** to find the determinant of that difference. Then I find the determinant.

$$A - \lambda \mathbf{I} = \begin{bmatrix} 1-\lambda & 4 & 2 \\ 0 & 2-\lambda & -1 \\ 0 & 0 & -3-\lambda \end{bmatrix}$$

$$\begin{aligned} \det \begin{bmatrix} 1-\lambda & 4 & 2 \\ 0 & 2-\lambda & -1 \\ 0 & 0 & -3-\lambda \end{bmatrix} &= (1-\lambda)(2-\lambda)(-3-\lambda)+0+0-(0+0+0) \\ &= (1-\lambda)(2-\lambda)(-3-\lambda) \end{aligned}$$

Because of all the 0 elements in *strategic* places, the only part of the determinant that isn't 0 is the product along the main diagonal. Set that product equal to 0, and you get eigenvalues equal to the elements along that diagonal. The property is true for all sizes of square matrices.

Powering up powers of matrices

In various applications involving probability models and manufacturing, you work with powers of square matrices. You solve problems by squaring the matrix or raising it to the third or fourth power or even higher. I show you, first, the relationship between the eigenvalues of a square matrix and the eigenvalues of powers of square matrices. Then, in "Getting It Straight with Diagonalization," I show you how to make powering up matrices a tad easier.

If A is an $n \times n$ matrix and λ is an eigenvalue of A, then λ^k is an eigenvalue of matrix A^k (for k = 2, 3, 4, . . .). For example, matrix A shown here has eigenvalues of 7 and -2.

$$A = \begin{bmatrix} 3 & -5 \\ -4 & 2 \end{bmatrix}, \lambda = 7, -2$$

I raise A to the third power and solve for the eigenvalues of the resulting matrix:

$$A^3 = \begin{bmatrix} 187 & -195 \\ -156 & 148 \end{bmatrix}$$

$$\begin{aligned} \det\begin{bmatrix} 187-\lambda & -195 \\ -156 & 148-\lambda \end{bmatrix} &= (187-\lambda)(148-\lambda) - 30{,}420 \\ &= 27{,}676 - 335\lambda + \lambda^2 - 30{,}420 \\ &= \lambda^2 - 335\lambda - 2{,}744 \\ &= (\lambda - 343)(\lambda + 8) \end{aligned}$$

The eigenvalues of A^3 are 343 and -8, which are the cubes of 7 and -2, respectively.

Thank goodness for technology. I was able to raise the matrix A to the third power with relative ease using my graphing calculator. You may be wondering if there's an alternative for those who aren't as lucky to own a handy-dandy calculating device. I'm so pleased that you're interested, because the answer is found in the next section on diagonalization!

Getting It Straight with Diagonalization

A *diagonal* matrix is a square matrix in which the only nonzero elements lie along the main diagonal. For example, here are three diagonal matrices in three different sizes:

$$A = \begin{bmatrix} 1 & 0 & 0 \\ 0 & 2 & 0 \\ 0 & 0 & -3 \end{bmatrix}, B = \begin{bmatrix} 7 & 0 \\ 0 & 4 \end{bmatrix}, C = \begin{bmatrix} 1 & 0 & 0 & 0 \\ 0 & 5 & 0 & 0 \\ 0 & 0 & -3 & 0 \\ 0 & 0 & 0 & 8 \end{bmatrix}$$

One of the nicest things about diagonal matrices is the ease of computations. In particular, raising diagonal matrices to various powers is achieved by just raising each element along the main diagonal to that power. So, raising matrix A to the fourth power,

$$A^4 = \begin{bmatrix} 1^4 & 0 & 0 \\ 0 & 2^4 & 0 \\ 0 & 0 & (-3)^4 \end{bmatrix} = \begin{bmatrix} 1 & 0 & 0 \\ 0 & 16 & 0 \\ 0 & 0 & 81 \end{bmatrix}$$

If you only had to raise diagonal matrices to higher powers in applications, then you'd be set. The next best thing, though, is that you can often change a square matrix into a diagonal matrix that's *similar* to that matrix and perform computations on the new matrix. The powers of a diagonal matrix are so much easier to find than powers of other matrices.

Also, as in a triangular matrix, the eigenvalues of a diagonal matrix lie along the main diagonal. So you can construct a matrix similar to a given matrix once you've found the eigenvalues of the original matrix.

Two matrices A and B are said to be *similar* if matrix B is equal to the product of the inverse of a nonsingular matrix S times matrix A times the matrix S: $B = S^{-1}AS$.

If two matrices are *similar* $n \times n$ matrices, then they have the same eigenvalues.

So, where all this business of diagonal matrices and similar matrices is going is that if you want a high power of a matrix, you try to find a similar matrix that's a diagonal matrix. You find the power of the diagonal matrix and then change the answer back into the original form.

For example, if A is an $n \times n$ matrix and if $B = S^{-1}AS$ where B is a diagonal matrix, then after raising B to the power k, you can use the same matrix S and its inverse and recover the power in terms of A with $A^k = SB^kS^{-1}$.

A matrix is *diagonalizable* (you can find a similar matrix that's diagonal) only if the matrix has eigenvectors that are linearly independent. I'll show you how to use the eigenvectors of a matrix to form a similar diagonal matrix.

The matrix A is diagonalizable, because its two eigenvalues, $\lambda_1 = 1$ and $\lambda_2 = 6$, have corresponding eigenvectors, $\mathbf{u}_1$ and $\mathbf{u}_2$, which are linearly independent.

$$A = \begin{bmatrix} 4 & -1 \\ -6 & 3 \end{bmatrix}$$

$$\lambda_1 = 1,\ \mathbf{u}_1 = \begin{bmatrix} 1 \\ 3 \end{bmatrix}$$

$$\lambda_2 = 6,\ \mathbf{u}_2 = \begin{bmatrix} 1 \\ -2 \end{bmatrix}$$

Letting the vectors $\mathbf{u}_1$ and $\mathbf{u}_2$ form the columns of vectors S, I find the inverse, S^{-1}.

$$S = \begin{bmatrix} 1 & 1 \\ 3 & -2 \end{bmatrix},\ S^{-1} = \begin{bmatrix} 0.4 & 0.2 \\ 0.6 & -0.2 \end{bmatrix}$$

If you need to review how to find the inverse of a matrix, go to Chapter 3. Now, finding the product $S^{-1}AS$, I first multiply matrix A by S and then the inverse of S by the product. I'll call the new matrix B.

$$S^{-1}(AS) = S^{-1}\left(\begin{bmatrix} 4 & -1 \\ -6 & 3 \end{bmatrix}\begin{bmatrix} 1 & 1 \\ 3 & -2 \end{bmatrix}\right)$$

$$= S^{-1}\left(\begin{bmatrix} 1 & 6 \\ 3 & -12 \end{bmatrix}\right)$$

$$= \begin{bmatrix} 0.4 & 0.2 \\ 0.6 & -0.2 \end{bmatrix}\begin{bmatrix} 1 & 6 \\ 3 & -12 \end{bmatrix}$$

$$= \begin{bmatrix} 1 & 0 \\ 0 & 6 \end{bmatrix} = B$$

The resulting vector should come as no surprise if you believed me that the two similar vectors, A and B, had the same eigenvalues. The eigenvalues of the diagonal matrix lie along that main diagonal. The goal now is to find the power of matrix A using matrix B.

So first I perform an operation on matrix B: I raise B to the fourth power.

$$B^4 = \begin{bmatrix} 1 & 0 \\ 0 & 6 \end{bmatrix}^4 = \begin{bmatrix} 1^4 & 0 \\ 0 & 6^4 \end{bmatrix} = \begin{bmatrix} 1 & 0 \\ 0 & 1{,}296 \end{bmatrix}$$

And next I recover the original matrix A, or rather, the fourth power of A, with $A^4 = SB^4S^{-1}$.

$$(SB^4)S^{-1} = \left(\begin{bmatrix} 1 & 1 \\ 3 & -2 \end{bmatrix}\begin{bmatrix} 1 & 0 \\ 0 & 1{,}296 \end{bmatrix}\right)S^{-1}$$

$$= \left(\begin{bmatrix} 1 & 1{,}296 \\ 3 & -2{,}592 \end{bmatrix}\right)S^{-1}$$

$$= \begin{bmatrix} 1 & 1{,}296 \\ 3 & -2{,}592 \end{bmatrix}\begin{bmatrix} 0.4 & 0.2 \\ 0.6 & -0.2 \end{bmatrix}$$

$$= \begin{bmatrix} 778 & -259 \\ -1{,}554 & 519 \end{bmatrix} = A^4$$

Here's a 3 × 3 matrix A with eigenvalues λ_1, λ_2, and λ_3 and the corresponding eigenvectors, $\mathbf{u}_1$, $\mathbf{u}_2$, and $\mathbf{u}_3$. I want to raise A to the sixth power.

$$A = \begin{bmatrix} 4 & -3 & 0 \\ 4 & -1 & -2 \\ 1 & -3 & 3 \end{bmatrix}$$

$$\lambda_1 = 1,\ \mathbf{u}_1 = \begin{bmatrix} 1 \\ 1 \\ 1 \end{bmatrix} \quad \lambda_2 = 2,\ \mathbf{u}_2 = \begin{bmatrix} 3 \\ 2 \\ 3 \end{bmatrix} \quad \lambda_3 = 3,\ \mathbf{u}_3 = \begin{bmatrix} 3 \\ 1 \\ 4 \end{bmatrix}$$

Using the eigenvalues of A, I write a matrix B, similar to A, by creating a diagonal matrix with the eigenvalues down the diagonal. Then I write matrix S using the eigenvectors and determine S^{-1}.

$$B = \begin{bmatrix} 1 & 0 & 0 \\ 0 & 2 & 0 \\ 0 & 0 & 3 \end{bmatrix},\ S = \begin{bmatrix} 1 & 3 & 3 \\ 1 & 2 & 1 \\ 1 & 3 & 4 \end{bmatrix},\ S^{-1} = \begin{bmatrix} -5 & 3 & 3 \\ 3 & -1 & -2 \\ -1 & 0 & 1 \end{bmatrix}$$

I raise matrix B to the sixth power and then find $A^6 = SB^6S^{-1}$.

$$B^6 = \begin{bmatrix} 1 & 0 & 0 \\ 0 & 2 & 0 \\ 0 & 0 & 3 \end{bmatrix}^6 = \begin{bmatrix} 1^6 & 0 & 0 \\ 0 & 2^6 & 0 \\ 0 & 0 & 3^6 \end{bmatrix} = \begin{bmatrix} 1 & 0 & 0 \\ 0 & 64 & 0 \\ 0 & 0 & 729 \end{bmatrix}$$

$$\begin{aligned}
(SB^6)S^{-1} &= \left(\begin{bmatrix} 1 & 3 & 3 \\ 1 & 2 & 1 \\ 1 & 3 & 4 \end{bmatrix} \begin{bmatrix} 1 & 0 & 0 \\ 0 & 64 & 0 \\ 0 & 0 & 729 \end{bmatrix} \right) S^{-1} \\
&= \left(\begin{bmatrix} 1 & 192 & 2{,}187 \\ 1 & 128 & 729 \\ 1 & 192 & 2{,}916 \end{bmatrix} \right) S^{-1} \\
&= \begin{bmatrix} 1 & 192 & 2{,}187 \\ 1 & 128 & 729 \\ 1 & 192 & 2{,}916 \end{bmatrix} \begin{bmatrix} -5 & 3 & 3 \\ 3 & -1 & -2 \\ -1 & 0 & 1 \end{bmatrix} \\
&= \begin{bmatrix} -1{,}616 & -189 & 1{,}806 \\ -350 & -125 & 476 \\ -2{,}345 & -189 & 2{,}535 \end{bmatrix} = A^6
\end{aligned}$$

Part V
The Part of Tens

In this part . . .

An *X* stands for ten and it also marks the spot. The Part of Tens includes lists of X related items, all given for you to put in your top X list. Whether it's calculators, Greek letters, or matrix applications, you can have your pick or take all X of them.

Chapter 17

Ten Real-World Applications Using Matrices

In This Chapter

- Working with biological and scientific models
- Tracking traffic and population movements
- Sending secret messages and playing games

In case you were wondering what in the world linear algebra is good for, in this chapter I offer ten examples of matrices at work. Of course, my pat answer should be that linear algebra is beautiful in its own right and needs no convincing for the true believer — but a little application never hurt anyone!

In this chapter, I show you some fairly common applications in very brief formats — just enough to convince you.

Eating Right

Diet and nutrition are concerns at every age. You need enough food to keep you healthy, but you don't want to overdo it. Older people need different amounts of the various nutrients than younger people and babies do.

Consider a situation where a cafeteria manager uses, on a daily basis, rice, macaroni, nonfat dry milk, and wheat bread as a part of the prepared meals. In each serving, the rice contains 5 milligrams of sodium, 21 milligrams of carbohydrates, and 2 grams of protein. The macaroni contains 1 gram of fat, 1 milligram of sodium, 28 grams of carbohydrates, and 5 grams of protein. The milk contains 1 gram of fat, 535 milligrams of sodium, 52 grams of carbohydrates, and 36 grams of protein. And the bread contains 4 grams of fat, 530 milligrams of sodium, 47 grams of carbohydrates, and 9 grams of protein. The manager has to provide a minimum of 12 grams of fat, 2 grams of sodium, 265 grams of carbohydrates, and 70 grams of protein for each person each day. She doesn't want to order too little or too much of the different food types.

So how does she solve this problem? With matrices, of course! She sets up an augmented matrix using the nutrients and amounts in each type of food. All the amounts are given in terms of grams, so the 530 milligrams becomes 0.530 grams.

	Mac	Nfm	Br	Rice	Tot
Fat	1	1	4	0	12
Sod	.001	.535	.530	.005	2
Carb	28	52	47	21	265
Prot	5	36	9	2	70

And now, solving for the amounts, the matrix becomes:

	Mac	Nfm	Br	Rice	Tot
Fat	1	0	0	0	.19
Sod	0	1	0	0	1.03
Carb	0	0	1	0	2.69
Prot	0	0	0	0	3.77

In order to supply the minimum daily requirements of the different nutrients, the manager needs to serve 0.19 grams of macaroni, 1.03 grams of nonfat dry milk, 2.69 grams of bread, and 3.77 grams of rice to each person.

Refer to Chapter 4 to find out how to simplify an augmented matrix to get the solutions of the system.

Controlling Traffic

Have you ever wondered who in the world programmed the traffic lights between your home and your business? During a two-month period when I was forced to use an alternate route to work, I kept track of the number of times I had to stop at the 23 traffic lights between home and work. On average, I was stopping at about 21 of the lights!

Anyway, I do have an appreciation for what traffic engineers must have to go through to keep the traffic moving. Consider Figure 17-1, depicting the four main streets in Aftermath City. Each street goes one way, and I show the number of cars entering and leaving the downtown area on the four streets during a one-hour period in the afternoon.

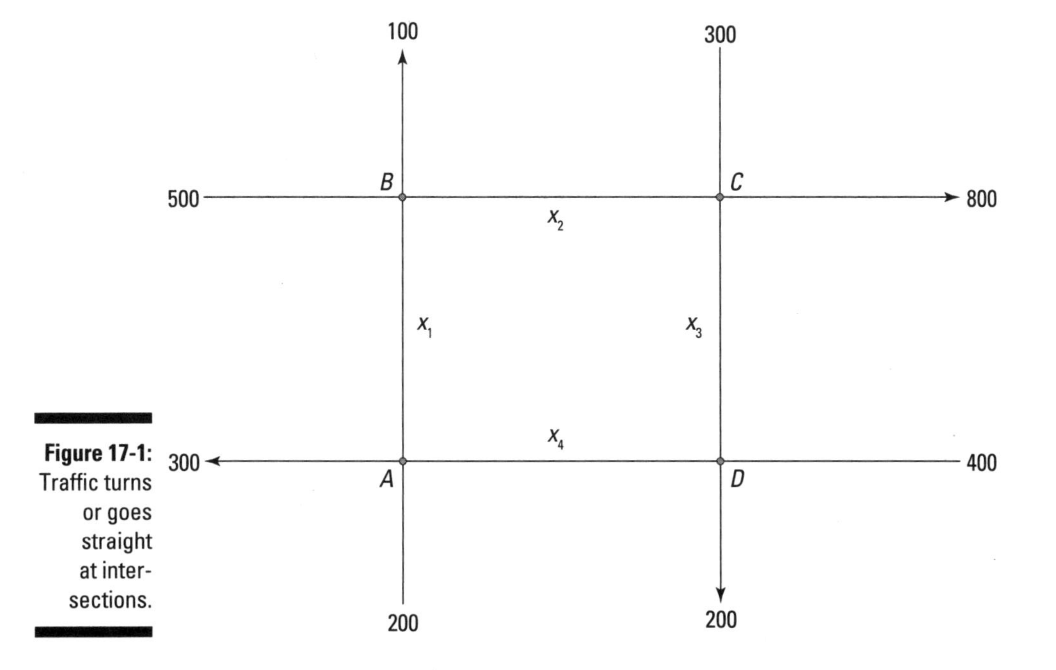

Figure 17-1: Traffic turns or goes straight at intersections.

One way to organize the information is to write equations about what's happening at each intersection:

A: $200 + x_4 = 300 + x_1$

B: $500 + x_1 = 100 + x_2$

C: $300 + x_2 = 800 + x_3$

D: $400 + x_3 = 200 + x_4$

Rewriting the equations and placing the coefficients in a matrix makes the relationships between the variables more evident.

$$\begin{array}{c} \\ A \\ B \\ C \\ D \end{array} \begin{array}{c} \begin{array}{cccc|c} x_1 & x_2 & x_3 & x_4 & \end{array} \\ \left[\begin{array}{cccc|c} 1 & 0 & 0 & -1 & -100 \\ 1 & -1 & 0 & 0 & -400 \\ 0 & 1 & -1 & 0 & 500 \\ 0 & 0 & 1 & -1 & -200 \end{array}\right] \end{array}$$

You find relationships between the amount of traffic on each street in between the intersections; the relationships can all be written in terms of what happens on the portion of the street x_4: $x_1 = x_4 - 100$, $x_2 = x_4 + 300$, and $x_3 = x_4 - 200$. One possible scenario is to let $x_4 = 200$, in which case, $x_1 = 100$, $x_2 = 500$, and $x_3 = 0$. It's all in the planning.

I cover systems of equations and using matrices to solve them in Chapter 4.

Catching Up with Predator-Prey

A common model for the interaction between one species, called the *predator,* and another species, called the *prey,* shows how the number in one category affects the other. For example, if there are lots of rabbits available, then foxes have plenty to eat and will *flourish* (grow in number). As the number of foxes grows, the number of rabbits will diminish (because they'll be eaten). When there are fewer rabbits, fewer foxes will be able to survive (or will move elsewhere). A fine balance serves to keep the fox and rabbit populations steady.

Just for the sake of argument, let's assume that rabbits provide up to 70 percent of the diet for the red fox population in a certain area. Also, the foxes eat 10 percent of the rabbits that are in the area in a one-month time period. If there were no rabbits available for food, then 40 percent of the foxes would die. And, if there were no foxes in the area, then the rabbit population would grow by 10 percent per month. When the rabbits are plentiful, the fox population grows by 30 percent of the rabbit population (in hundreds). The transition matrix shown here describes the interplay between foxes and rabbits.

$$\begin{array}{c} \\ F \\ R \end{array} \begin{array}{c} \begin{array}{cc} F & R \end{array} \\ \begin{bmatrix} 0.60 & 0.30 \\ -0.10 & 1.10 \end{bmatrix} \end{array} * \begin{bmatrix} \text{number of foxes} \\ \text{number of rabbits (in hundreds)} \end{bmatrix}$$

Here's how the elements in the matrix are obtained:

> Number of Foxes Next Month = 0.60 (Foxes This Month If No Rabbits) + 0.30 (Rabbits This Month)
>
> Number of Rabbits Next Month = –0.10 (Rabbits This Month If Plentiful Foxes) + 1.10 (Rabbits This Month If No Foxes)

Working through the problem, you find more on the evolution of this system by solving for eigenvalues and corresponding eigenvalues for the matrix. Then, observing what happens over repeated generations, you find that both populations tend to grow at 3 percent per month with a ratio of 100 foxes per 4,600 rabbits. (To find more on eigenvalues and eigenvectors — see Chapter 16.)

Creating a Secret Message

If your hero is James Bond, then you've probably dabbled in secret codes rather than exploding fountain pens. The simplest way to encrypt a message is to let each letter of the alphabet be represented by one of the other letters. These codes are fairly easy to decipher (and I practice my skills on these Cryptoquotes in the daily newspaper). Matrices add a twist to secret messages — making most of them unbreakable.

For example, I want to send a message to my fellow spy saying: *Meet me at the bridge.* I begin by assigning each letter in the message its numerical place in the alphabet.

M	E	E	T	M	E	A	T	T	H	E	B	R	I	D	G	E
13	5	5	20	13	5	1	20	20	8	5	2	18	9	4	7	5

I now break up the message into 3×1 vectors, filling in the last element with a 27 (because the message has only 17 letters, so you can't fill the last vector completely). The number 27 is also equivalent to the number 1, so it'll come out as an A. Another option is to use the number 24 (X) to fill in any blank spaces.

$$\begin{bmatrix} 13 \\ 5 \\ 5 \end{bmatrix}, \begin{bmatrix} 20 \\ 13 \\ 5 \end{bmatrix}, \begin{bmatrix} 1 \\ 20 \\ 20 \end{bmatrix}, \begin{bmatrix} 8 \\ 5 \\ 2 \end{bmatrix}, \begin{bmatrix} 18 \\ 9 \\ 4 \end{bmatrix}, \begin{bmatrix} 7 \\ 5 \\ 27 \end{bmatrix}$$

Now I encrypt my message by multiplying each vector by a 3×3 matrix, A. I pick a matrix with a relatively nice inverse, because my fellow spy may have to compute the results by hand.

$$A = \begin{bmatrix} 2 & -1 & 5 \\ -2 & 3 & -8 \\ 1 & -2 & 5 \end{bmatrix}$$

$$A^* \begin{bmatrix} 13 \\ 5 \\ 5 \end{bmatrix} = \begin{bmatrix} 46 \\ -51 \\ 28 \end{bmatrix}, A^* \begin{bmatrix} 20 \\ 13 \\ 5 \end{bmatrix} = \begin{bmatrix} 52 \\ -41 \\ 19 \end{bmatrix},$$

$$A^* \begin{bmatrix} 1 \\ 20 \\ 20 \end{bmatrix} = \begin{bmatrix} 82 \\ -102 \\ 61 \end{bmatrix}, A^* \begin{bmatrix} 8 \\ 5 \\ 2 \end{bmatrix} = \begin{bmatrix} 21 \\ -17 \\ 8 \end{bmatrix},$$

$$A^* \begin{bmatrix} 18 \\ 9 \\ 4 \end{bmatrix} = \begin{bmatrix} 47 \\ -41 \\ 20 \end{bmatrix}, A^* \begin{bmatrix} 7 \\ 5 \\ 27 \end{bmatrix} = \begin{bmatrix} 144 \\ -215 \\ 132 \end{bmatrix}$$

My fellow spy is expecting a code consisting of positive, two-digit numbers. Some of the elements in the vector products are negative, and some have three digits. I add or subtract multiples of 26 to each of the elements until I get two-digit numbers. When I end up with a one-digit number, such as 8, I just put a 0 in front. Then I list all the elements from the vectors in a row.

$$\begin{bmatrix} 46 \\ -51 \\ 28 \end{bmatrix} \rightarrow \begin{bmatrix} 46 \\ 01 \\ 28 \end{bmatrix}, \begin{bmatrix} 52 \\ -41 \\ 19 \end{bmatrix} \rightarrow \begin{bmatrix} 52 \\ 11 \\ 19 \end{bmatrix},$$

$$\begin{bmatrix} 82 \\ -102 \\ 61 \end{bmatrix} \rightarrow \begin{bmatrix} 82 \\ 02 \\ 61 \end{bmatrix}, \begin{bmatrix} 21 \\ -17 \\ 8 \end{bmatrix} \rightarrow \begin{bmatrix} 21 \\ 09 \\ 08 \end{bmatrix},$$

$$\begin{bmatrix} 47 \\ -41 \\ 20 \end{bmatrix} \rightarrow \begin{bmatrix} 47 \\ 11 \\ 20 \end{bmatrix}, \begin{bmatrix} 144 \\ -215 \\ 132 \end{bmatrix} \rightarrow \begin{bmatrix} 92 \\ 19 \\ 80 \end{bmatrix}$$

460128521119820261210908471120921980

I send the sequence of numbers to my fellow spy. She has the code-breaker, the inverse of the matrix I used to do the encrypting. She breaks up the numbers into sets of three two-digit numbers for the vectors. Then she multiplies each vector by the inverse matrix.

$$A^{-1} = \begin{bmatrix} -1 & -5 & -7 \\ 2 & 5 & 6 \\ 1 & 3 & 4 \end{bmatrix}$$

$$A^{-1} * \begin{bmatrix} 46 \\ 01 \\ 28 \end{bmatrix} = \begin{bmatrix} -247 \\ 265 \\ 161 \end{bmatrix}, A^{-1} * \begin{bmatrix} 52 \\ 11 \\ 19 \end{bmatrix} = \begin{bmatrix} -240 \\ 273 \\ 161 \end{bmatrix},$$

$$A^{-1} * \begin{bmatrix} 82 \\ 02 \\ 61 \end{bmatrix} = \begin{bmatrix} -519 \\ 540 \\ 332 \end{bmatrix}, A^{-1} * \begin{bmatrix} 21 \\ 09 \\ 08 \end{bmatrix} = \begin{bmatrix} -122 \\ 135 \\ 80 \end{bmatrix},$$

$$A^{-1} * \begin{bmatrix} 47 \\ 11 \\ 20 \end{bmatrix} = \begin{bmatrix} -242 \\ 269 \\ 160 \end{bmatrix}, A^{-1} * \begin{bmatrix} 92 \\ 19 \\ 80 \end{bmatrix} = \begin{bmatrix} -747 \\ 759 \\ 469 \end{bmatrix}$$

Adding or subtracting multiples of 26 to each element until the number is between 1 and 26, my fellow spy reads:

$$\begin{bmatrix}-247\\265\\161\end{bmatrix} \to \begin{bmatrix}13\\5\\5\end{bmatrix}, \begin{bmatrix}-240\\273\\161\end{bmatrix} \to \begin{bmatrix}20\\13\\5\end{bmatrix},$$

$$\begin{bmatrix}-519\\540\\332\end{bmatrix} \to \begin{bmatrix}1\\20\\20\end{bmatrix}, \begin{bmatrix}-122\\135\\80\end{bmatrix} \to \begin{bmatrix}8\\5\\2\end{bmatrix},$$

$$\begin{bmatrix}-242\\269\\160\end{bmatrix} \to \begin{bmatrix}18\\9\\4\end{bmatrix}, \begin{bmatrix}-747\\759\\469\end{bmatrix} \to \begin{bmatrix}7\\5\\1\end{bmatrix}$$

13	5	5	20	13	5	1	20	20	8	5	2	18	9	4	7	5	1
M	E	E	T	M	E	A	T	T	H	E	B	R	I	D	G	E	A

The original vectors are re-created (with the last number showing up as a 1 rather than a 27 for the missing letter).

Saving the Spotted Owl

Several years ago, the news was full of stories of the conflict between the loggers in the Pacific Northwest and the threatened extinction of the spotted owl. Studies were done regarding the life cycle of the spotted owl and how the loss of timberland would affect the number of owls in the area.

The life cycle of spotted owls (which are expected to live about 20 years) is divided into three stages: juvenile, sub-adult, and adult. R. Lamberson is responsible for one study in which it was determined that, among female owls counted each year, the number of new juvenile owls was expected to be about 33 percent of the number of adult owls, only 18 percent of the juveniles are expected to survive to become sub-adults, 71 percent of the sub-adults are expected to become adults, and 94 percent of the adults are expected to survive. The transition matrix that models these figures is as follows:

$$\begin{array}{ccc}\text{juv} & \text{sub} & \text{adu}\end{array}$$
$$\begin{bmatrix}0 & 0 & 0.33\\0.18 & 0 & 0\\0 & 0.71 & 0.94\end{bmatrix} * \begin{bmatrix}\text{number juvenile}\\\text{number sub-adult}\\\text{number adult}\end{bmatrix}$$

Multiplying the transition matrix by a vector whose elements are the number of juvenile, sub-adult, and adult owls, the expected number of each stage owl is determined and long-range predictions are made.

TIP

Matrix multiplication is explained in Chapter 3.

Migrating Populations

Birds migrate, fish migrate, and people migrate from community to community or from one part of the country to another. City and community planners need to take into account the projected migration of the population as they make plans for roads and shopping centers and deal with other infrastructure issues.

In a particular metropolitan area, it's been observed that, each year, 4 percent of the city's population moves to the suburbs and 2 percent of the population of the suburbs moves to the city. Here's a transition matrix reflecting these changes:

$$\begin{array}{c} \\ \text{city} \\ \text{burbs} \end{array} \begin{array}{c} \begin{array}{cc} \text{city} & \text{burbs} \end{array} \\ \begin{bmatrix} 0.96 & 0.04 \\ 0.02 & 0.98 \end{bmatrix} \end{array}$$

From the matrix, you see that 96 percent of the people who were in the city stayed in the city the next year, and 98 percent of the people who were in the suburbs stayed in the suburbs.

You can determine the number of people expected to populate the different areas by multiplying by a vector with the current populations. The long-range projections are found by raising the transition matrix to various powers. According to this model, if the trends stay the same, the populations will become pretty stable in about 60 years, with 33 percent of the population living in the city and 67 percent living in the suburbs.

Plotting Genetic Code

Genetics affects everyone. From hair color to eye color to height, the genes passed along to you from your parents determine what you look like. My parents both had brown eyes, but I have two blue-eyed brothers. How does that work?

Just a brief genetics lesson: Brown eyes are *dominant,* and blue eyes are *recessive,* which means that brown wins over blue. But blue genes can come from brown-eyed parents. Eyes come in many different shades, too, so I'm going to work with a genetics model that doesn't have quite so many choices of color. (If you're interested in genetics, you can find all kinds of information in *Genetics For Dummies,* by Tara Rodden Robinson, PhD [Wiley].)

A particular type of flower comes in red, white, or pink. Assume that the flowers get only two genes for color. The flowers have the genes RR (red), Rr (pink), or rr (white). If both parent flowers are RR (red), then the offspring are also red. If one parent is RR (red) and the other Rr (pink), then the resulting flowers have half a chance of being RR (red) and half a chance of being Rr (pink). In a biology class, you get to map all these out. What I present here is a transition matrix showing parent and offspring and the probability that the offspring will have a particular color:

$$\begin{array}{cc} & \text{Offspring} \\ & \begin{array}{ccc} \text{RR} & \text{Rr} & \text{rr} \end{array} \\ \text{Parent} \begin{array}{c} \text{RR} \\ \text{Rr} \\ \text{rr} \end{array} & \begin{bmatrix} .5 & .5 & 0 \\ .25 & .5 & .25 \\ 0 & .5 & .5 \end{bmatrix} \end{array}$$

According to the matrix, a red flower has a 50 percent chance of having red offspring and a 50 percent chance of having pink offspring, if the other parent is Rr. A red flower can't have white offspring. A pink flower has a 50 percent chance of having pink offspring, a 25 percent chance of having white offspring, and a 25 percent chance of having red offspring.

What about future generations? What is the probability that a red flower will have a red grandflower (Sure! Isn't the offspring of the offspring the *grand*spring?) I show you here the next generation and, also, the generation after 20 more times.

$$\begin{array}{cc} & \text{Offspring}^2 \\ & \begin{array}{ccc} \text{RR} & \text{Rr} & \text{rr} \end{array} \\ \begin{array}{c} \text{RR} \\ \text{Rr} \\ \text{rr} \end{array} & \begin{bmatrix} .375 & .5 & .125 \\ .25 & .5 & .25 \\ .125 & .5 & .375 \end{bmatrix} \end{array} \qquad \begin{array}{cc} & \text{Offspring}^{20} \\ & \begin{array}{ccc} \text{RR} & \text{Rr} & \text{rr} \end{array} \\ \begin{array}{c} \text{RR} \\ \text{Rr} \\ \text{rr} \end{array} & \begin{bmatrix} .25 & .5 & .25 \\ .25 & .5 & .25 \\ .25 & .5 & .25 \end{bmatrix} \end{array}$$

According to the probability matrix, there's a 12.5 percent chance that the offspring of the offspring of a red flower will be white. And look at 20 generations down the line. The probabilities of the colors all line up at 25 percent for red or white and 50 percent for pink.

Distributing the Heat

Heat sources can come from several different directions. When an object is heated on one side, that side tends to be warmer, with cooling temperatures observed as you move away from the heat source. Consider the object in Figure 17-2 to be a square metal plate or a square room or anything with a relatively uniform interior. The numbers on each side indicate the heat being applied to that side in degrees Fahrenheit.

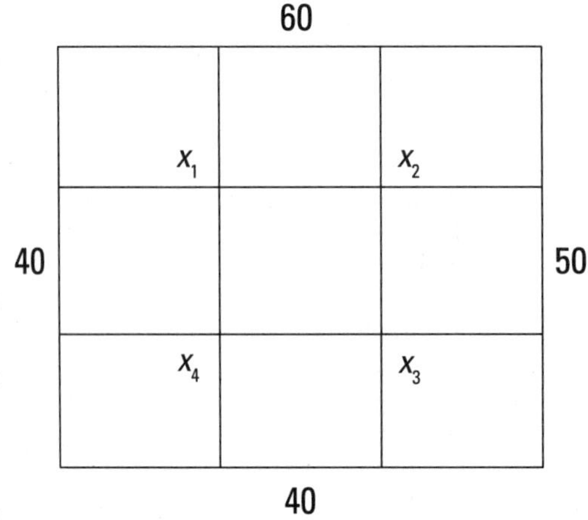

Figure 17-2: Different heats are applied to each side.

Now, using a *mean value property* (where you find the temperature at a particular point by adding the adjacent temperatures and dividing by the number of values used), I write the four equations representing the heat at each of the central points:

$$x_1 = \frac{40 + 60 + x_2 + x_4}{4},\ x_2 = \frac{60 + 50 + x_3 + x_1}{4}$$
$$x_3 = \frac{50 + 40 + x_2 + x_4}{4},\ x_4 = \frac{40 + 40 + x_3 + x_1}{4}$$

Simplifying the equations and setting up an augmented matrix, I use row operations to solve for the temperatures at the central points:

$$\left[\begin{array}{cccc|c} 4 & -1 & 0 & -1 & 100 \\ -1 & 4 & -1 & 0 & 110 \\ 0 & -1 & 4 & -1 & 90 \\ -1 & 0 & -1 & 4 & 80 \end{array}\right] \rightarrow \left[\begin{array}{cccc|c} 1 & 0 & 0 & 0 & 48.75 \\ 0 & 1 & 0 & 0 & 51.25 \\ 0 & 0 & 1 & 0 & 46.25 \\ 0 & 0 & 0 & 1 & 43.75 \end{array}\right]$$

The temperatures at x_1, x_2, x_3, and x_4, respectively, are 48.75, 51.25, 46.25, and 43.75. Each central point is affected by the two sides it's closest to and the other two central points it's adjacent to.

You can find information on solving systems of equations using augmented matrices in Chapter 4.

Making Economical Plans

A well-known model for an economy is the Leontief production model. The model takes into account the different production arenas of a country, city, or area and how they interact with one another. Here I show you a simplified version of how the model works.

Consider an economy in which you find three major production sectors: agriculture, manufacturing, and service. A production target is set for the month: 18 units of agriculture, 20 units of manufacturing, and 24 units of service. When producing these units, though, some of the sectors need the very products being produced to create this output. In fact, for each unit of output, agriculture needs 0.10 units of agriculture, 0.50 units of manufacturing, and 0.15 units of service. The manufacturing sector needs 0.05 units of agriculture and 0.15 units of service for each unit produced. And the service sector needs 0.10 units of service and 0.40 units of manufacturing.

To find the total number of units of each — the amounts needed for the production goals and the amounts needed by each sector to meet these goals, I set up an input matrix, A, and a demand vector, **d.**

$$\begin{array}{c} \\ \\ \text{Agr} \\ \text{A=Man} \\ \text{Ser} \end{array} \begin{array}{c} \text{Used in production} \\ \begin{array}{ccc} \text{Agr} & \text{Man} & \text{Ser} \end{array} \\ \begin{bmatrix} .10 & .50 & .15 \\ .05 & 0 & .15 \\ 0 & .40 & .10 \end{bmatrix} \end{array} \qquad \mathbf{d} = \begin{bmatrix} 18 \\ 20 \\ 24 \end{bmatrix}$$

To find the total amount of each type needed, **x,** I solve $\mathbf{x} = A\mathbf{x} + \mathbf{d}$, which simplifies to $\mathbf{x} = (I - A)^{-1}\mathbf{d}$.

$$\mathbf{x} = (I - A)^{-1}\mathbf{d} = \left(I - \begin{bmatrix} .10 & .50 & .15 \\ .05 & 0 & .15 \\ 0 & .40 & .10 \end{bmatrix} \right)^{-1} * \begin{bmatrix} 18 \\ 20 \\ 24 \end{bmatrix} = \begin{bmatrix} 42.05 \\ 27.97 \\ 39.10 \end{bmatrix}$$

So it takes 42.05 units of agriculture to create the 18 units in demand and what's necessary to create the units of agriculture, manufacturing, and service. Likewise with the other sectors.

Playing Games with Matrices

When you were a kid, you probably played lots of paper games — starting with tic-tac-toe — and now you've graduated to computer games that tax your wits. One particular game can be played on either paper or computer — although using the computer is much faster and more accurate. The game is called Squares or Magic Squares. It can be played with any size square, but it seems challenging enough to me with just a three-by-three grid.

You start with some random pattern of white and black markers on the squares and win by changing them to a pattern in which every marker is black except the center marker. I show you a possible starting pattern and the target pattern in Figure 17-3.

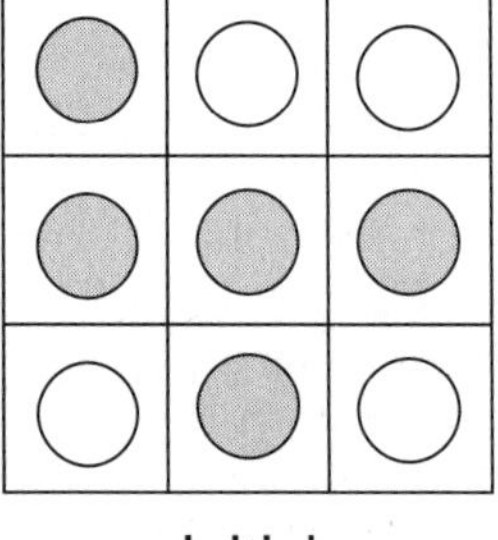

Initial

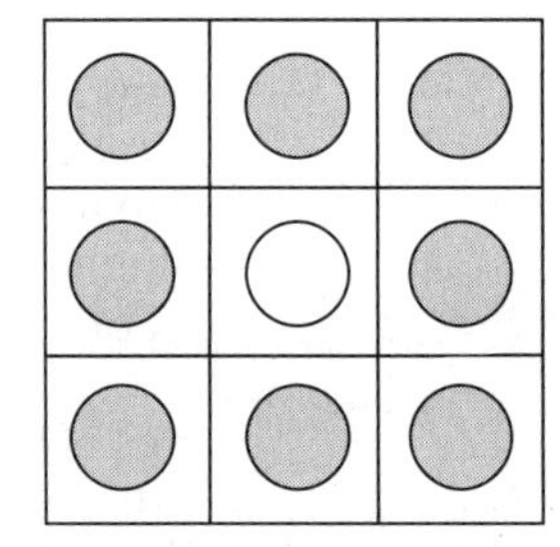

Final Target

Figure 17-3: Change the markers until you create the final pattern.

You might wonder why this game is considered such a challenge. After all, you can just change the five markers — four to black and one to white. Here's the catch: When you change one marker, you also have to change two, three, or four others. I've labeled the squares to describe the legal moves:

a	b	c
d	e	f
g	h	i

The rules are

- When you change any corner (*a, c, g,* or *i*) you also change each square adjacent to that corner:
  - *a: a, b, d, e*
  - *c: b, c, e, f*
  - *g: d, e, g, h*
  - *i: e, f, h, i*
- When you change the middle marker on any side, you change all the markers on that side.
  - *b: a, b, c*
  - *d: a, d, g*
  - *f: c, f, i*
  - *h: g, h, i*
- When you change the center marker, you change the markers above, below, left, and right of it, too: *b, e, h, d, f.*

Now you see the challenge — and why a computer game makes this much more efficient. But you can do this without a computer! In fact, any (doable) target pattern can always be reached in no more than nine moves. (But there are some initial patterns that have no solution — so you'd be switching markers forever if you stumbled on one of those. Computer games avoid those nasty types.)

The solution comes from a series of 9×1 vectors — the first nine of which are put into a 9×9 matrix. Each of the nine squares has its own vector based on the other squares that are affected when a switch is made on that square. The vectors contain 1s for the squares that are switched and 0s for the squares that are changed. I show you the general vector **v** and the vectors for *a, d,* and *e* here:

$$\mathbf{v} = \begin{bmatrix} a \\ b \\ c \\ d \\ e \\ f \\ g \\ h \\ i \end{bmatrix} \quad \mathbf{v}_a = \begin{bmatrix} 1 \\ 1 \\ 0 \\ 1 \\ 1 \\ 0 \\ 0 \\ 0 \\ 0 \end{bmatrix} \quad \mathbf{v}_d = \begin{bmatrix} 1 \\ 0 \\ 0 \\ 1 \\ 0 \\ 0 \\ 1 \\ 0 \\ 0 \end{bmatrix} \quad \mathbf{v}_e = \begin{bmatrix} 0 \\ 1 \\ 0 \\ 1 \\ 1 \\ 1 \\ 0 \\ 1 \\ 0 \end{bmatrix}$$

Now picture all nine vectors — v_a, v_b, . . . v_i — put into a huge 9 × 9 matrix. And then, imagine finding the inverse of that matrix. Here's the inverse matrix, though, in case you're in a hurry:

$$A^{-1} = \begin{bmatrix} 1 & 0 & 1 & 0 & 0 & 1 & 1 & 1 & 0 \\ 1 & 1 & 1 & 1 & 1 & 1 & 0 & 0 & 0 \\ 1 & 0 & 1 & 1 & 0 & 0 & 0 & 1 & 1 \\ 1 & 1 & 0 & 1 & 1 & 0 & 1 & 1 & 0 \\ 1 & 0 & 1 & 0 & 1 & 0 & 1 & 0 & 1 \\ 0 & 1 & 1 & 0 & 1 & 1 & 0 & 1 & 1 \\ 1 & 1 & 0 & 0 & 0 & 1 & 1 & 0 & 1 \\ 0 & 0 & 0 & 1 & 1 & 1 & 1 & 1 & 1 \\ 0 & 1 & 1 & 1 & 0 & 0 & 1 & 0 & 1 \end{bmatrix}$$

To find the solution — which squares to switch — you create a vector representing the initial setup and add it to a vector representing the target pattern. Then you multiply that inverse matrix times the sum. Using the initial and target patterns I showed you at the beginning of this section, here's the addition. The 2s mean that you switched twice and are back where you started, so they become 0s.

$$\text{initial} + \text{target} = \begin{bmatrix} 1 \\ 0 \\ 0 \\ 1 \\ 1 \\ 1 \\ 0 \\ 1 \\ 0 \end{bmatrix} + \begin{bmatrix} 1 \\ 1 \\ 1 \\ 1 \\ 0 \\ 1 \\ 1 \\ 1 \\ 1 \end{bmatrix} = \begin{bmatrix} 2 \\ 1 \\ 1 \\ 2 \\ 1 \\ 2 \\ 1 \\ 2 \\ 1 \end{bmatrix} \rightarrow \begin{bmatrix} 0 \\ 1 \\ 1 \\ 0 \\ 1 \\ 0 \\ 1 \\ 0 \\ 1 \end{bmatrix}$$

Now multiply the inverse matrix times the sum:

$$A^{-1} * \begin{bmatrix} 0 \\ 1 \\ 1 \\ 0 \\ 1 \\ 0 \\ 1 \\ 0 \\ 1 \end{bmatrix} = \begin{bmatrix} 2 \\ 3 \\ 2 \\ 3 \\ 4 \\ 4 \\ 3 \\ 4 \\ 4 \end{bmatrix} \rightarrow \begin{bmatrix} 0 \\ 1 \\ 0 \\ 1 \\ 0 \\ 0 \\ 1 \\ 1 \\ 0 \end{bmatrix} \rightarrow \begin{bmatrix} 0 \\ b \\ 0 \\ d \\ 0 \\ 0 \\ g \\ h \\ 0 \end{bmatrix}$$

The 3s that appeared in the product mean three switches, which revert back to 1s. The 4s in the product go back to 0s. (In general, odd numbers become 1s and even numbers become 0s.) It will take just four moves to change the initial pattern to the target pattern. Switch *b, d, g,* and *h* (and all the squares that come with each).

Chapter 18

Ten (Or So) Linear Algebra Processes You Can Do on Your Calculator

In This Chapter

- Graphing lines and finding intersections
- Taking the opportunity to do matrix operations
- Performing row operations on matrices
- Making the calculator round off numbers

Graphing calculators are wonderful instruments. You carry them around in your backpack, briefcase, or purse, and you whip them out to do all sorts of mundane tasks — things that you just don't want to waste your brain power on.

In this chapter, I show you some graphing capabilities, including how to solve a system of linear equations by graphing lines. I also show you several processes you can do involving matrices. Finally, I show you how to take a firm grip on large numbers and let your calculator round them down to size.

Note: There are way too many calculators on the market for me to be able to tell you exactly which button to push on your specific calculator, so I give more general instructions. If you're not familiar with your graphing calculator, have the user's manual handy — especially the part showing all the different functions in the drop-down menu of a particular button.

If you don't have your manual anymore (just like most of my students!), go online. Most calculator manufacturers have helpful Web sites to assist you.

Letting the Graph of Lines Solve a System of Equations

In Chapter 4, I talk about systems of equations and their solutions. I also show you how to graph systems of two linear equations on the coordinate axes. Here, I show you how to take one of the systems of equations and put it into your calculator for a solution.

To solve a system of two linear equations in two unknowns, do the following:

1. **Write each equation in *slope-intercept* form, $y = mx + b$.**

   This is also the *function form* where you show that the variable y is a function of the variable x.

2. **Enter the two equations in the y-menu of your calculator.**

3. **Graph the lines.**

   You may have to adjust your screen to make both of the lines appear. In particular, you want to see where the two lines cross.

   Some calculators have a Fit button that adjusts the height of the graphs to a particular width so you can view both lines at the same time.

4. **Use the Intersection tool.**

   You usually have to select the lines, one at a time, as prompted. Then you're invited to make a "guess." You can use the left-right arrows to move the pointer closer to the intersection, if you like.

5. **Press Enter to get the answer.**

   You'll be given both coordinates of the point (both variables x and y).

Now for an example. I want to solve the following system of equations:

$$\begin{cases} 2x + 3y = 9 \\ 4x - y = 11 \end{cases}$$

Writing the equations in slope-intercept form (Step 1), I get $y = -\frac{2}{3}x + 3$ and $y = 4x - 11$. I put the first equation in the first position of the y-menu and the second equation in the second position (Step 2). Graphing the lines, I find that the standard setting (from -10 to $+10$ in both directions) works just fine (Step 3). Figure 18-1 shows you what the graph might look like.

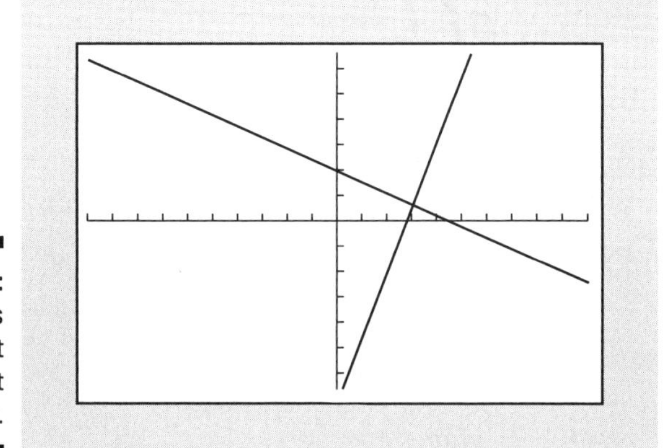

Figure 18-1: The lines intersect in the first quadrant.

Using the Intersection tool (Step 4), I'm prompted by the calculator to choose the two lines. Some calculators have you check off the equations; other calculators have you click on the actual graphs of the lines. I press Enter (Step 5) and initiate the finding of the solution. Some calculators give the answer as an ordered pair (3,1), and others show $x = 3$, $y = 1$. In either case, I've found the point of intersection of the lines — which corresponds to the solution of the system. If my lines have no intersection, then I'll see two parallel lines that never touch. If I try to make the calculator solve the parallel system, anyway, I'll get some sort of error message.

Making the Most of Matrices

Matrix operations are quick and accurate when performed with a graphing calculator. You do have some downsides, though: The matrices have to be entered into the calculator one element at a time. And then, if you're working with multiple matrices, you have to keep track of what you put where — which matrix is which and what you named them. The computations performed on matrices often result in numbers with many decimal places, making it hard to see all the elements at one time. You either have to change the numbers to fractions or round them off. I show you how to quickly and easily round the numbers down for the display in the "Adjusting for a Particular Place Value" section, later in this chapter.

To enter a matrix into the calculator, you choose a name. No, Henry isn't an appropriate name in this case. Usually the calculator gives you a choice from A, B, C, D, and so on. (How dull.) After choosing a name, you enter the dimension of the matrix, such as 2×3. Then you enter each element in its appropriate place. Now the fun begins.

Adding and subtracting matrices

You can add or subtract matrices that have the same dimensions. If you try to make the calculator add or subtract matrices with different dimensions, you'll get a rude message such as "Are you kidding me?" No, not really. You'll get something more polite such as "ERROR."

To add or subtract two matrices with the same dimensions, you do the following:

1. **Enter the elements into the two matrices.**

   Keep track of which matrix is which, in terms of the elements.

2. **With a new screen, enter the name of the first matrix, insert the operation symbol (+ or –), and then insert the name of the second matrix.**
3. **Press Enter.**

Did you think it was going to be something complicated and difficult? Sorry. Not this time.

A calculator that I use shows the following for the operation: [A] + [B]. The brackets indicate a matrix.

Multiplying by a scalar

When you multiply a matrix by a scalar, you essentially multiply each element in the matrix by the same number. Using your calculator to do this task saves you from having to perform the same multiplication over and over again. And the process is very simple. If you want to multiply matrix A by 2, in my calculator you just type in **2*[A]** and — *voilà!* — you get the answer.

Multiplying two matrices together

The hardest part of multiplying two matrices together has to do with the Law of Order. No, you don't call on Sam Waterston. It's just that when multiplying matrices, order matters, and that's the law! You can multiply a 2×3 matrix times a 3×5 matrix, in that order, but you can't multiply in the opposite order. (In Chapter 3, I go through all the reasons for the rule.) For now, just keep track of which matrix is which when multiplying. Your calculator won't let you get away with any switch — you'll get an error message if you try. But prevention is the best way to go.

To multiply a 2×3 matrix A times a 3×5 matrix B in my calculator, I just type in **[A]*[B]** and press Enter to get the resulting 2×5 matrix.

Performing Row Operations

Many tasks in linear algebra involve vectors and matrices. And, quite often, you need to change the format of the matrix to solve a system of equations or write a relationship between variables or multipliers. Graphing calculators are great for performing operations on matrices, especially when the elements in the matrices are large and cumbersome or when you have to divide by numbers that don't result in nice integers.

I list many of the most common operations in this section. ***Note:*** Not all operations may be on every calculator, and some calculators may have more operations than these. Your user's manual is your best guide.

Switching rows

You have a matrix with four rows and would prefer that the fourth row be where the first row is. (Maybe a 1 is the first element of the fourth row.) Your calculator should be able to do the switch (often called *row swap*). You'll have a specific script to follow when entering the commands for the row-switching operation. You'll have to identify what matrix is to be operated on, because you may have several matrices loaded in your calculator. Then you'll have to designate the rows.

For example, on my calculator, I'd type in **rowSwap([A],1,4),** meaning that I want to switch rows 1 and 4 in my matrix A. Of course, I first have to enter the matrix and save it as matrix A. Then I have to find the *rowSwap* command. In most calculators, you can't just type in the letters using the alphabetical key, you have to find the built-in operation indicating a matrix. Check under the matrix button for a drop-down menu involving operations.

Adding two rows together

You add two rows of a matrix to create a different format. You're usually targeting just one of the elements in the matrix, and all the other elements in the rows have to just follow along. For example, in matrix A, I want to add rows 1 and 3, because the sum of the first elements in those two rows is equal to 0.

$$A = \begin{bmatrix} 1 & 2 & 3 & 4 & 5 \\ 0 & -2 & 5 & 6 & 4 \\ -1 & 3 & 6 & 7 & 2 \\ 3 & 1 & 5 & 3 & 6 \end{bmatrix}$$

Once you have the sum, you need to determine where you're going to put it. If my goal is to get all 0s below the 1 in the first row, first column, then I want the sum of rows 1 and 3 to be placed in row 3.

The calculator entry has to include which matrix you're working with, which rows need to be added together, and where to put the sum. In my calculator, I find the matrix operation that adds rows together, and the process I type in is ***row*+([A],1,3)**. The calculator automatically places the sum in row 3, the second row named in the script. If you want the sum to be placed in row 1, then you use *row*+([A],3,1).

Adding the multiple of one row to another

If your goal is to have all 0s above and below a particular element in your matrix, then it's peachy if all the elements are opposites of the target element; all you'd have to do is add the rows to the target row. You usually aren't that lucky — it's more common that you'd have to add a multiple of one row to another row.

For example, matrix A in the preceding section has a 1 in the first row, first column, and a 3 in the fourth row, first column. To change the 3 to a 0, I need to multiply row 1 by -3 and add this alteration of row 1 to those in row 4. I want all the results to be placed in row 4. The commands for this operation include

- Naming the matrix
- Naming the multiplier and which row gets multiplied
- Naming the row that the multiple is getting added to
- Indicating where the result gets placed

Whew! This is what I have to put in my calculator: ***row*+(–3,[A],1,4)**. I type in the multiplier first, then the matrix, then the row that gets multiplied, and then the row that the product gets added to. The result goes in that fourth row.

Right now, I have to tell you a secret: I use a mailing label, stuck to the inside cover of my calculator, to write down all these commands. I can never remember what goes where, and my user's manual is never handy. So these little scripts get written down for quick reference. You may want to do the same.

Multiplying a row by a scalar

You can multiply an entire matrix by a scalar — each element in the matrix gets multiplied by that same number. But sometimes you just want one of the rows changed or multiplied by a particular number. In matrix A, I'd like to get rid of the fractions in row 2, so I choose to multiply row 2 by 6. In matrix B, I want a 1 for the third element in row 3, so I multiply every element in row 3 by $^1/_2$.

$$A = \begin{bmatrix} 1 & 3 & 4 & -2 & 5 \\ 0 & \frac{1}{2} & -3 & \frac{2}{3} & \frac{5}{6} \\ 0 & 3 & 5 & 6 & -4 \end{bmatrix} \quad B = \begin{bmatrix} 1 & 0 & 3 & 4 & -9 & 5 & 6 \\ 0 & 1 & 5 & 6 & 3 & 9 & 0 \\ 0 & 0 & 2 & 5 & 4 & 3 & -1 \end{bmatrix}$$

The operation for multiplying a single row by a scalar is relatively simple. You just have to name the matrix, the multiplier, and the row. For example, in my calculator, I enter ***row*(6,[A],2),** putting the multiplier first, the matrix second, and the row third. The resulting elements are all returned to the same row that got multiplied.

Creating an echelon form

Not all calculators have this capability, but some will take a matrix and change it to the *row echelon form* or the *reduced row echelon form.* You use the echelon forms to determine the solution of a system of equations or to write the relationship between variables in equations or other relationships. (Refer to Chapter 4 for more on the echelon forms.) The calculator usually has a restriction on the types of matrices that the echelon form operations are performed upon: The number of rows in the matrix can't be greater than the number of columns.

The row-echelon form changes your matrix into one that has 1s in the *aii* positions (row and column are the same) whenever the row is not all 0s, and 0s below each 1. Rows of all 0s appear as the last rows of the matrix. The reduced row echelon form has the 1s in the a_{ii} positions (when the row isn't all 0s) and 0s both below and above each 1. The matrix E is shown in its original form, row-echelon form, and reduced row-echelon form:

$$E = \begin{bmatrix} 1 & 4 & -3 & 18 \\ 3 & -1 & 6 & -12 \\ 1 & 1 & 3 & -2 \end{bmatrix}$$

$$\rightarrow \begin{bmatrix} 1 & -\frac{1}{3} & 2 & -4 \\ 0 & 1 & -\frac{15}{13} & \frac{66}{13} \\ 0 & 0 & 1 & -\frac{62}{33} \end{bmatrix} \rightarrow \begin{bmatrix} 1 & 0 & 0 & \frac{8}{11} \\ 0 & 1 & 0 & \frac{32}{11} \\ 0 & 0 & 1 & -\frac{62}{33} \end{bmatrix}$$

To perform these operations with your calculator, you usually just have to access the operation and then type in the name of the matrix.

Raising to Powers and Finding Inverses

Using your calculator, you can raise a matrix to a really high power and even use the reciprocal to find a matrix's inverse. One restriction is, of course, that you're dealing with a square matrix. Only square matrices can multiply times themselves. And only square matrices have inverses (although not all square matrices do have inverses).

Raising matrices to powers

Raising the matrix A to the 20th power is horrible to have to do with paper and pencil. With a graphing calculator, though, all you have to do is enter the elements of the matrix, save the matrix, and then raise the matrix to that power. Raising matrix A to the 20th power, I enter **[A]^20** in my calculator.

Inviting inverses

The inverse of a matrix is another matrix of the same size as the original matrix. The inverse of matrix A times the matrix A, itself, is an identity matrix. Even multiplying in the opposite direction gives you the same answer. Symbolically: $[A]^{-1}*[A] = [A]*[A]^{-1} = I$.

To find the inverse of a square matrix, you must use the built-in reciprocal button. In some calculators, it looks like x^{-1}. You can't use the caret button, ^-1, to get the inverse in most calculators. The calculator will tell you if the matrix is singular (has no inverse).

Determining the Results of a Markov Chain

Some mathematical models of real-life situations involve the change of the state of the situation based on a probability. A *Markov chain* is a process in which the outcome of an experiment or the makeup of the next generation is completely dependent on the previous situation or generation. Basic to studying a Markov chain is a *transition matrix.* A transition matrix gives all the probabilities that something moves from one state to another. For example, in the matrix shown here, you see the probabilities that people who bought items A, B, or C will buy A, B, or C the next time.

$$\begin{array}{c} \\ A \\ B \\ C \end{array}\begin{array}{c} \begin{array}{ccc} A & B & C \end{array} \\ \begin{bmatrix} .65 & .10 & .25 \\ .25 & .05 & .70 \\ .35 & .15 & .50 \end{bmatrix} \end{array}$$

You read the entries from the left side to the top. For example, you see that if a person buys A now, the probability is 0.65, or 65 percent, that he'll buy A next time; 10 percent that he'll buy B next time; and 25 percent that he'll buy C next time. The probability is only 5 percent that a person buying B this time will buy B next time, and so on.

You use the transition matrix to determine what percentage of people will buy A, B, and C over a period of time. If you square the matrix, then you see that the percentage of people who bought A the third time after buying it the second time has gone down to 53.5 percent and the number who bought C after buying A has gone up to 35.75 percent.

$$\begin{bmatrix} .65 & .10 & .25 \\ .25 & .05 & .70 \\ .35 & .15 & .50 \end{bmatrix}^2 = \begin{bmatrix} .535 & .1075 & .3575 \\ .42 & .1325 & .4475 \\ .44 & .1175 & .4425 \end{bmatrix}$$

What's most telling, though is what happens in the long run. You can raise the matrix to a high power — in this case, I only had to do the eighth power — to see what the long-term trends are. The values are all rounded to the nearest thousandth.

$$\begin{bmatrix} .65 & .10 & .25 \\ .25 & .05 & .70 \\ .35 & .15 & .50 \end{bmatrix}^8 = \begin{bmatrix} .484 & .114 & .402 \\ .484 & .114 & .402 \\ .484 & .114 & .402 \end{bmatrix}$$

You see from the Markov chain that 48.4 percent of the people will be buying A, 11.4 percent will be buying B, and 40.2 percent will be buying C.

Solving Systems Using $A^{-1}*B$

In Chapter 4, I show you how to solve systems of linear equations in a multitude of ways (well, maybe *multitude* is overstating it a bit). But here I present a way to solve systems of equations that have a single solution. You use the coefficient matrix, A; use the constant matrix, B; and multiply the inverse of the coefficient matrix times the constant matrix.

For example, I want to solve the following system of equations:

$$\begin{cases} 3x_1 - 2x_2 + x_3 + 4x_4 - 9x_5 = -27 \\ 2x_1 + 3x_2 + 3x_3 - x_4 - x_5 = 8 \\ -4x_1 - x_2 - 3x_3 + 2x_4 + 6x_5 = 23 \\ 5x_1 - 7x_2 + x_3 + x_4 + x_5 = 3 \\ x_1 + x_2 + 7x_3 + 4x_4 + 2x_5 = 50 \end{cases}$$

What a task! Five equations and five unknowns. But, as long as the matrix made up of the coefficients isn't singular (that is, has no inverse), we're in business. Create a matrix A, made up of the coefficients of the variables, and a matrix B, made up of the constants:

$$A = \begin{bmatrix} 3 & -2 & 1 & 4 & -9 \\ 2 & 3 & 3 & -1 & -1 \\ -4 & -1 & -3 & 2 & 6 \\ 5 & -7 & 1 & 1 & 1 \\ 1 & 1 & 7 & 4 & 2 \end{bmatrix} \quad B = \begin{bmatrix} -27 \\ 8 \\ 23 \\ 3 \\ 50 \end{bmatrix}$$

The system of equations is now represented by A*X = B, where the vector X contains five elements representing the values of the five variables.

Now tell the calculator to multiply the inverse of matrix A times matrix B. The resulting vector has the values of the variables.

$$[A]^{-1}*[B] = \begin{bmatrix} 1 \\ 2 \\ 3 \\ 4 \\ 5 \end{bmatrix} = [X] = \begin{bmatrix} x_1 \\ x_2 \\ x_3 \\ x_4 \\ x_5 \end{bmatrix}$$

You see that $x_1 = 1$, $x_2 = 2$, $x_3 = 3$, $x_4 = 4$, and $x_5 = 5$.

Adjusting for a Particular Place Value

You're concerned with place value when you're trying to rein in all those decimal places that result from computations. When you divide 4 by 9, you get 0.44444444 . . . with 4s going on forever. You round to thousandths by keeping the first four 4s and lopping off the rest. You round up if what's being removed is bigger than 5 and just truncate (chop off) if it's smaller than 5. If what's to be removed is exactly 5, you either round up or down to whichever number is even. Sounds like too much hassle? Then let your calculator do the work for you.

One technique is to just change the *mode,* or settings, of your calculator. You can choose whether to let the calculator show all the decimal places in its capacity or to limit the decimal places to one, two, three, or however many places you want. This is especially nice when working with matrices. By limiting the size of the numbers resulting from computations, you have an easier time seeing the values of the different elements.

For example, in Figure 18-2, I show you what the screen view looks like after raising a 3 × 3 matrix to the eighth power. (This is the matrix I used in the earlier section, "Determining the Results of a Markov Chain.") The decimal numbers are so long, that you only see the first column.

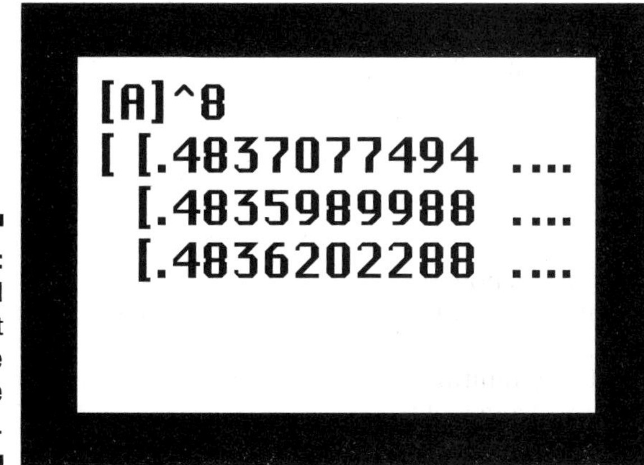

Figure 18-2: You scroll to the right to see the rest of the matrix.

Some calculators even have a *round* function that is invoked after computations are completed. You can type in instructions to have the calculator round all elements of a particular matrix to your chosen number of decimal places. For example, in my calculator, if I use the *round* function, my instruction *round* ([A], 2) tells the calculator to take a look at matrix A and round each element to two decimal places. Sweet! I'd be able to see all the digits of all the elements in the matrix in Figure 18-2.

Chapter 19

Ten Mathematical Meanings of Greek Letters

In This Chapter

- Taking a look at some familiar and not so familiar meanings for Greek letters
- Getting the historical scoop on the use of some Greek letters

If you ever belonged to a college fraternity or sorority, then you know all about having to recite the Greek letters in order and recognizing their capital and lowercase symbols. Most fraternities and sororities use three Greek letters in their names, but some only use two. And many honorary societies have adopted Greek letters to give their name a bit of traditional and historical flavor.

Mathematicians, statisticians, scientists, and engineers have all adopted one Greek letter or more as their own, special symbol to indicate a value or ratio. The Greek letters represent the standard for some value or computation that's used often in each particular field.

In this chapter, I acquaint you with some of the more popular uses of the letters of the Greek alphabet.

Insisting That π Are Round

Most schoolchildren know that "pi are squared," so you have to correct them and say that they usually aren't — at least in bakeries. The area of a circle, of course, is found with the famous formula: $A = \pi r^2$.

You may think that the Greek letter π has *always* represented the number 3.14159 . . . , ever since the relationship between the circumference and diameter of a circle was discovered, but that isn't so. The mathematician Leonhard Euler (1707–1783) is credited with popularizing the use of π when writing mathematical discourses, although there is some record of π also being used shortly before Euler was born. The Greek letter π also is referred to as *Archimedes's constant,* giving you a better idea of how long the ratio between a circle's circumference and diameter has been recognized.

Determining the Difference with Δ

The capital form of the Greek letter delta, Δ, signifies a change or a *difference.* In mathematics, Δx represents "the change in x" — how much the variable changes under certain circumstances. When using the lowercase version of delta, δ, the letter still represents a change or variation in mathematics. And attach a + or – to the δ and you have a partial charge on an ion in chemistry.

Summing It Up with Σ

The capital letter sigma, Σ, represents the sum terms in a list or sequence. Σ is often called the *summation operator,* indicating that you add up all the terms in a certain format for however long you want.

Change the letter to a lowercase sigma, σ, and you enter the world of statistics. The letter σ represents *standard deviation;* square sigma, σ^2, and you have *variation,* or spread.

Row, Row, Row Your Boat with ρ

The lowercase Greek letter rho, ρ, takes on different meanings in different settings. In statistics, ρ is a correlation coefficient, indicating how much one entity determines another — sort of like predictability. In physics, ρ refers to density. Density is measured in mass per unit volume, such as 40 pounds per cubic foot. And in finance, you find ρ having to do with sensitivity to interest rate. (I must be sensitive and not hurt interest's feelings!) As you can see, ρ is a pretty popular symbol, spanning statistics, physics, and banking.

Taking on Angles with θ

The lowercase Greek letter theta, θ, is a familiar sight to geometry students. Theta is a standard symbol for an angle. Rather than name an angle with three points, such as $\angle$ ABC, you insert the θ symbol right into the interior of the angle and refer to the angle with that one Greek letter. You also see θ used as the representative angle measure in many trigonometric identities and computations.

In finance, you find θ as a symbol representing sensitivity to passage of time. In thermodynamics, θ is used as a potential temperature.

Looking for a Little Variation with ε

The lowercase Greek letter epsilon, ε, in mathematics signifies *just a little different.* You might hear one mathematician say to another, "He's within epsilon of proving it." (Okay, so we're an odd bunch.) Within epsilon, or $\pm\varepsilon$ indicates that you're not exactly there; you're above, below, to the right, to the left, or otherwise askew of the target.

In computer science, ε indicates an empty string, in code theory. And in astrometry, the symbol ε indicates the Earth's axial tilt — we're just a bit off.

Taking a Moment with μ

The lowercase Greek letter mu, μ, is rather a*mu*sing. (Sorry, I couldn't help myself.) Actually, μ is one of the most popular of the Greek letters — used by many mathematical and scientific fields.

In number theory, μ represents the *Möbius function* (a function from number theory that outputs 0, 1, or –1, depending on the prime factorization of the input number). In statistics and probability, μ represents the mean or expected value of a population. And then μ is the coefficient of friction in physics and the service rate in queuing theory. The letter also denotes 10^{-6}, or one-millionth.

So, how do you know, if you happen upon a μ just lying around, what it's supposed to represent? It all depends on who you're with and what you're talking about!

Looking for Mary's Little λ

The letter lambda, λ, is another all-purpose lowercase Greek letter. You see it earlier in this book, in Chapter 16, when I discuss eigenvalues and eigenvectors. The λs represent any eigenvalues of the particular matrix.

λ also represents 1 microliter — a measure of volume that's equivalent to one cubic millimeter (not very big). When measuring radio waves, λ represents one wavelength. In physics and engineering, λ represents the average lifetime of an exponential distribution. And in probability theory, you see λ as the expected number of occurrences in a *Poisson distribution* (which expresses the probability of events occurring based on averages from past observations).

Wearing Your ΦBK Key

A high personal honor is being chosen as a member of Phi Beta Kappa, ΦBK, an honor society whose members are the best of the best achievers. The society was founded in 1776 at the College of William & Mary. It's the oldest honor society in the United States. The letters ΦBK represent "Love of learning is the guide of life."

Coming to the End with ω

You hear *the alpha and the omega* in reference to the beginning and the end — the first and last letters of the Greek alphabet. The lowercase letter omega, ω, is used in other situations, of course. In probability theory, ω represents a possible outcome of an experiment. And, although the capital letter N is usually the choice for representing the set of natural numbers, you also see ω used for this designation in set theory.

The capital omega, K, represents the ohm for electricians and the rotation of a planet for astronomers. Mathematicians also have an omega function, designated K, which counts how many prime factors a number has.

Glossary

absolute value: The numerical value or worth of a number without regard to its sign; the number's distance from zero. The absolute value of both +3 and –3 is 3.

adjacent: Next to; sharing a side or vertex; having nothing between. The numbers 2 and 3 are adjacent whole numbers; in rectangle *ABCD,* the sides *AB* and *BC* are adjacent.

adjoint: The transpose of a square matrix obtained by replacing each element with its cofactor. Each element a_{ij} is replaced with the cofactor of the element in a_{ji}. Also called *adjugate.*

associativity: The property of addition and multiplication that preserves the final result when the elements in the operation are regrouped. Adding (8 + 3) + 4 yields the same result as adding 8 + (3 + 4).

augmented matrix: A matrix consisting of the coefficients of a system of linear equations and a final column containing the constants. Each column contains coefficients of the same variable from the system.

basis: A set of vectors, $\{\mathbf{v}_1, \mathbf{v}_2, \mathbf{v}_3, \ldots\}$, which spans a vector space, *V,* and has the property that the vectors are linearly independent.

characteristic equation: The linear equation obtained by setting the determinant of the matrix formed by $A - \lambda\mathbf{I}$ equal to 0. The symbol λ represents the eigenvalues of the matrix A.

characteristic polynomial: The determinant of the square matrix defined by subtracting a multiple of an identity matrix from a transformation matrix, $A - \lambda\mathbf{I}$.

closure: The property of number systems and vector spaces in which particular operations used in the system produce only elements of that system or vector space. The whole numbers are closed to addition, but the whole numbers are *not* closed to division (dividing 4 by 5 does not result in a whole number).

coefficient: A constant multiplier of a variable, matrix, or vector. In the term $2x$, the number 2 is the coefficient.

coefficient matrix: A matrix consisting of all the coefficients of the variables in a system of linear equations.

cofactors of matrix: Created from a matrix by multiplying the determinants of the minors of the matrix by powers of –1. The cofactor A_{ij} of matrix A is equal to $(-1)^{i+j}\det(M_{ij})$, where M_{ij} is a minor of the matrix.

collinear: Lying on the same line.

column matrix (vector): An $m \times 1$ matrix; a matrix with one column and m rows.

commutativity: The property of addition and multiplication that preserves the final result when the elements in the operation are reversed in order. Multiplying 4×5 yields the same result as multiplying 5×4.

consecutive: Items or elements that follow one another in some sequence. The integers 4, 5, and 6 are three consecutive integers.

consistent system of equations: A system of equations that has at least one solution.

constant: A value that doesn't change. In the expression $x^2 + x + 4$, the number 4 is a constant, and the xs are variables.

constraint: A qualification or limiting rule that affects the values that a variable may have. The variable x may be limited to only numbers greater than 2: $x > 2$.

contraction: A function or transformation in which distances are shortened. The function f performed on the element u is a contraction if $f(u) = ru$, where $0 < r < 1$.

coordinate axes: Perpendicular lines dividing the plane into four quadrants. Traditionally, the x-axis is an horizontal line, and the y-axis is a vertical line.

coordinate vector: A column or row vector of real numbers $a_1, a_2, \ldots, a_k$, which are used to express the ordered basis $S = \mathbf{v}_1, \mathbf{v}_2, \ldots, \mathbf{v}_k$ in the form $a_1\mathbf{v}_1, a_2\mathbf{v}_2, \ldots, a_k\mathbf{v}_k$.

dependent system of equations: A system of equations in which at least one of the equations is a linear combination of one or more of the other equations in the system.

determinant: The sum of all the products of all the possible permutations of the elements of a square matrix. Each product is multiplied by either +1 or –1.

diagonal matrix: A square matrix in which all the elements not on the main diagonal are equal to zero.

diagonalizable matrix: A matrix for which you can construct another matrix similar to that original matrix that is a diagonal matrix.

dilation: A function or transformation in which distances are lengthened. The function f performed on the element u is a dilation if $f(u) = ru$, where $r > 1$.

dimension of matrix: The number of rows and columns of a matrix, expressed in the format $m \times n$. The dimension 2×3 indicates a matrix with two rows and three columns.

dimension of a vector space: The number of vectors in the basis of the vector space.

distributive property: The property of multiplication over addition in which each element in a grouping is multiplied by another element outside the grouping and the values before and after the process are the same. When distributing the number 2 over the sum of 4 and 5, the results are the same on each side of the equation: $2(4 + 5) = 2(4) + 2(5)$.

dot product (inner product) of two vectors: The sum of the products of the corresponding elements of the vectors.

eigenvalue: A number associated with a matrix in which multiplying a vector by the matrix is the same as multiplying that vector by the eigenvalue; the resulting vector is a multiple of the original vector multiplied by the eigenvalue. Given the square matrix A, vector $\mathbf{x}$, and eigenvalue λ, then $A\mathbf{x} = \lambda\mathbf{x}$.

eigenvector: A vector $\mathbf{x}$ associated with a square matrix A such that $A\mathbf{x} = \lambda\mathbf{x}$, where λ is some scalar. The scalar λ is called an *eigenvalue* of A; you say that $\mathbf{x}$ is an *eigenvector* corresponding to λ.

equivalent matrices: Two $m \times n$ matrices in which one is obtained from the other by performing elementary row operations.

free variable: A variable in a system of equations for which the other variables can be written in terms of that variable. The variable x_3 is a free variable if $x_1 = 2x_3$ and $x_2 = 3x_3$.

geometric mean: The number between two other numbers that is the positive square root of the two numbers. The geometric mean of 4 and 9 is 6, because 6 is the square root of $4 \times 9 = 36$.

homogeneous system of equations: A system of equations, all set equal to zero. The system always has a solution; the solution is *trivial* if all the variables are equal to zero and *nontrivial* if some of the variables are not equal to zero.

identity: An element in a set associated with an operation such that performing the operation with another element and the identity does not change the value or state of that other element. In the real number system, the additive identity is 0 and the multiplicative identity is 1.

identity matrix: A square matrix with its main diagonal (from upper left to lower right) consisting of 1s and the rest of the elements 0s.

image: The resulting matrix or vector after performing a function operation on a matrix or vector.

inconsistent system of equations: A linear system that has no solution.

independent equations: A system of equations for which none of the equations is a linear combination of the other equations in the system.

index of element in matrix: The row and column designation or position of the element a, indicated by a subscript, a_{ij}. The element a_{24} is in the second row and fourth column of the matrix.

inverse matrix: A square matrix associated with another square matrix such that the product of the matrix and its inverse is an identity matrix.

inversion (of a permutation): A permutation in which a larger integer precedes a smaller integer. When considering the permutations of the first three positive integers, the permutation 132 has one inversion (3 comes before 2), and the permutation 321 has three inversions (3 comes before 2, 3 comes before 1, 2 comes before 1).

invertible matrix: A square matrix that has an inverse. Also called a *nonsingular matrix.*

kernel of linear transformation: The subset of a vector space for which a linear transformation takes all the vectors to a 0 vector.

linear combination: The sum of the products of two or more quantities, each multiplied by some constant value. The linear combination of the elements of a set of vectors and some scalars is written $a_1\mathbf{v}_1 + a_2\mathbf{v}_2 + \ldots + a_k\mathbf{v}_k$, where each a_i represents a real number and each $\mathbf{v}_i$ represents a vector.

linear equation: An equation in the form $a_1x_1 + a_2x_2 + \ldots + a_kx_k = b$, where the as are constants, the xs are unknowns, and the b is a constant.

linear independence: When the linear combination of a set of elements is equal to 0 only when each of the scalar multiples is 0. The set of vectors $\{\mathbf{v}_1, \mathbf{v}_2, \ldots, \mathbf{v}_k\}$ is linearly independent when $a_1\mathbf{v}_1 + a_2\mathbf{v}_2 + \ldots + a_k\mathbf{v}_k = 0$ only if $a_1 = a_2 = \ldots = a_k = 0$.

linear transformation (operator): A transformation or operation performed on elements from a set in which addition and scalar multiplication are preserved. Letting T represent the transformation, **u** and **v** represent vectors, and a and b represent scalars, then $T(a\mathbf{u} + b\mathbf{v}) = aT(\mathbf{u}) + bT(\mathbf{v})$.

magnitude of vector: The length of a vector. The value obtained by computing the square root of the sum of the squares of the elements of the vector.

main diagonal of matrix: The elements of a square matrix in which the two elements of the index are the same: $a_{11}, a_{22}, \ldots, a_{kk}$. These elements start in the upper left-hand corner and run down to the lower right-hand corner of the matrix.

matrix: A rectangular array of numbers or elements with m horizontal rows and n vertical columns.

minor of matrix: A subset of a matrix delineated by removing a row and column associated with a particular element. A sub-matrix is related to the a_{ij} element of the original matrix in that the ith row and jth column are eliminated.

natural basis of vectors: The basis of a set of vectors where one element in each vector is equal to 1 and the other elements are equal to 0, and no vector has the 1 in the same position. See also *standard basis of matrices.*

noninvertible matrix: A square matrix that does not have an inverse. Also called a *singular matrix.*

nonsingular matrix: A square matrix that has an inverse. Also called an *invertible matrix.*

non-trivial solution: A solution of a homogeneous system of equations in which the values of the scalars are not all zero.

null space: The set of all the solutions of a system of homogeneous equations.

ordered pair (triple, quadruple): The listing of the coordinates of points or elements of a vector, in parentheses and separated by commas, in which the order is natural and sequential. The point on the coordinate system (2,3) is an ordered pair where $x = 2$ and $y = 3$.

orthogonal: Vectors whose inner product is 0. An orthogonal basis of a vector space is one in which all the vectors are orthogonal to one another.

orthonormal: An orthogonal basis of a vector set in which the vectors all have a magnitude of 1.

parallelepiped: A polyhedron in which all the faces are parallelograms.

parallelogram: A four-sided polygon in which the opposite sides are both parallel and congruent.

parameter: A variable used to express a relationship and whose value distinguishes the various cases.

parametric equation: An equation in which a parameter is used to distinguish between different cases. In the slope-intercept form of the linear equation, $y = mx + b$, the m and b are parameters, affecting the slant and y-intercept of the graph of the line.

permutation: A rearrangement of an ordered listing of elements. The six permutations of the first three letters of the alphabet are: abc, acb, bac, bca, cab, and cba.

perpendicular: At right angles to one another.

polyhedron: A multi-surface figure.

polynomial: A function represented by $f(x) = a_n x^n + a_{n-1} x^{n-1} + a_{n-2} x^{n-2} + \ldots + a_1 x^1 + a_0$, where as are real numbers and ns are whole numbers.

range of a function *f*: All the results of performing a function operation on all the elements in a set.

range of linear transformation: All the vectors that are results of performing a linear transformation on a vector space.

ray: A geometric figure starting with an endpoint and including all the points extending along a line in one direction from that endpoint.

real number: Any rational or irrational number.

reciprocal: Any real number (except 0) raised to the –1 power. The reciprocal of 2 is $^1/_2$. The reciprocal of $^4/_3$ is $^3/_4$. The product of a number and its reciprocal is 1.

reflection: A transformation in which all reflected points are transported to the opposite side of the line of reflection at an equal distance to the line of reflection and on a line perpendicular to the line of reflection.

reverse diagonal: The diagonal of a square matrix that runs from the upper-right corner to the lower-left corner of the matrix.

right angle: An angle measuring 90 degrees with rays that are perpendicular to one another.

rotation: A transformation in which points are transported to positions by a degree measure. The degree measure is determined by the angle formed from the original point to the center of the rotation to the rotated or image point.

row matrix: A $1 \times n$ matrix; a matrix with one row and n columns.

scalar: A constant or constant multiple.

semiperimeter: Half the perimeter of a polygon.

singular matrix: A square matrix that does not have an inverse. Also called a *noninvertible matrix.*

skew-symmetric matrix: A matrix in which each element $a_{ij} = -a_{ji}$. The matrix is equal to the negative of its transpose. If the element in the second row, third column is 4, then the element in the third row, second column is –4.

solution of system linear equations: Sequence of numbers *satisfying* (making true statements) when substituted into the equations of the system.

span: All the vectors in a vector space, V, that are linear combinations of a set of vectors, $\{\mathbf{v}_1, \mathbf{v}_2, \ldots, \mathbf{v}_k\}$, called its *spanning set.*

spanning set: A set of vectors, $\{\mathbf{v}_1, \mathbf{v}_2, \ldots, \mathbf{v}_k\}$, whose linear combinations produce the vectors in a vector space.

square matrix: A matrix with the same number of rows and columns.

standard basis of matrices: A basis consisting of vectors that are columns of an identity matrix. Each vector has one nonzero element, which is a 1.

standard position of a vector: A vector positioned with its endpoint at the origin.

subspace: A non-empty vector space that is a subset of a vector space with the same operations as the vector space.

transformation: A passage from one expression, figure, or format to another through a defined process or rule.

transition matrix: A matrix used in the rule or process defined by a transformation.

translation: A transformation in which points are transported to positions by a length or distance along a vector placed at a particular angle.

transpose of matrix: A matrix constructed by changing all rows of a matrix to columns and all columns to rows. Each element a_{ij} becomes element a_{ji}.

triangular matrix: A matrix in which all the elements either above or below the main diagonal are 0. An *upper triangular* matrix has all zeros below the main diagonal, and a *lower triangular* matrix has all zeros above the main diagonal.

trivial solution: When the solution of a homogeneous system has each variable equal to 0.

unit vector: A vector whose magnitude is equal to 1.

variable: An unknown quantity that can take on any of the values from a particular set. In most situations, the variable x in the expression $2x + 3x^2$ takes on the value of any real number.

vector: A member of a vector space.

vector in two-space: A measurable quantity that is described by both its magnitude and its direction. A directed line segment in the plane.

vector space: A set of elements subject to properties involving two defined operations, $\oplus$ and $\otimes$, and including identities and inverses.

zero matrix: An $m \times n$ matrix in which each element is 0.

Index

Symbols and Numerics

• A •

• B •

• C •

• D •

• E •

• F •

• G •

• H •

• I •

• K •

• L •

• M •

• O •

• P •

• Q •

• R •

• S •

• T •

• U •

• V •

• W •

• Z •

USINESS, CAREERS & PERSONAL FINANCE

ccounting For Dummies, 4th Edition*
78-0-470-24600-9

ookkeeping Workbook For Dummies†
78-0-470-16983-4

ommodities For Dummies
78-0-470-04928-0

oing Business in China For Dummies
78-0-470-04929-7

E-Mail Marketing For Dummies
978-0-470-19087-6

Job Interviews For Dummies, 3rd Edition*†
978-0-470-17748-8

Personal Finance Workbook For Dummies*†
978-0-470-09933-9

Real Estate License Exams For Dummies
978-0-7645-7623-2

Six Sigma For Dummies
978-0-7645-6798-8

Small Business Kit For Dummies, 2nd Edition*†
978-0-7645-5984-6

Telephone Sales For Dummies
978-0-470-16836-3

USINESS PRODUCTIVITY & MICROSOFT OFFICE

ccess 2007 For Dummies
78-0-470-03649-5

xcel 2007 For Dummies
78-0-470-03737-9

ffice 2007 For Dummies
78-0-470-00923-9

utlook 2007 For Dummies
78-0-470-03830-7

PowerPoint 2007 For Dummies
978-0-470-04059-1

Project 2007 For Dummies
978-0-470-03651-8

QuickBooks 2008 For Dummies
978-0-470-18470-7

Quicken 2008 For Dummies
978-0-470-17473-9

Salesforce.com For Dummies, 2nd Edition
978-0-470-04893-1

Word 2007 For Dummies
978-0-470-03658-7

DUCATION, HISTORY, REFERENCE & TEST PREPARATION

frican American History For Dummies
78-0-7645-5469-8

lgebra For Dummies
78-0-7645-5325-7

lgebra Workbook For Dummies
78-0-7645-8467-1

rt History For Dummies
78-0-470-09910-0

ASVAB For Dummies, 2nd Edition
978-0-470-10671-6

British Military History For Dummies
978-0-470-03213-8

Calculus For Dummies
978-0-7645-2498-1

Canadian History For Dummies, 2nd Edition
978-0-470-83656-9

Geometry Workbook For Dummies
978-0-471-79940-5

The SAT I For Dummies, 6th Edition
978-0-7645-7193-0

Series 7 Exam For Dummies
978-0-470-09932-2

World History For Dummies
978-0-7645-5242-7

OOD, GARDEN, HOBBIES & HOME

ridge For Dummies, 2nd Edition
78-0-471-92426-5

oin Collecting For Dummies, 2nd Edition
78-0-470-22275-1

ooking Basics For Dummies, 3rd Edition
78-0-7645-7206-7

Drawing For Dummies
978-0-7645-5476-6

Etiquette For Dummies, 2nd Edition
978-0-470-10672-3

Gardening Basics For Dummies*†
978-0-470-03749-2

Knitting Patterns For Dummies
978-0-470-04556-5

Living Gluten-Free For Dummies†
978-0-471-77383-2

Painting Do-It-Yourself For Dummies
978-0-470-17533-0

IEALTH, SELF HELP, PARENTING & PETS

nger Management For Dummies
78-0-470-03715-7

nxiety & Depression Workbook or Dummies
78-0-7645-9793-0

ieting For Dummies, 2nd Edition
78-0-7645-4149-0

og Training For Dummies, 2nd Edition
78-0-7645-8418-3

Horseback Riding For Dummies
978-0-470-09719-9

Infertility For Dummies†
978-0-470-11518-3

Meditation For Dummies with CD-ROM, 2nd Edition
978-0-471-77774-8

Post-Traumatic Stress Disorder For Dummies
978-0-470-04922-8

Puppies For Dummies, 2nd Edition
978-0-470-03717-1

Thyroid For Dummies, 2nd Edition†
978-0-471-78755-6

Type 1 Diabetes For Dummies*†
978-0-470-17811-9

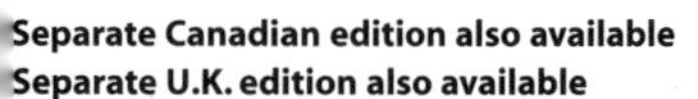

Separate Canadian edition also available
Separate U.K. edition also available

)PERATING SYSTEMS & COMPUTER BASICS

Mac For Dummies, 5th Edition
78-0-7645-8458-9

aptops For Dummies, 2nd Edition
78-0-470-05432-1

inux For Dummies, 8th Edition
78-0-470-11649-4

lacBook For Dummies
78-0-470-04859-7

lac OS X Leopard All-in-One esk Reference For Dummies
78-0-470-05434-5

Mac OS X Leopard For Dummies
978-0-470-05433-8

Macs For Dummies, 9th Edition
978-0-470-04849-8

PCs For Dummies, 11th Edition
978-0-470-13728-4

Windows® Home Server For Dummies
978-0-470-18592-6

Windows Server 2008 For Dummies
978-0-470-18043-3

Windows Vista All-in-One Desk Reference For Dummies
978-0-471-74941-7

Windows Vista For Dummies
978-0-471-75421-3

Windows Vista Security For Dummies
978-0-470-11805-4

PORTS, FITNESS & MUSIC

oaching Hockey For Dummies
78-0-470-83685-9

oaching Soccer For Dummies
78-0-471-77381-8

itness For Dummies, 3rd Edition
78-0-7645-7851-9

ootball For Dummies, 3rd Edition
78-0-470-12536-6

GarageBand For Dummies
978-0-7645-7323-1

Golf For Dummies, 3rd Edition
978-0-471-76871-5

Guitar For Dummies, 2nd Edition
978-0-7645-9904-0

Home Recording For Musicians For Dummies, 2nd Edition
978-0-7645-8884-6

iPod & iTunes For Dummies, 5th Edition
978-0-470-17474-6

Music Theory For Dummies
978-0-7645-7838-0

Stretching For Dummies
978-0-470-06741-3

Get smart @ dummies.com®

- Find a full list of Dummies titles
- Look into loads of FREE on-site articles
- Sign up for FREE eTips e-mailed to you weekly
- See what other products carry the Dummies name
- Shop directly from the Dummies bookstore
- Enter to win new prizes every month!

Separate Canadian edition also available
Separate U.K. edition also available

vailable wherever books are sold. For more information or to order direct: U.S. customers visit www.dummies.com or call 1-877-762-2974.
K. customers visit www.wileyeurope.com or call (0) 1243 843291. Canadian customers visit www.wiley.ca or call 1-800-567-4797.